Statistical Modelling using GENSTAT®

This book is based on the Open University course M346 *Linear Statistical Modelling*. Details of this and any other Open University courses can be obtained from the Course Reservations Centre, PO Box 724, The Open University, Milton Keynes, MK7 6ZS, United Kingdom, tel. +44 (0)1908 653231.

For availability of other course components, contact Open University Worldwide Ltd, The Berrill Building, Walton Hall, Milton Keynes, MK7 6AA, United Kingdom, tel. +44 (0)1908 858785, e-mail ouwenq@open.ac.uk

Alternatively, much useful information can be obtained from the Open University's website, http://www.open.ac.uk

Statistical Modelling using GENSTAT®

K. J. McConway and M. C. Jones

The Open University

P. C. Taylor

University of Hertfordshire

John Wiley & Sons, Ltd

First published in Great Britain in 1999 by
Arnold, a member of the Hodder Headline Group,
338 Euston Road, London NW1 3BH

John Wiley & Sons Ltd, The Atrium, Southern Gate, Chichester, West
Sussex, PO19 8SQ, United Kingdom

For details of our global editorial offices, for customer services and
for information about how to apply for permission to reuse the copyright
material in this book please see our website at www.wiley.com.

GENSTAT is a registered trademark of Lawes Agricultural Trust (Rothamsted Experimental Station)

Windows is a registered trademark of Microsoft Corporation in the United Kingdom and/or other countries

British Library Cataloguing-in-Publication Data
A catalogue record for this book is available from the British Library

A Library of Congress Cataloging-in-Publication Data
A catalogue record for this book is available from the Library of Congress

ISBN-13: 978-0-470-68568-6

4 5 6 7 8 9 10

Typeset in 10/12 pt Times by Focal Image Ltd, London

Contents

Preface

This book is a self-contained second course in statistics, concerning itself with a variety of important aspects of both the general linear model (covering regression and analysis of variance) and the generalized linear model, all with a strongly applied slant, and with use of the software package GENSTAT for Windows integrated into the text.

As described in more detail in Section 1.3, after an initial review of prerequisite material, the basics of GENSTAT are introduced and first used in conjunction with a review of simple linear regression. This makes up the first four chapters of the book. The next four chapters concern the linear model, ranging over multiple linear regression and the analysis of variance for a variety of basic designed experiments. The following four chapters look at various aspects of the generalized linear model, including logistic regression and loglinear modelling amongst other things. The final chapter consists of four data analysis case studies using techniques from the rest of the book. This chapter is called 'Further data analyses' since the whole book revolves around a large number of real datasets, their analysis using GENSTAT, and interpretation of the results.

As well as its obvious role in the classroom study of these topics, the book is eminently suitable for solitary study, based as it is on the Open University's distance teaching module M346, *Linear Statistical Modelling*.

The book has copious quantities of exercises, mostly requiring the use of GENSTAT, interspersed throughout, and solutions are given at the end; as students learn by doing, the exercises often ask you to 'have a go', and the solutions therefore include quite a lot of teaching material by way of explanation of what you have done! At the end of each chapter (after the first) there is also a summary of the methodology that has been introduced in that chapter.

A number of standard statistical methods and ideas are presumed to have been learnt – but not to a particularly high theoretical standard – in a preceding course in statistics. These include: histograms, boxplots and scatterplots; the normal, Poisson and binomial distributions; the central limit theorem; maximum likelihood estimation; confidence intervals; hypothesis testing (including t testing); t, χ^2 and F distributions; correlation; and some methods for examining the assumptions of models. Most of these topics are, in any case, reviewed in the early stages of the book, particularly Chapter 2. You would be in a better position also if you have a grounding in the basic ideas of linear regression, although Chapter 4 provides a review of this topic.

The mathematical prerequisites for using this book are not especially high: you will need to be able to appreciate mathematical formulae and graphs, but will not have to follow complicated algebraic arguments in the text. Calculus and linear algebra are not necessary, although one or two formulae using the latter are given in an optional manner to indicate the links that there are with more mathematical treatments of the subject.

Some familiarity with using a computer is expected, and some experience of using a statistical software package would be advantageous.

The statistical software used throughout the book is the following version of GENSTAT for Windows:

Genstat 5, Release 4.1, Third Edition, Student Version.

This version of GENSTAT was current at the time of writing the book, although much of what is written here applies also to earlier Windows versions of GENSTAT, and it is expected that virtually all of it will remain essentially current for some years to come.

As well as installing your copy of GENSTAT, you should create on your PC a directory called c:\lmgen and its subdirectory c:\lmgen\data (except, of course, in the unlikely event of your already having a directory called lmgen used for something else, in which case you should use a different name!).

The book proceeds through the analysis of some 78 real datasets. These are stored on the publisher's website and you should download them into c:\lmgen\data from the following address:

www.wiley.com/go/genstat

The datasets are given there both in the '.gsh' form that you will find is the form used throughout the book, and also in plain text format.

Finally, we would like to record our particular appreciation of the excellent editorial work done on our M346 manuscript by Open University Publishing Editor Roger Lowry. Our thanks also go to Open University colleagues Jane Williams, Andrew Bertie and Roberta Cheriyan for their reading, writing and organizing contributions, respectively; to other members of the OU Statistics Department for their support and input; to other members of the University who were involved in the production of the course and the book; and to Professors Phil Brown (University of Kent) and Cliff Lunneborg (University of Washington) for their comments on course drafts.

We wish you the best of luck with your studies and hope that you will gain maximum benefit, knowledge and enjoyment from this book.

Kevin McConway, Chris Jones and Paul Taylor

The Open University

March 1999.

A free 28 day trial version of Genstat is available to download from:

http://www.nag.co.uk/stats/TT/stmodbk.html

1

Introduction

The science of *statistics* is about solving problems, in a very wide range of areas, by analysing *data*. Whether one's interest is in medicine or engineering or the social sciences or astronomy or almost any other subject, there will always be important practical questions that can only be approached by the collection of suitable data and the analysis of those data by statistical methods. The emphasis of this book will, therefore, be on looking at real datasets that were collected as part of someone's research, and on developing and understanding statistical tools that will then be applied to answer the question(s) that the researcher set out to address.

The *theory* of statistical methods is a mathematical subject that will not be to the fore in this book, although you should gain an awareness that the methodology that is covered is not *ad hoc* but is firmly underpinned by theoretical understanding. The *implementation* of statistical methods, i.e. the detailed way in which a method produces its answers, could also be approached in a rather mathematical way, but it will not be here. Modern advances in computer technology mean that the computer can perform any long and tedious manipulations, and hence take the implementational strain off our hands; you will, therefore, be using ready-made statistical software – GENSTAT, introduced in Chapter 3 – to analyse data and hence help to answer the questions of interest. What *you* will be asked to do in this book is to understand the methods (but not necessarily in every detail), to decide which method(s) are appropriate to answer a given question, to apply the methods, and to interpret the answers that you get.

The authors hope that your first course in statistics also had the orientation described above[1], although you will not be greatly disadvantaged if it did not. There is a range of basic statistical methods and ideas that will be called upon as the book progresses; Chapter 2 of the book comprises a brief review of many of these. The ideas include the normal, Poisson and binomial distributions, the central limit theorem, maximum likelihood estimation, confidence intervals, hypothesis testing, t, χ^2 and F distributions, and some methods for examining the assumptions of models. The book builds on the basic ideas of linear regression, a topic reviewed in Chapter 4. Chapter 4 also briefly mentions the related topic of correlation.

---

[1] See, for example, Daly, F., Hand, D. J., Jones, M. C., Lunn, A. D. and McConway, K. J. (1995) *Elements of Statistics*, Wokingham, Addison-Wesley.

1.1 What methods will this book cover?

Data of interest in this book will involve at least two measurements on each of a number of individuals. Note that 'individuals' sometimes means individual people but could also mean individual animals or blocks of wood or stars or even groups of people, depending on the particular context. One of these measurements will be treated as a *response* variable; the questions that will be answered will concern how the response variable depends on the other variables, the *explanatory* variables, that have been measured. Examples might include: analysing the effects of different concentrations of glue (the explanatory variable is the concentration) on the strength of plywood (the response variable); or predicting the votes that the Conservatives might receive (the response variable) in each constituency in the next general election based on the votes they received in the constituency at the previous election and on other factors such as whether the constituency is rural or urban and whether or not the same candidate is standing (these are several explanatory variables); or analysing the effects of several drug treatments on alleviating the symptoms of asthma (measured in some appropriate way to give a response variable), taking into account other factors such as age, sex and weather conditions (the treatments together with the other factors are all explanatory variables).

The questions asked are of two fundamental types. It may be that a better *understanding* of the effects of the explanatory variables on the response variable is required. Alternatively, it may simply be desired to use the relationship to *predict* responses for future individuals from values of their explanatory variables. The first of these makes some attempt to mimic the mechanism by which explanatory variables influence the response variable; the second is less concerned with the meaning of the model, and concentrates instead on the usefulness of the results.

The kind of *statistical modelling* approach that will be taken to such problems is to develop a simplified *model* for the relationship between the response and the explanatory variables based on the data. This is an *empirical* approach to model building, which is distinct from using context-driven theoretical models such as might arise from, for example, physical or biological or economic theory. (However, methods of model fitting are common to both modelling situations.) If you find yourself in a situation where well-understood models are available, and if it can be shown that such a model fits well to the data, then there is no need for the empirical approach. But, in most subject areas, all too rarely is this the case, and then empirical models are very useful.

You should already have met a first example of *linear* statistical modelling, namely the basic *linear regression* model. In that model, often called *simple* linear regression, a single response variable, assumed to be normally distributed, is related to a single, often continuous, explanatory variable by a straight line that models the *mean* value of the response variable for each value of the explanatory variable. (So, if the data exhibit an overall trend in the response/explanatory relationship, the line reflects that trend. This may be an increasing dependence of the response variable on the explanatory variable, or a decreasing one; and, if there is no trend, the line will be horizontal.) The essentials of simple linear regression are reviewed in Chapter 4.

Linear regression can be extended in many practically important ways. There may be many explanatory variables rather than just one. The asthma drug trial scenario mentioned above provides an example: the drug type, patient's age, patient's sex and weather conditions could all be explanatory variables. Also the explanatory variables may be of different types. Age is essentially a continuous variable. Sex, on the other hand, is a *binary* variable, i.e. it takes just two values, perhaps coded 0 for male and 1 for female. How 'weather conditions' is measured has not been specified: perhaps the researchers would use several weather-related measurements like temperature and/or humidity and/or rainfall together with pollen count and/or some measure of pollution. What if the pollution measure were only defined on a three-point scale: high pollution, medium pollution, low pollution? You will learn how to cope with all these types of explanatory variable.

Some studies are *observational*, in that the data are those observed on some appropriate kind of random sample from the population of interest. In such a case, the researcher gets whatever values of explanatory variables happen to turn up. But in *experimental* studies, the researcher may be able to *assign* certain explanatory variable values to individuals. In the trial above, asthma researchers can control how many patients are given each treatment (although randomization is still needed to specify precisely which patient gets which drug). You will learn how such an experiment should be designed and analysed.

Linear regression extensions of the type mentioned so far provide the content of the book up to the end of Chapter 8. There is a general umbrella term for such models: the *general linear model*. However, the book will not explicitly deal with such a grand overarching model, although you should notice many similarities in what is done in various contexts as the book proceeds.

If we can have all sorts of different types of explanatory variables, we can have all sorts of different response variables too. For instance, there has not been any definition of how alleviation of asthma symptoms might be defined, or over what kind of timescale. Perhaps some continuous clinical measurement of ease of breathing might be measured; perhaps it might even be modelled by a normal distribution. But perhaps the researchers simply count the number of asthma episodes over, say, a month. Or perhaps the patient simply says, yes, I feel better, or, no, I don't. How do we cope with different types of response like these?

For general forms of response, it will prove useful to introduce a general framework within which to operate. This is the *generalized linear model*. We shall work within this framework from Chapter 9 onwards. Apologies, on behalf of the statistical community, for the confusing similarity in names!

For all the 'generals' and 'generalizeds', you may still be concerned about the apparently limiting nature of the word 'linear'. But, in fact, 'linear modelling' is much wider, and more generally useful, than it seems at first glance. In the simple linear regression model, the mean of the response variable is modelled by a function of the form $\alpha + \beta x$, where x represents the explanatory variable and α and β are *parameters* of the model. This is indeed the equation of a straight line in x. However, the word 'linear' actually refers to such a function being linear *in the parameters* and not (necessarily) in the explanatory variable(s). That is, a linear function in this context should be a sum of terms each of which are of the form: a parameter times a function of x (where 1 is included as a function of x). So, $\alpha + \beta x^2$ is also a linear model while $\alpha + \beta x/(\gamma + x)$ is not because of the way γ enters the model.

Exercise 1.1

Which of the following are linear functions in the parameters α, β and γ, and which are not? (x, x_1 and x_2 are explanatory variables.)

(a) $\alpha + \beta \log x + \gamma x^{1/4}$;

(b) $\alpha/x + \beta + \gamma x$;

(c) $\alpha + \exp(\beta x)$;

(d) $\alpha + \beta x_1 + \gamma x_2$.

As well as allowing functions of the explanatory variables, and several of them at that, in the linear part of the model, the generalized linear model actually allows the mean of the response variable to be modelled by some function of a linear function.[2] The point is that by the time linear functions, in all their generality, are used via some further general function, the models that can be handled include many that don't appear to have anything to do with straight-line fitting at all! Of course, there is also a major place in statistical methodology for non-linear models, but a good knowledge of how to apply linear modelling ideas will stand you in good stead in a surprisingly wide variety of applications.

A brief outline of the book's content will be given in Section 1.3. Before that, let us look at one particular dataset and, with the assistance of simple graphical and numerical techniques that you should already be familiar with, explore the data with a view to understanding the kinds of problems that this book aims to equip you to solve.

1.2 Exploring an interesting dataset

The subject of this subsection is a study of bone marrow transplantation in leukaemia patients. The researchers looked at data from 37 patients who received a 'non-depleted allogeneic bone marrow transplant' as treatment for their leukaemia. Many such patients unfortunately develop a condition called 'acute graft-versus-host disease', or GvHD for short. The main aim of this study was to understand what characteristics of a bone marrow graft recipient and his/her donor were most closely linked with the development or otherwise of GvHD, and hence to be able to predict for any new patient whether he or she was at high or low risk of developing GvHD after bone marrow transplantation (and, if at high risk, to do something about it).

The researchers studied the records of 20 transplant patients who did not develop GvHD and 17 who did, in their search for explanatory characteristics. The variables on which measurements were reported are:

(a) the recipient's age (in years) [recage, for short];

(b) the recipient's sex (coded 0 for male, 1 for female) [recsex];

(c) the age of the donor of the transplanted material (years) [donage];

(d) whether the donor was male (coded 0) or female, and if female whether the donor had ever been pregnant (2) or not (1) [donmfp];

(e) the type of leukaemia that the patient had (coded 1 for 'acute myeloid leukaemia', 2 for 'acute lymphocytic leukaemia' and 3 for 'chronic myeloid leukaemia') [type];

---

[2]Don't worry if this means little to you now; it will mean much more by the end of the book.

(f) the ratio of two clinical measurements, the mean counts per minute in 'mixed epidermal cell lymphocyte reactions' and the mean counts per minute in 'mixed lymphocyte reactions', a dimensionless quantity which will simply be referred to as the 'index' [indx];

(g) whether or not the recipient developed GvHD (coded 0 for 'no', 1 for 'yes') [gvhd].

Note to readers: you are not expected to understand all the medical terms used in association with these data, nor indeed all the non-statistical technical terms associated with any of the examples in this book. An understanding on the level of 'a certain kind of transplant patient and whether or not they develop an undesirable complication' will suffice. Of course, if you were the statistician involved in the original research project, the more you understood on the medical side, the better.

The data are given in Table 1.1 (overleaf). Also shown is a useful 'patient number', which is not a potential explanatory factor for predicting GvHD.

Measurements (a)–(f) are all available prior to the bone marrow transplantation. Predictions based on any or all of them could thus be very valuable. They form six (potential) explanatory variables. Measurement (g), the binary variable GvHD, is the response variable.

You should already be familiar with the notion of a *scatterplot* of one continuous variable against another. For example, Figure 1.1 shows a scatterplot of donor age versus recipient age. The scatterplot shows quite a clear relationship between the two: as recipient age increases, so does donor age. This, of course, reflects an attempt by the clinicians to match transplant donor and recipient as closely as possible in terms of age.

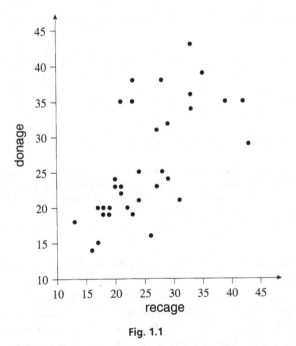

Fig. 1.1

However, the relationships of real interest to the researchers here are those involving GvHD as the response variable, which we would like to plot on the vertical axis (the *y*-axis), and the explanatory variables, each of which we would like to plot in turn along

Table 1.1 The data on bone marrow transplantation and GvHD

Patient	Recipient age	Recipient sex	Donor age	Donor M/F/Preg	Type	Index	GvHD
1	27	0	23	1	2	0.27	0
2	13	0	18	1	2	0.31	0
3	19	0	19	1	1	0.39	0
4	21	0	22	1	2	0.48	0
5	28	0	38	0	2	0.49	0
6	22	1	20	1	2	0.50	0
7	19	0	19	1	2	0.81	0
8	20	0	23	1	2	0.82	0
9	33	1	36	0	1	0.86	0
10	18	1	19	0	1	0.92	0
11	17	0	20	0	2	1.10	0
12	31	1	21	0	3	1.52	0
13	23	1	38	0	2	1.88	0
14	17	0	15	1	2	2.01	0
15	26	1	16	1	2	2.40	0
16	28	0	25	0	1	2.45	0
17	24	0	21	2	1	2.60	0
18	18	1	20	0	2	2.64	0
19	24	1	25	2	1	3.78	0
20	20	1	24	0	3	4.72	0
21	23	0	35	2	1	1.10	1
22	21	0	35	2	2	1.16	1
23	21	1	23	0	3	1.45	1
24	33	0	43	0	3	1.50	1
25	29	0	24	2	3	1.85	1
26	42	0	35	2	2	2.30	1
27	27	1	31	0	3	2.34	1
28	43	1	29	2	2	2.44	1
29	22	1	20	0	1	3.70	1
30	35	0	39	2	1	3.73	1
31	16	0	14	1	1	4.13	1
32	39	0	35	2	2	4.52	1
33	28	0	25	2	3	4.52	1
34	29	0	32	0	3	4.71	1
35	23	1	19	1	3	5.07	1
36	33	0	34	0	3	9.00	1
37	19	0	20	0	1	10.11	1

Source: Bagot, M., Mary, J. Y., Heslan, M., Keuntz, M., Cordonnier, C., Vernant, J. P., Dubertret, L. and Levy, J. P. (1988) 'The mixed epidermal cell lymphocyte-reaction is the most predictive factor of acute graft-versus-host disease in bone marrow graft recipients', *British Journal of Haematology*, **70**, 403–9.
Dataset name: gvhd.

the horizontal axis (the x-axis). In Figure 1.2, this has been done with recipient age as explanatory variable.

Straight away, we have found that the scatterplot – which tells us so much when, in particular, the response variable is on a continuous scale – is less adequate when the response variable is binary. The dominant feature of Figure 1.2 is basically that the response variable takes only values 0 and 1! However, we can still see something on closer inspection: those patients who do not contract GvHD (**gvhd** = 0) all have 'lowish' ages (33 or under in this dataset) while those who do contract GvHD (**gvhd** = 1) spread out more across ages and, in particular, include some older recipients.

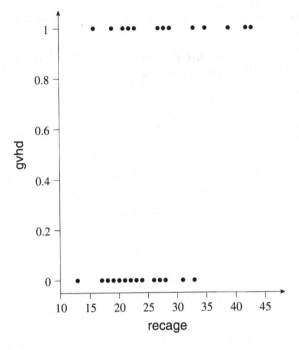

Fig. 1.2

Also, in Figure 1.2, no account is taken of coincident datapoints, which are simply plotted in the same way as if only appearing singly in the dataset. This problem would get worse if the dataset were bigger, in which case it would become hard to glean much from such a plot at all. This 'overplotting' problem is common to many computer packages, although some have ways round it.

However, within each of the two groups defined by the value of GvHD, the collection of recipient ages is just like an ordinary single sample. An alternative way of presenting these data might therefore be to separate the data into two parts corresponding to gvhd = 0 and gvhd = 1 and then make *boxplots* of each group separately, drawing the boxplots on the same scale. This is done for GvHD and recipient age in Figure 1.3.

It is clearer from Figure 1.3 than from Figure 1.2 that there might be some relationship between GvHD and recipient age.

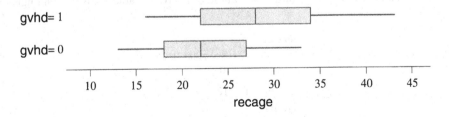

Fig. 1.3

Exercise 1.2

Figure 1.4 gives pairs of comparative boxplots for the two GvHD groupings against (a) donor age and (b) index. Describe what you see.

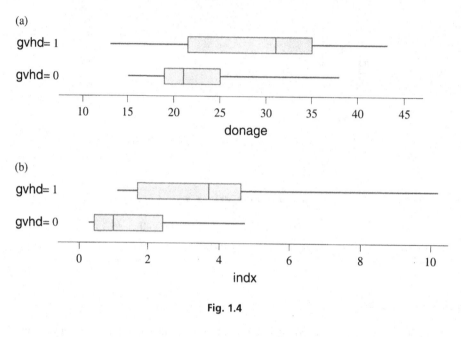

Fig. 1.4

So, GvHD occurrence seems to have some dependence on each of the explanatory variables looked at so far (in fact, GvHD appears more likely for older recipients and donors and for larger index values).

What about the other three explanatory variables? None of these, recipient sex, **donmfp** or **type**, is measured on anything like a continuous scale. Indeed, recipient sex is binary. With both a binary response variable and a binary explanatory variable it may not be worth going to the trouble of producing graphical representations since the same information can quite readily be obtained from the raw numbers themselves. So, when considering the GvHD and recipient sex variables only, the information can be laid out in the form of a *contingency table*, i.e. a table in which the counts of patients in each 'cell' corresponding to each combination of the values of the two variables is given. Table 1.2 is the contingency

Table 1.2

		Recipient sex	
		0	1
GvHD	1	12	5
	0	11	9

table for GvHD and recipient sex, with the response variable GvHD continuing to be portrayed in the vertical direction. This is obtained simply by counting patients in Table 1.1.

Since interest is in how GvHD depends on recipient sex, it might also be useful to calculate the proportions in each cell, these proportions being calculated with respect to the *column totals*. For example, 12 out of $12 + 11 = 23$, i.e. $12/23 = 0.52$ of male recipients (coded 0) suffered from GvHD while $11/23 = 0.48$ of male recipients did not develop GvHD. The complete table of proportions is given in Table 1.3.

Table 1.3 Proportions of male and female recipients developing and not developing GvHD.

		Recipient sex	
		0	1
GvHD	1	0.52	0.36
	0	0.48	0.64

While, for male recipients, about half developed GvHD, a slightly smaller percentage of females developed it. But the differences do not seem very great, particularly given the small sample sizes. Perhaps there is little or no dependence of GvHD on recipient sex. (No endeavour is made here to formalize this comparison.)

The other two explanatory variables, **type** and **donmfp**, are neither continuous nor binary; in fact they each take three different values only. The fact that **donmfp** is coded 0, 1 or 2 while **type** is coded 1, 2 or 3 is irrelevant. (You may have seen the word 'categorical' used for such a variable before.) We can still produce a contingency table, only now we have three columns corresponding to the explanatory variable rather than two. Table 1.4 is such a contingency table for the variable **type**.

Table 1.4 Counts and proportions for GvHD as a function of **type**.

		type					type		
		1	2	3			1	2	3
GvHD	1	5	4	8	GvHD	1	0.45	0.25	0.80
	0	6	12	2		0	0.55	0.75	0.20

In the type 1 leukaemia group, around a half seem to develop GvHD. Fewer seem to develop GvHD in the type 2 leukaemia group, while rather more than half of type 3 leukaemia sufferers develop GvHD. Again, we must not read too much into these results because of the small sample sizes, but first indications are that **type**, too, could be an important explanatory variable affecting GvHD.

Exercise 1.3
Produce contingency tables in both count and proportion form to investigate whether there is any apparent dependence of GvHD on **donmfp**. Describe what you see.

Our graphical/numerical exploration of these data suggests that several of the explanatory variables have some value with respect to predicting GvHD. However, scatterplots, boxplots and contingency tables only allow us to look at the dependence of the response variable on explanatory variables one at a time, each plot or table ignoring the other explanatory variables available. But, in combination, the explanatory variables may be able to tell us much more about the response variable. For example, it *could* be that low recipient age, low donor age and low index together, and in conjunction perhaps with one particular type of leukaemia and donors who have not been pregnant, is particularly indicative of not contracting GvHD. If so, this would be very useful clinical information. And what of patients who are, say, young and with a low index value but who have a different type of leukaemia and whose donor has been pregnant? How good is the prognosis for such people?

Further questions ensue. Could equally good predictions of the development or otherwise of GvHD be made based on a small subset of the explanatory variables? For example, given values of the other explanatory variables, is the value of the variable donmfp really important to GvHD? Given the relationship between recipient age and donor age (Figure 1.1), is it best to include just one of these in a model and not the other? Do we in fact gain anything by combining explanatory variables at all: might the single best explanatory variable tell us as much about GvHD as anything else? Are there complicated 'interactions' between explanatory variables, e.g. might recipient age, say, have a substantial effect on GvHD if the recipient is female but have little or no effect if the recipient is male? Is there more than one model which is (roughly) equally plausible given the data? Are the predictions from all the plausible models essentially the same, or are there important differences depending on which model we choose?

It is clear that there are many new aspects to regression modelling once several explanatory variables are involved. Also, the capability to deal with explanatory variables, and indeed response variables, of various different measurement types is an important ingredient of this kind of modelling. Providing appropriate modelling methodology is the central theme of this book.

Further analysis of the GvHD dataset is postponed until Chapter 9. However, as a foretaste of what can be done for such data, here is *one* of the several reasonable models for these data that more sophisticated analysis comes up with. This model says that, for a patient with any given values of the three explanatory variables donor age, donmfp and index, the probability p that this patient develops GvHD is modelled as

$$p = \frac{e^{q_1}}{1 + e^{q_1}}$$

where

$$q_1 = -6.72 + 0.1633 \text{ donage} + 2.271 \log(\text{indx})$$
$$+ 1.23 \text{ (if donmfp} = 1)$$
$$+ 1.53 \text{ (if donmfp} = 2). \tag{1.1}$$

Here, and unless otherwise stated, logs are taken to base e. If this model is a good one (and in Chapter 9, we shall see that the donmfp terms are probably not necessary), it says that the probability of developing GvHD increases, in an apparently rather complicated way, with values of each of donor age, donmfp and index. Notice that neither of the sex variables nor recipient age appears in this model; the latter would probably not improve matters because of its association with donor age, which does appear in the model. The variable type is not included in this model either, although it does appear in other reasonable ones. Notice too that logs have been taken of the index explanatory variable. A second model for the GvHD data that also seems fairly plausible is the simpler model

$$p = \frac{e^{q_2}}{1 + e^{q_2}}$$

where

$$q_2 = -1.293 + 1.738 \log(\mathsf{indx}). \tag{1.2}$$

This model is responsible for the title of the source article for these data, which claims that the index 'is the most predictive [single] factor of acute graft-versus-host disease in bone marrow graft recipients'.

The form of these models should help illustrate the comments towards the end of Section 1.1 that linear statistical modelling, of which this is an example, is not all that restrictive a kind of modelling!

1.3   A brief outline of the book

This introductory chapter is completed by a very brief outline of the book. Its aim is to help you keep track of the many ways in which the basic linear regression model is extended in the chapters to come; you might find it useful to refer back to this section if ever you find yourself immersed in the details and losing track of the wider scheme of things.

As already mentioned, Chapters 2 and 4 largely comprise review material reminding you of various statistical ideas and methods that will be assumed in the rest of the book. Chapter 2 is quite general in this regard; Chapter 4 concentrates on linear regression. The main situation in Chapter 4 is of a normally distributed response variable and a single quantitative explanatory variable. The terms 'quantitative' and 'categorical' signify different types of explanatory variable. They will be fully explained in Section 2.8.

Chapter 3 introduces the GENSTAT statistical software, which forms a driving force for the majority of the book. No prior knowledge of how to use your copy of GENSTAT is assumed. As part of its introduction it is applied in Chapter 3 to the topics of Chapter 2. GENSTAT is used for providing appropriate graphical output, both for initial data examination and later for model checking (i.e. assessing the adequacy and appropriateness of any model that is proposed), and also for implementing the statistical methods and providing results.

Chapters 5–13 all concern extensions to simple linear regression, all continuing to deal with a single response variable. Chapters 5–8 retain the normality assumption for this response variable; Chapters 9–13 relax this assumption. Figure 1.5 indicates where various combinations of types of response and explanatory variables are studied in detail.

Three of the later chapters are not mentioned in Figure 1.5, namely Chapters 8, 11 and 13.

The distinction between observational studies and designed experiments was made briefly in Section 1.1 (and will be reinforced in Chapter 5). The analysis of designed experiments is considered, in the context of a normally distributed response variable, in

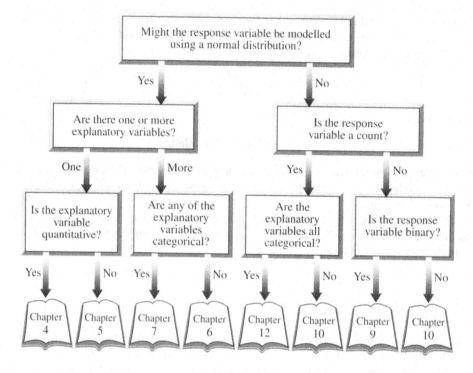

Fig. 1.5

Chapters 5 and 7 and also – with emphasis on the design as well as the analysis – in Chapter 8.

The question of model checking is addressed throughout the book. However, some of the more advanced 'diagnostic' techniques are considered more fully, in a quite general context, in Chapter 11.

Finally, Chapter 13 discusses the analysis of some substantial datasets and the many questions asked of them, drawing on many of the methods and approaches that have appeared earlier in the book.

Even though this book will allow you to approach a very wide variety of modelling situations, there will remain yet more tools, under the generalized linear modelling framework, that are beyond the scope of this book. The very brief postscript that follows Chapter 13 indicates what some of these are.

2

Review of statistical concepts

Before you can be equipped to tackle the more advanced statistical methodology contained in this book, it will be useful to remind you of some of the more basic statistical methods and ideas that form a solid grounding on which to build further work. This chapter, together with Chapter 4, therefore provides a review of the elements of statistics that will be assumed in later chapters.

The review will be fairly brief and speedy; most of the concepts and techniques are covered in greater depth in introductory textbooks, such as *Elements of Statistics* referred to in Chapter 1. The topics in this chapter include: histograms, probability plots and transformations; the normal, t, χ^2, F, Bernoulli, binomial and Poisson distributions; confidence intervals and hypothesis testing in general, including t testing in particular; maximum likelihood estimation; the central limit theorem and normal-based confidence intervals; and quantitative and categorical variables. The relative emphases placed on each here reflect the relevance of each in the current book.

What is missing from this chapter is any review of methodology for linear regression. This topic is the one that is most central to the current book, in the sense that one could think of the entire book as consisting of extensions to it! Its importance therefore warrants a separate review chapter, Chapter 4.

All of the exercises in this chapter bar two (Exercises 2.15 and 2.17) can be done without the help of a computer. This means you need pen, paper and a calculator, and in many cases you will also be expected to refer to a book of statistical tables. If you already have access to a statistical software package with which you are familiar, you are welcome to use this to carry out the exercises (in which case you might also like to reproduce the figures and summary statistics given in some exercises). This is why the datasets in this chapter are given, in ASCII form, on the data disk. But you are discouraged from trying to use GENSTAT at this stage. You will learn how to perform most of these tasks in GENSTAT as part of the introduction to that software package in Chapter 3.

2.1 The normal distribution

2.1.1 Basic attributes

Many measurements, e.g. leaf lengths or heights of adult males or blood plasma nicotine levels, can be thought of as being on a *continuous* scale, i.e. taking real numbers as values. (Measurements are actually always taken in *discrete* form, e.g. height to the nearest millimetre, but since height itself can take any, continuous, value, it remains very useful to model such measurements on their underlying continuous scale.)

Probability models, being models for the way measurements vary, are a fundamental tool of the statistician. They describe various features of the *distribution* of a *random variable*. Features include measures of *location* – whereabouts the measurements tend to congregate – and of *spread* or *scale* – how tightly they cluster around their central location. The basic descriptor of the distribution is the *probability density function*. 'Density function', 'density' or 'p.d.f.' are often used as shorthand for this.

Although many other models for continuous data exist, one with a very major role in statistical practice, and in this book, is the *normal distribution*. The normal distribution is sometimes called the *Gaussian distribution*. This has probability density function

$$f(x) = \frac{1}{\sigma\sqrt{2\pi}} \exp\left[-\frac{1}{2}\left(\frac{x-\mu}{\sigma}\right)^2\right], \quad -\infty < x < \infty.$$

For a random variable X with this distribution, we write

$$X \sim N(\mu, \sigma^2)$$

to signify that X is distributed as the normal distribution with two parameters μ and σ^2.

The normal distribution has a characteristic symmetric bell shape, shown in Figure 2.1 in the special case of the *standard* normal distribution – for which $\mu = 0$ and $\sigma^2 = 1$. In

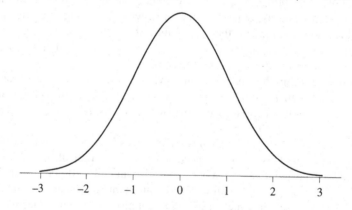

Fig. 2.1   The probability density function of the $N(0, 1)$ distribution.

the general case, the parameter μ determines the location of the normal distribution, and is in fact the population *mean*.[1] The parameter σ^2 determines the spread of the normal distribution, and is the population *variance*. It is equally valid to parametrize the spread by the population *standard deviation* $\sigma = \sqrt{\sigma^2}$. The way the normal density varies with μ and σ^2 is illustrated in Figure 2.2.

The importance of the standard normal distribution lies in the fact that if $X \sim N(\mu, \sigma^2)$ then

$$Z = \frac{X-\mu}{\sigma} \sim N(0, 1).$$

---

[1] μ is also the point about which the density is symmetric.

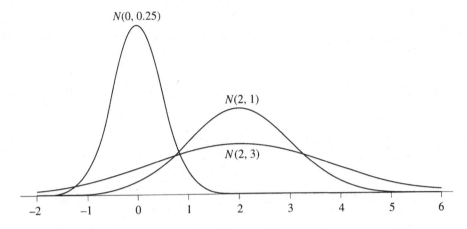

Fig. 2.2 Some normal probability density functions.

The inverse relationship is that if $Z \sim N(0, 1)$ then

$$X = \sigma Z + \mu \sim N(\mu, \sigma^2).$$

2.1.2 Data and the normal distribution

As in Chapter 1, plotting data by, for example, using histograms or boxplots or scatterplots as appropriate, is an important first step in analysing data.

Exercise 2.1

Files hald, forearm and magsus contain the following datasets. In hald are 50 measurements of the breaking strengths (kg) of samples of linen thread. In forearm are the lengths of the forearms (inches) of 140 randomly selected males. In magsus are 101 measurements of magnetic susceptibility (SI units) of small areas in the Roman quarry of Mons Claudianus in the Egyptian desert. Table 2.1 shows the first five values in each of these datasets.

Table 2.1

hald	1.40	1.52	1.63	1.69	1.73	⋯
forearm	17.3	18.4	20.9	16.8	18.7	⋯
magsus	4.76	3.68	5.55	2.95	2.79	⋯

The sources are as follows.

Dataset name: hald;
Source: Hald, A. (1952) *Statistical Theory with Engineering Applications*, New York, John Wiley.

Dataset name: forearm;
Source: Pearson, K. and Lee, A. (1903) 'On the laws of inheritance in man: I. Inheritance of physical characters', *Biometrika*, **2**, 357–462.

Dataset name: magsus;
Source: Williams-Thorpe, O., Jones, M. C., Tindle, A. G. and Thorpe, R. S. (1996) 'Magnetic susceptibility variations at Mons Claudianus and in Roman columns: a method of provenancing to within a single quarry', *Archaeometry*, **38**, 15–41.

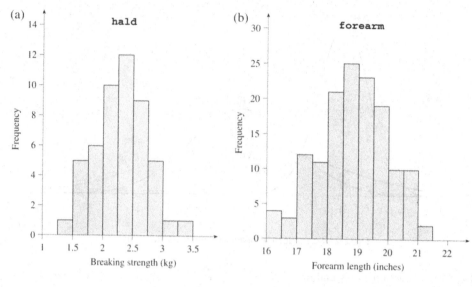

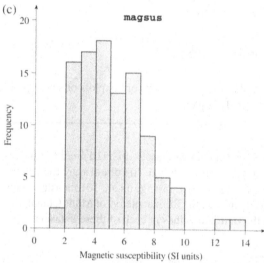

Fig. 2.3

Figure 2.3 gives *histograms* of each of the datasets in (a) hald, (b) forearm and (c) magsus. Which of these would you say has approximately the symmetric bell shape characteristic of the normal distribution and which not?

You were reminded of the notion of *skewness* of a distribution in the solution to Exercise 2.1. In fact, a distribution in which there are (relatively) many small values and (relatively) few large values is said to be *positively skewed*; conversely, a distribution in which there are (relatively) many large values and (relatively) few small values is said to be *negatively skewed*.

The histograms in Exercise 2.1 indicate that the normal distribution might provide a reasonable model for the breaking strength and forearm length datasets. The precise form of normal distribution that best fits such data is given by replacing the unknown parameter values by estimates of them obtained from the data. Let the data be written $x_1, x_2, \ldots, x_n$ for a random sample of size n. The usual estimates of μ and σ^2 are the *sample mean*

$$\bar{x} = \frac{1}{n} \sum_{i=1}^{n} x_i$$

and the *sample variance*

$$s^2 = \frac{1}{n-1} \sum_{i=1}^{n} (x_i - \bar{x})^2,$$

respectively. The *sample standard deviation* $s = \sqrt{s^2}$ estimates σ.

Exercise 2.2
Histograms of each of the datasets in (a) hald and (b) forearm are given again in Figure 2.4, but this time with fitted normal curves added. Do these fitted curves support your conclusions in Exercise 2.1? The normal distribution fitted in Figure 2.4(a) has parameter values 2.299 and 0.1689 respectively. What must be the sample mean and sample standard deviation of the hald dataset?

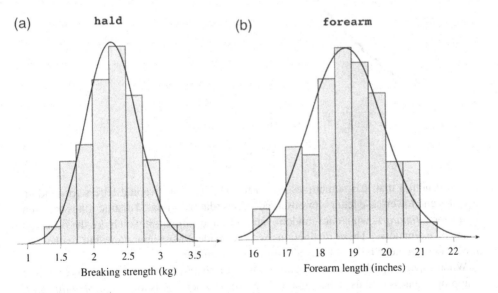

Fig. 2.4

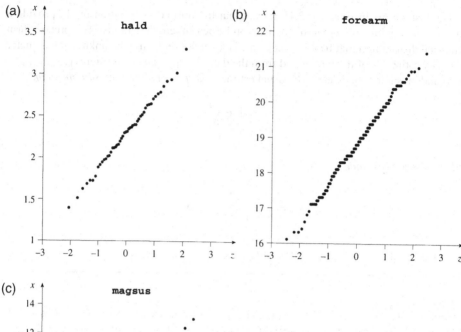

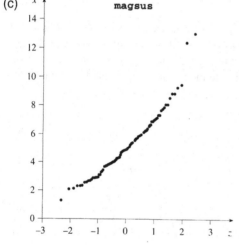

Fig. 2.5

A distinction that it is sometimes useful to make is to use capital letters for random variables and lower-case letters for observed values (data). A related distinction is between parameter *estimators*, which are random variables, and *estimates*, which are the observed values of estimators. These distinctions will be made where important, but neither the authors nor you need be very fussy about these conventions.

When a specific probability model is proposed for a dataset, one way of visually checking its appropriateness is to use a histogram, as above; another way is to use a *probability plot*. Rearrange the data into ascending order, calling the ordered data $x_{(1)} \le x_{(2)} \le \ldots \le x_{(n)}$. Also, define the normal quantiles, or *normal scores*, $z_1, z_2, \ldots, z_n$, by the equation

$$\Phi(z_i) = \frac{i}{n+1}, \quad i = 1, 2, \ldots, n.$$

Here, $\Phi(z) = P(Z \leq z)$, where Z is a standard normal random variable, is the *distribution function* of the standard normal distribution. The definition of the distribution function for other distributions is analogous. The distribution function is also referred to as the 'cumulative distribution function' or 'c.d.f.'. The probability plot consists of plotting the ordered xs against the zs. If what you see approximates to a straight line, then the data may plausibly be assumed to be modelled by a normal distribution.

Exercise 2.3
Figure 2.5 gives normal probability plots of each of the datasets in (a) `hald`, (b) `forearm` and (c) `magsus`. Do these plots support your conclusions from Exercises 2.1 and 2.2?

Later in the book, you will see examples of *half-normal plots* being used in place of the full normal probability plots above. Half-normal plots are simply full normal plots folded over, i.e. plots in which absolute values only are considered, negative and positive values being treated as the same. These are intended for use only when the distribution's symmetry about zero is not in question, since information on symmetry, or lack of it, is lost. The idea is that other detailed aspects of the distribution might show up better. You will first meet examples of this in Chapter 5.

2.1.3 Transforming to normality

The dataset `magsus` was seen in Section 2.1.2 to be too skewed to be reasonably modelled by a normal distribution. An alternative distribution is needed. One possibility is to try to transform the data $x_1, x_2, \ldots, x_n$ in such a way that the transformed data $w_1, w_2, \ldots, w_n$ can be approximated by a normal distribution. Transformation is a popular and useful technique, which works quite often, but it is not a panacea.

There is an enormous number of different possible transformations. If the original data are positive and highly positively skewed, then transformations such as $w_i = x_i^{1/2}$ or, most popularly of all, $w_i = \log x_i$, tend to reduce high x values much more than low ones, and hence decrease the skewness of the data.

Exercise 2.4
For the `magsus` data, many analysts' first reaction would be to take logs of the data. In Figure 2.6, this has been done and appropriate plots of the transformed data made. In Figure 2.7 the exercise has been repeated for the square root transformation. What do you conclude?

The transformations considered in Exercise 2.4 are part of the following useful *ladder of powers*, which lists possible transformations:

$$\ldots, x^{-2}, x^{-1}, x^{-1/2}, \log x, x^{1/2}, x^1, x^2, \ldots.$$

The transformation x^1 leaves the value of x as it is and $\log x$ takes the place of x^0. This is not an arbitrary insertion. There are good reasons for the position of log in the ladder, but

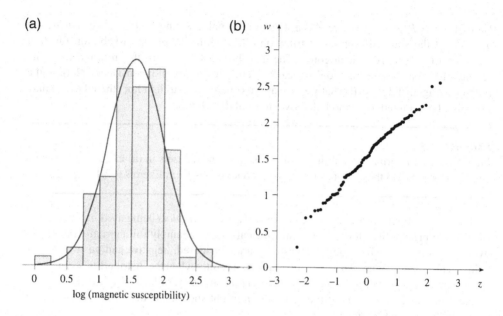

Fig. 2.6

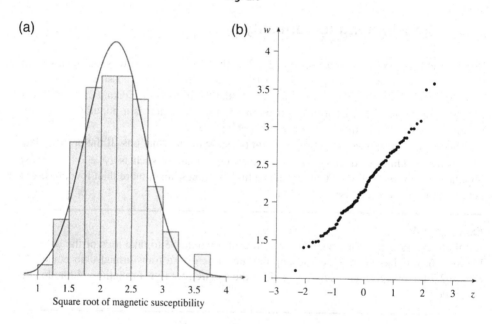

Fig. 2.7

you need not worry about them. The important property to remember (as will be reinforced in Section 4.5) is that, provided $x > 1$, powers below 1 reduce the high values in a dataset relative to the low values, while powers above 1 have the opposite effect of stretching out high values relative to low ones. The further up or down the ladder from x^1, the greater the effect.

2.1.4   Some distributional properties

Some basic properties of means and variances underlie some of the useful results that will be taken advantage of later in the book.

For any distribution whatsoever, and not just the normal nor even just continuous distributions, if X is a random variable with mean μ and variance σ^2 and if a and b are constants, then

$$E(aX + b) = a\mu + b \quad \text{and} \quad V(aX + b) = a^2\sigma^2,$$

where E and V stand for (population) *expectation* and *variance*, respectively. Also, if $X_1, X_2, \ldots, X_n$ are random variables with means $\mu_1, \mu_2, \ldots, \mu_n$ and variances σ_1^2, $\sigma_2^2, \ldots, \sigma_n^2$, then

$$E\left(\sum_{i=1}^{n} X_i\right) = \sum_{i=1}^{n} \mu_i$$

and, if $X_1, X_2, \ldots, X_n$ are independent,

$$V\left(\sum_{i=1}^{n} X_i\right) = \sum_{i=1}^{n} \sigma_i^2.$$

When X and $X_1, X_2, \ldots, X_n$ additionally are normally distributed, the end-products of the above are also normally distributed:

$$aX + b \sim N(a\mu + b, a^2\sigma^2) \tag{2.1}$$

and

$$\sum_{i=1}^{n} X_i \sim N\left(\sum_{i=1}^{n} \mu_i, \sum_{i=1}^{n} \sigma_i^2\right).$$

The first of these results can be used to prove the relationship between general and standard normal distributions.

2.2   Confidence intervals

By estimating the values of model parameters, we attempt to infer properties of the population from which the data are taken. The estimating of the values of parameters and the understanding of the behaviour, especially the variability, of estimators collectively form the topic of *statistical inference*, or just *inference* for short. Statistical inference covers point estimation, interval estimation, hypothesis testing and more.

Instead of stating a single number, a *point estimate*, as our estimate of a parameter, it is usually much better to acknowledge the uncertainty in our estimate by providing a range of plausible values for the parameter: this is a *confidence interval*. Suppose that the limits of a confidence interval obtained from data for a parameter θ, the *confidence limits*, are (θ_-, θ_+) and that it is declared that this is a 95% confidence interval. It is important to appreciate the meaning of this statement. The confidence interval is a random interval that would have been different if the data had been different. The confidence statement says that, in 95% of hypothetical replications of the current situation (provided the model is good), the confidence intervals obtained would include the true value of θ. Any particular confidence interval may or may not include θ. See Figure 2.8 for an example of this.

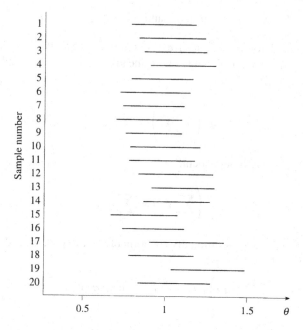

Fig. 2.8 Twenty datasets of size $n = 100$ were generated from a $N(\theta, 1)$ distribution with $\theta = 1$. Each dataset yielded a 95% confidence interval for θ as shown. On average, all but $0.05 \times 20 = 1$ of these confidence intervals will include $\theta = 1$; in this case, indeed, all but one (sample 19) did.

2.2.1 Confidence interval for the mean of a normal distribution; Student's *t* distribution

If $X_1, X_2, \ldots, X_n$ is a random sample from a $N(\mu, \sigma^2)$ distribution, the point estimator of μ is

$$\widehat{\mu} = \overline{X} = \frac{1}{n} \sum_{i=1}^{n} X_i,$$

which has the property of *unbiasedness*:

$$E(\widehat{\mu}) = \mu.$$

Also,

$$V(\widehat{\mu}) = \frac{\sigma^2}{n}.$$

In fact,

$$\overline{X} \sim N\left(\mu, \frac{\sigma^2}{n}\right).$$

These results are simple consequences of properties in Section 2.1.4.

A confidence interval for μ is of the form $\widehat{\mu} \pm$ multiple of s.e.$(\widehat{\mu})$, where s.e.$(\widehat{\mu})$ is the *standard error* of $\widehat{\mu}$, which is defined to be the square root of $V(\widehat{\mu})$. The standard error is a name for a standard deviation when applied to an estimator. This standard error is not known because σ^2 is not known, but a natural estimator, replacing σ^2 by s^2, is available. The resulting $100(1 - \alpha)\%$ confidence interval for μ is

$$\overline{X} \pm [t_{1-\alpha/2}(n - 1)]\frac{s}{\sqrt{n}}.$$

The multiplier $t_{1-\alpha/2}(n - 1)$ is the $100(1 - \alpha/2)$ *percentage point* (i.e. the $100(1 - \alpha/2)\%$ point) of *Student's t distribution*, or the *t distribution* for short, with $n - 1$ *degrees of freedom* (d.f.). In general, $t_\gamma(v)$ denotes the 100γ *percentage point* or γ *quantile* of the t distribution with v degrees of freedom; that is, if $T \sim t(v)$, then $P(T \leq t_\gamma(v)) = \gamma$. The t distribution is a symmetric continuous distribution that looks like the normal distribution only with more weight on more extreme observations: see Figure 2.9. The distributional result driving the formula above is that, under normality,

$$\frac{\overline{X} - \mu}{S/\sqrt{n}} \sim t(n - 1). \tag{2.2}$$

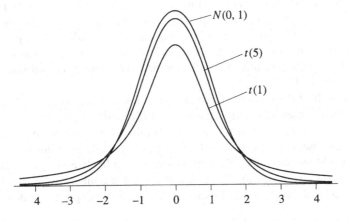

Fig. 2.9 Probability density functions of the $N(0, 1)$, $t(5)$ and $t(1)$ distributions.

Exercise 2.5

Obtain (a) a 95% confidence interval for the mean linen thread breaking strength using the data in `hald`, and (b) a 90% confidence interval for the mean male forearm length using the data in `forearm`. You are given that, for `hald`, $\bar{x} = 2.299$, $s^2 = 0.1689$, and for `forearm`, $\bar{x} = 18.80$, $s^2 = 1.255$. In each case, report the values of the appropriate t percentage points (quantiles) used. Assume that normal models are appropriate in each case.

2.3 Hypothesis testing

Sometimes, research questions are framed not as 'What is a plausible range of values for such and such a parameter?' but rather 'Are the data consistent with this particular value for the parameter?'. A *hypothesis test* is a test of such a hypothesized value for a parameter.

The hypothesized value, θ_0, of the parameter θ defines a *null hypothesis*, H_0, of the form H_0: $\theta = \theta_0$. The opposite of the null hypothesis is the *alternative hypothesis*, H_1. This may be of the form H_1: $\theta \neq \theta_0$, a *two-sided* alternative leading to a *two-sided test*. On the other hand, if a natural 'direction' in which θ may differ from θ_0 is apparent before the test, then H_1 may be *one-sided*, e.g. for the case where it is known beforehand that $\theta \geq \theta_0$ then H_1: $\theta > \theta_0$, leading to a *one-sided test*.[2]

In this book, an approach to hypothesis testing often called *significance testing* will be taken. In this approach, a *significance probability*, or *SP* for short, is obtained. The SP is often called the *p-value*. The SP is the probability, *assuming the null hypothesis is true*, that an event as extreme or more extreme than that observed would occur. To measure this, we need a *test statistic*, a function of the data, which in the current context would be an estimator of θ. Denote the test statistic, thought of as a random variable, as T and its observed value from the data as t. Then, the SP would be the probability of obtaining as extreme or more extreme a value as the observed t, in both 'big' and 'small' directions in the two-sided case and in just one direction in the one-sided case, the probabilities being evaluated under the null hypothesis. The set of extreme values of T containing the observed value t provides an SP 'in the obtained direction', and the other set of extreme values of T contributes an SP 'in the opposite direction'. The two-sided SP is the sum of these, and (unless something very unexpected is going on) the one-sided SP is just the SP in the obtained direction. See Figure 2.10 for clarification. Often, but not always, there is symmetry in the distribution under the null hypothesis resulting in the two-sided SP being twice the one-sided SP.

The value of the SP should be interpreted as follows. The smaller the SP value, the less likely it is that, under the null hypothesis (i.e. if H_0 were true), we would have obtained the observed value of the test statistic. So, a small value of SP is likely to mean that H_0 is untrue, in which case we should prefer H_1. However, we need to bear in mind that a small value of SP could just mean that H_0 holds but a rather unlikely event has taken place. A small SP is therefore taken, with the proviso that a rather unlikely event may have taken place, to give evidence against H_0: the smaller the SP, the stronger the evidence.

---

[2] In some cases one may wish to set up the hypotheses as, for example, $H_0 : \theta \leq \theta_0$, $H_1 : \theta > \theta_0$. These are dealt with in exactly the same way as $H_0 : \theta = \theta_0$, $H_1 : \theta > \theta_0$.

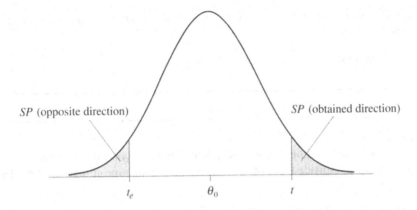

Fig. 2.10 The density function shown is that implied by the null hypothesis; t_e is the point as extreme as t but in the opposite direction.

There are no hard and fast rules for 'when is small small', but: values greater than about 0.10 are usually taken as negligible evidence against H_0; values around 0.05 or so are taken to provide marginal evidence against H_0; and rather smaller values, particularly beyond 0.01, are taken as providing strong evidence against H_0.

There is a strong link between hypothesis testing and confidence intervals. Let us consider the two-sided testing situation for simplicity. If a $100(1-\alpha)\%$ confidence interval for θ excludes θ_0, then the SP must be less than α: both approaches say that θ_0 seems unlikely on the basis of the data. Likewise, if a $100(1-\alpha)\%$ confidence interval for θ includes θ_0, then the SP must be greater than or equal to α. There is the same equivalence for one-sided tests involving one-sided confidence intervals.

2.3.1   The one-sample *t* test

Consider testing $H_0 : \mu = \mu_0$ where the data $X_1, X_2, \ldots, X_n$ can be assumed to come from the $N(\mu, \sigma^2)$ distribution. An appropriate test statistic is the *t statistic*

$$T = \frac{\overline{X} - \mu_0}{S/\sqrt{n}}.$$

SPs can be evaluated using the t distribution on $n - 1$ degrees of freedom since, under the null hypothesis, result (2.2) applies. This is the *one-sample t test*.

Exercise 2.6
It is required to test the null hypothesis $H_0 : \mu = 2$ where μ is the mean breaking strength of the linen thread on which data are provided in `hald`. Assume normality and suppose that no particular alternative to $\mu = 2$ has been suggested.

(a) Perform the appropriate t test and report and interpret the result.

Suppose now that it is suspected that, if anything, the mean breaking strength might be larger than 2 kg.

(b) Repeat part (a) given this extra information.

Exercise 2.7

(a) Using only the solution to Exercise 2.5(b), what can you say about the SP when performing a two-sided test of the null hypothesis that mean male forearm length is 18.5?

(b) Obtain better information about the SP for the test in part (a) and report your conclusion.

Assume normality throughout.

The one-sample t test is also the appropriate tool to use when wishing to test for a zero mean difference in *matched pairs* data (provided normality can be assumed). For such data, values of two quantities whose comparison is of interest are both made on the same individual (and both measurements are repeated on many individuals to produce the data). Examples might involve the effects of two different drugs on some human ailment or the growth of two different types of plant when one plant of each type is grown in a pot. (This drugs example only applies to situations in which the same ailment resurfaces once one treatment is discontinued. More permanent cures demand different data and statistical approaches; see Section 2.3.2.)

Claiming that the mean values of the two responses are equal is equivalent to claiming that the mean difference, μ_D, between the values is zero. Once differences are calculated, the one-sample t test can be applied to them.

Exercise 2.8

The datafile weantoil contains paired data for 25 subjects. The first five data pairs are shown in Table 2.2.

Table 2.2

Society	Age at weaning	Age at toilet training
Alorese	2.3	1.8
Balinese	2.7	3.0
Bena	1.8	4.7
Chagga	2.7	0.8
Comanche	1.5	1.2
⋮	⋮	⋮

Source: Whiting, J. M. and Child, I. L. (1962) *Child Training and Personality*. New Haven (CT), Yale University Press.
Dataset name: weantoil.

The data concern ages (in years) of weaning and toilet training in young children. In Western society, it is the norm that children are toilet trained substantially later than they are weaned. In 'primitive' societies, this may not be the case. The subjects in `weantoil` are in fact 25 different primitive societies. A glance at Table 2.2 shows that in only two of the five cases shown does weaning precede toilet training. The question is, on average, does toilet training occur after weaning in primitive societies? To investigate this, the differences between the toilet-training and weaning ages were calculated (as toilet-training age minus weaning age) and are as follows.

```
−0.5    0.3    2.9   −1.9   −0.3   −0.5   −0.5   −0.5   0.5
−0.6   −0.2   −1.3    0.9    0.5    0.2   −1.8   −0.4   0.2
 0.5   −0.3   −1.4    0.2   −0.5    1.4   −0.8
```

A normal probability plot of these differences is given in Figure 2.11.

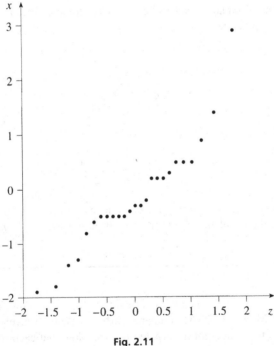

Fig. 2.11

(a) Make a rough assessment of the appropriateness of a normality assumption for the distribution of these differences.

(b) Assuming normality, test the null hypothesis that $\mu_D = 0$, i.e. that the mean difference is zero, against the alternative that $\mu_D > 0$, i.e. that the toilet-training ages are greater than the weaning ages. Why is such a one-sided test appropriate?

2.3.2   The two-sample t test

Suppose that we have two independent random samples $X_{1,1}, X_{1,2}, \ldots, X_{1,n_1}$ and $X_{2,1}$, $X_{2,2}, \ldots, X_{2,n_2}$, of sizes n_1 and n_2, respectively, from two distinct populations with means μ_1 and μ_2, and suppose that we wish to test the null hypothesis $H_0 : \mu_1 = \mu_2$. The difference from the matched pairs situation is that each individual provides a data value for only one of the two circumstances under comparison (e.g. each person receives only one of two possible treatments). The *two-sample t test* is the analogue of the one-sample t test if it is assumed that, independently:

$$X_{1,i} \sim N(\mu_1, \sigma^2), \quad i = 1, 2, \ldots, n_1;$$

$$X_{2,j} \sim N(\mu_2, \sigma^2), \quad j = 1, 2, \ldots, n_2.$$

Note the assumption of equal variances for the two groups, which should be at least approximately true in practice.

The two-sample test statistic is

$$T = \frac{\overline{X}_1 - \overline{X}_2}{S_p \sqrt{\frac{1}{n_1} + \frac{1}{n_2}}}$$

where

$$S_p^2 = \frac{(n_1 - 1)S_1^2 + (n_2 - 1)S_2^2}{n_1 + n_2 - 2}$$

and where $\overline{X}_i$ and S_i^2 are the sample mean and variance of group i, $i = 1, 2$. The reference distribution for calculation of the SP is the t distribution with $n_1 + n_2 - 2$ degrees of freedom.

Exercise 2.9

The data in Table 2.3 concern the effects of carpeting in hospital rooms. The study, carried out in a hospital in Montana, USA, compared airborne bacteria levels in each of eight carpeted rooms with those in eight uncarpeted rooms. Note that each room appears only once in either carpeted or uncarpeted state, so this is not a matched pairs situation. Take the normality of each group as read.

Table 2.3   Levels of airborne bacteria (number of colonies per ft^3 of air).

Carpeted	11.8	8.2	7.1	13.0	10.8	10.1	14.6	14.0
Uncarpeted	12.1	8.3	3.8	7.2	12.0	11.1	10.1	13.7

Source: Walter, W. G. and Stobie, A. (1968) 'Microbial air sampling in a carpeted hospital', *Journal of Environmental Health*, **30**, 405.

Dataset name: carpet.

(a) Calculate the sample means. Which mean is bigger?

(b) Calculate the sample variances. Are the variances roughly equal? Is it valid to perform a two-sample t test? (A rule of thumb that has been suggested is that the t test is applicable if the larger sample variance is no more than about three times the smaller.)

(c) If appropriate, perform a two-sided t test and report your conclusions.

Exercise 2.10

For 12 patients suffering chronic renal failure, measurements were taken before and after they underwent haemodialysis. The measurements involved a ratio of heparin and protein plasma levels, which shall be referred to here simply as plasma measurements. For another 12 patients, also with chronic renal failure and undergoing haemodialysis, the same before and after measurements were made. The two groups of patients are distinguished by having either low (Group 1) or normal (Group 2) plasma heparin cofactor II (HCII) levels. Interest lies in whether the plasma measurements increase with haemodialysis by a different amount in low and normal HCII groups.

All the data are reproduced in Table 2.4.

Table 2.4

Group 1 (low HCII)			Group 2 (normal HCII)		
Patient	Before	After	Patient	Before	After
1	1.41	1.47	13	2.11	2.15
2	1.37	1.45	14	1.85	2.11
3	1.33	1.50	15	1.82	1.93
4	1.13	1.25	16	1.75	1.83
5	1.09	1.01	17	1.54	1.90
6	1.03	1.14	18	1.52	1.56
7	0.89	0.98	19	1.49	1.44
8	0.86	0.89	20	1.44	1.43
9	0.75	0.95	21	1.38	1.28
10	0.75	0.83	22	1.30	1.30
11	0.70	0.75	23	1.20	1.21
12	0.69	0.71	24	1.19	1.30

Source: Toulon, P., Jacquot, C., Capron, L., Frydman, M. O., Vignon, D. and Aiach, M. (1987) 'Antithrombin III and heparin cofactor II inpatients with chronic renal failure undergoing regular haemodialysis', *Thrombosis and Haemostasis*, **57**, 263–8.
Dataset name: hcii.

Within each group, there are paired data that yield differences. Is the mean difference in Group 1 different from the mean difference in Group 2? To investigate this, proceed through parts (a) and (b) below.

(a) Obtain after–before differences for each group. A comparative pair of boxplots for the two sets of differences is given in Figure 2.12. Does normality for each set of differences seem a tenable assumption? Are the variances roughly equal?

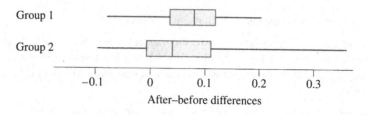

Fig. 2.12

(b) Assuming that the necessary assumptions are tenable, perform a two-sample t test and report your conclusion.

2.4   Chi-squared and F distributions

2.4.1   The χ^2 distribution

Continue with the assumption that a random sample $X_1, X_2, \ldots, X_n$ comes from a $N(\mu, \sigma^2)$ distribution. The sample variance

$$S^2 = \frac{1}{n-1} \sum_{i=1}^{n} (X_i - \overline{X})^2$$

has the properties that

$$E(S^2) = \sigma^2,$$

so that S^2 is an unbiased estimator of σ^2, and

$$V(S^2) = \frac{2\sigma^4}{(n-1)}.$$

Moreover, the distribution of the random variable

$$\frac{(n-1)S^2}{\sigma^2} = \sum_{i=1}^{n} \left(\frac{X_i - \overline{X}}{\sigma} \right)^2$$

is given by the *chi-squared distribution*, i.e. the χ^2 distribution, with $\nu = n - 1$ degrees of freedom. This is written

$$\frac{(n-1)S^2}{\sigma^2} \sim \chi^2(n-1).$$

χ^2 random variables are continuous, taking positive values only. Their distributions are positively skewed: the smaller the d.f. the more skewed the distribution; see Figure 2.13. Its genesis is as the distribution of a sum of squares of independent standard normal variables. That is, if $Z_1, Z_2, \ldots, Z_\nu$ are independent $N(0, 1)$ random variables, $W = Z_1^2 + Z_2^2 + \cdots + Z_\nu^2 \sim \chi^2(\nu)$. The first two moments of W are $E(W) = \nu$ and $V(W) = 2\nu$.

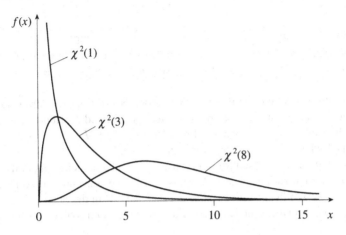

Fig. 2.13 $\chi^2(1)$, $\chi^2(3)$ and $\chi^2(8)$ densities.

The χ^2 distribution will crop up regularly in this book, usually as a reference distribution for a hypothesis test. Here, let us note its role in providing confidence intervals for the variance σ^2 of a normal distribution $N(\mu, \sigma^2)$. If $c_L = c_{\alpha/2}(n-1)$ and $c_U = c_{1-\alpha/2}(n-1)$ are the appropriate percentage points (quantiles) of the $\chi^2(n-1)$ distribution, a $100(1-\alpha)\%$ confidence interval for σ^2 is given by

$$\left(\frac{(n-1)S^2}{c_U}, \frac{(n-1)S^2}{c_L} \right).$$

Exercise 2.11
Obtain a 95% confidence interval for the variance of the distribution of linen thread breaking strengths using hald. What values of c_L and c_U are used to obtain this confidence interval?

2.4.2   The *F* distribution

A further distribution will also occur regularly in this book in hypothesis testing situations. This, the *F distribution* is defined as the distribution of the ratio of two independent χ^2 random variables each divided by their degrees of freedom: i.e. if the random variable F is defined by

$$F = \frac{X_{\nu_1}/\nu_1}{X_{\nu_2}/\nu_2},$$

where independently $X_{\nu_i} \sim \chi^2(\nu_i)$, $i = 1, 2$, then we say that F has the F distribution with ν_1 and ν_2 degrees of freedom, written

$$F \sim F(\nu_1, \nu_2).$$

An important example of an F random variable is the ratio

$$\frac{S_1^2/\sigma_1^2}{S_2^2/\sigma_2^2}$$

where S_1^2 and S_2^2 are sample variances from independent samples from $N(\mu_1, \sigma_1^2)$ and $N(\mu_2, \sigma_2^2)$ distributions, since $(n_i - 1)S_i^2/\sigma_i^2$ follows a χ^2 distribution with $n_i - 1$ degrees of freedom, $i = 1, 2$. This ratio has the F distribution on $n_1 - 1$ and $n_2 - 1$ degrees of freedom.

Notice that this distribution is indexed by two degrees of freedom parameters (and the order in which they are written is important). It, too, is a skewed distribution for continuous random variables taking positive values. Provided $v_2 > 2$, $E(F) = v_1/(v_2 - 2)$, which is approximately 1 when $v_1 = v_2$ (except for small v_1, v_2).

The F distribution can be used for hypothesis tests and confidence intervals concerning the ratio of two normal variances, as in Exercise 2.12. The convention used in this book is that a two-sided F significance probability is twice the SP in the obtained direction. The latter is $P(F > s_1^2/s_2^2)$ where the sample variances are labelled so that their ratio is greater than one.

Exercise 2.12

Test the hypothesis that the variances are equal for the two groups of differences in the hcii data (assuming normality). How would the result of this test have affected your answer to Exercise 2.10(a) had you performed it then?

2.5 Bernoulli, binomial and Poisson distributions

In this section, you are reminded of some basic distributions for discrete data. These will also play an important role in this book.

2.5.1 The Bernoulli distribution

A binary random variable is one that takes values 0 or 1 only. The zeros and ones are usually codes for no or yes, failed or succeeded, heads or tails, lived or died, etc. The values 0 and 1 could also be a recoding of 1 and 2 or any other pair of numbers. It is immaterial which is called zero and which is called one, although 0 is often associated with 'failure' and 1 with 'success'.

There is only one possible distribution for a binary random variable and this is the *Bernoulli distribution* with parameter p, written Bernoulli(p):

$$P(X = 1) = p, \quad P(X = 0) = 1 - p, \quad 0 \le p \le 1.$$

Exercise 2.13

Table 2.5 gives the results of an experiment to see whether bees have an affinity for flowers whose petals have lines. Before the experiment, a number of bees were trained to feed off sugar syrup located on a glass table under which were a number of irregularly shaped flowers. After the bees had become accustomed to this, the original flowers were replaced by 16 new flowers, 8 with lines (which in nature might be serving as 'nectar guides') and 8 without. The data are coded 1 for each bee feeding off a lined flower, 0 for each feeding off a plain flower. A total of 107 bees fed off the test table.

Table 2.5   For each bee, 0/1 denotes feeding off plain/lined flower.

1	1	1	0	0	1	1	0	1	0	0	1	1	1	0	1	0	1	1	0	1	1	0			
1	1	0	0	1	1	0	0	1	1	1	1	1	0	0	1	1	0	1	0	1	0	1			
1	0	1	0	0	0	0	0	0	1	0	1	1	1	1	0	1	1	0	1	1	1	1			
0	0	1	1	1	1	0	0	1	0	1	1	1	1	1	0	1	1	1	1	1	1	1			
0	1	0	0	1	1	1	1	0	0	0	0	1	1	1											

Source: Free, J. B. (1970) 'Effect of flower shapes and nectar guides on the behavior of foraging honeybees', *Behavior*, **37**, 269–85.
Dataset name: honeybee.

The natural estimator of p is the proportion of ones that occur in the dataset. What is your estimate of the probability that a randomly chosen bee feeds off a lined flower?

2.5.2   The binomial distribution

If the outcomes of *Bernoulli trials* (like those for bees above) are independent of one another and the value of p is the same for all trials, then the random variable X that denotes the number of ones, usually called the number of *successes*, follows a *binomial distribution*. This has a known parameter n, the number of trials, and an unknown parameter p, the success probability. We write $X \sim B(n, p)$, where X can take values $0, 1, \ldots, n$.

The *probability mass function* of the $B(n, p)$ distribution is

$$p(x) = P(X = x) = \binom{n}{x} p^x (1 - p)^{n-x}, \quad x = 0, 1, \ldots, n$$

where

$$\binom{n}{x} = \frac{n!}{x!(n - x)!}.$$

The mean and variance of this distribution are

$$E(X) = np \quad \text{and} \quad V(X) = np(1 - p).$$

If $p = \frac{1}{2}$, the binomial distribution is symmetric; otherwise it is not.

The Bernoulli(p) distribution for a single Bernoulli trial is the same as the $B(1, p)$ distribution.

Example 2.1
Given that it may be reasonable to assume that the bees in Table 2.5 act independently, the more usual way of presenting these data would simply be to say that, of the 107 bees in the experiment, 66 fed off the lined flowers. That is, the data yield a *single* observation $x = 66$ from a binomial distribution with $n = 107$ and unknown p. ■

It is also common to obtain a random sample $X_1, X_2, \ldots, X_k$ say, where each X_i is taken independently from the $B(n, p)$ distribution, but we shall not consider this here. Later in the book we shall be concerned with situations where each X_i comes independently from a different binomial distribution, with parameters n_i (known) and p_i (unknown).

For a single realization X from the $B(n, p)$ distribution, the natural estimator of p is $\widehat{p} = X/n$.

Exercise 2.14
What are the formulae for the mean and variance of $\widehat{p}$?

A $100(1 - \alpha)\%$ confidence interval for p is obtained by solving (by computer) the pair of equations

$$P(X \leq x) = \tfrac{1}{2}\alpha, \quad P(X \geq x) = \tfrac{1}{2}\alpha. \tag{2.3}$$

Part (a) of the following exercise should not be attempted by hand, but you may be able to answer it if you are using appropriate statistical software. Any reader should be able to do part (b) if they look at the solution to part (a) first.

Exercise 2.15
(a) Obtain a 95% confidence interval for the proportion of bees preferring lined flowers.

(b) If the bees were choosing flowers at random, p would be 0.5. Does your confidence interval give any evidence that the bees are not selecting at random in this case?

2.5.3   The Poisson distribution

While by no means the only distribution for discrete data with a fixed upper limit (given by n), the binomial distribution is none the less an extremely common model for such situations and a first port of call for modellers in many such cases. Likewise, for *count* data, in which the data are discrete but can take any non-negative integer value without upper limit, the 'default' modelling distribution is the *Poisson distribution*. The Poisson distribution has parameter $\lambda > 0$ and probability mass function

$$p(x) = P(X = x) = \frac{\lambda^x e^{-\lambda}}{x!}, \quad x = 0, 1, \ldots .$$

If $X \sim \text{Poisson}(\lambda)$, $E(X) = V(X) = \lambda$.

The usual estimator for λ from a random sample $X_1, X_2, \ldots, X_n$ is the sample mean:

$$\widehat{\lambda} = \frac{1}{n} \sum_{i=1}^{n} X_i.$$

Exercise 2.16
For each of the 365 days in 1971, Table 2.6 gives the observed frequency distribution of deaths in one Montreal hospital, and Figure 2.14 is a bar chart giving the data in graphical form. (Notice that it is much more convenient to give the data this way than as a string of 365 individual numbers.)

Table 2.6 Deaths per day in a Montreal hospital.

Number of deaths	Number of days on which this many deaths were observed
0	220
1	113
2	23
3	8
4	1
>4	0
Total	365

Source: Zweig, J. P. and Csank, J. Z. (1978) 'The application of a Poisson model to the annual distribution of daily mortality at six Montreal hospitals', *Journal of Epidemiology and Community Health*, **32**, 206–11. Dataset name: hospdth.

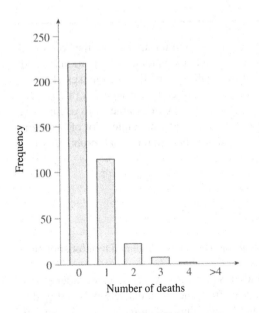

Fig. 2.14

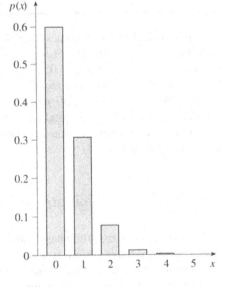

Fig. 2.15

The simplest type of model for $X_1, X_2, \ldots, X_{365}$, the numbers of deaths on days $1, 2, \ldots, 365$, treats them as being realizations of independent random variables. If we also assume that these 365 days are typical of other days, at least in the same hospital at around the same time, we can treat the data as a random sample of count data. So we might reasonably consider modelling the data with a Poisson distribution. (Note that the assumptions here may not be valid, as there could be structure to the data that is not evident from the frequency distribution. For example, perhaps the deaths occur in clusters, for instance in the coldest part of the winter. We would need the full string of 365 numbers in order to look into this further.)

(a) Calculate the sample mean. This is your estimate of λ.

(b) Figure 2.15 is a plot of the Poisson distribution corresponding to $\lambda = \overline{X}$. From a comparison of this plot with the bar chart in Figure 2.14 does the fitted Poisson distribution seem to be a good model for these data?

(c) A further (partial) check of the Poisson model is to calculate the sample variance. Recalling that, under the Poisson model, the mean and variance are the same, what does the sample variance tell you about the Poisson model for the data?

Confidence intervals for the Poisson parameter can also be obtained, by computer, via equations (2.3). Exercise 2.17 should be attempted only if you have appropriate software to hand.

Exercise 2.17

Assuming a Poisson model, obtain a 90% confidence interval for the mean number of deaths per day in this Montreal hospital.

If p is small, then, as n becomes large, the $B(n, p)$ distribution approaches the Poisson(λ) distribution with $\lambda = np$. This Poisson approximation to the binomial distribution is good, roughly speaking, when $p \leq 0.05$ even for quite small n. This link also makes clear the Poisson distribution's role as a model for *rare events*: a rare event is an event that is itself rather unlikely to happen (small p) but has a lot of opportunities (large n) to occur. One can think of the deaths in hospital example in this way: there are quite a lot of people in the hospital on any one day, but each person individually has a rather small probability (we hope!) of dying that day.

2.6   Maximum likelihood estimation

In this book, as is the case for much of statistical science, the fundamental tool for statistical inference will be the *likelihood function*.

Suppose for a moment that we have a random sample $X_1, X_2, \ldots, X_n$ of independent discrete data, where each X_i is assumed to arise from the same discrete probability distribution with probability mass function $p(x; \theta)$ depending on an unknown parameter θ.

For each given value θ_0 of θ, the probability of the observed data $x_1, x_2, \ldots, x_n$ under the model $p(x; \theta_0)$ is given by the *likelihood*

$$L(\theta_0) = p(x_1; \theta_0)p(x_2; \theta_0) \ldots p(x_n; \theta_0).$$

L is the likelihood function (for the discrete case), a function of θ treating the data as given. If θ_0 is an inappropriate parameter value, the probability of the observed data under the probability model $p(x; \theta_0)$ – i.e. the value of the likelihood function – will be small; conversely, if θ_0 is close to a value for θ under which the data do indeed appear to have been generated, then the value of the likelihood function will be (relatively) large. In this way, the likelihood is a measure of how reasonable is (the assumed probability model using) each candidate value of θ as a model for the data.

The continuous analogue of the above involves the model density function $f(x; \theta)$. For each given value θ_0 of θ, the likelihood is then

$$L(\theta_0) = f(x_1; \theta_0)f(x_2; \theta_0) \ldots f(x_n; \theta_0).$$

If we now employ f to denote both the continuous density and discrete mass functions, we can talk of

$$L(\theta) = f(x_1; \theta)f(x_2; \theta) \ldots f(x_n; \theta)$$

as defining the likelihood function for either case.

The best point estimate of θ from the likelihood perspective is therefore that value of θ that maximizes the likelihood. This gives the *maximum likelihood estimator*, denoted $\widehat{\theta}$.

It is very often useful to transform L by taking logs, so that the *log-likelihood function* is given by

$$l(\theta) = \log L(\theta) = \sum_{i=1}^{n} \log f(x_i; \theta).$$

The value of θ that maximizes the log-likelihood function $l(\theta)$ is the same as the value that maximizes the likelihood function $L(\theta)$, and so is also given by $\widehat{\theta}$. Taking logs does not change the point of maximization, but it usually makes calculations simpler.

It turns out that, with one exception, each of the point estimators that has been mentioned in this chapter is also a maximum likelihood estimator, or mle for short. This goes for $\overline{X}$ as estimator of μ in the $N(\mu, \sigma^2)$ model; for the sample proportion X/n as estimator of p in the $B(n, p)$ model; and for $\overline{X}$ as estimator of λ in the Poisson(λ) model. The only estimator that we have used that is not the maximum likelihood estimator is S^2 as estimator of σ^2 in the $N(\mu, \sigma^2)$ model.

Exercise 2.18

The mle of σ^2 is

$$\widehat{\sigma}^2 = \frac{1}{n} \sum_{i=1}^{n} (X_i - \overline{X})^2.$$

Calculate the value of $\widehat{\sigma}^2$ for the hald and forearm datasets, using the values of s^2 given in Exercise 2.5. Is there much difference between the values of $\widehat{\sigma}^2$ and s^2 for these datasets? How does any difference between $\widehat{\sigma}^2$ and s^2 depend on n?

Maximum likelihood estimators have desirable theoretical properties. First, they are generally *asymptotically unbiased*. That is, [3]

$$E(\widehat{\theta}) \rightarrow \theta \quad \text{as} \quad n \rightarrow \infty.$$

What is more, the mle is often exactly unbiased (i.e. $E(\widehat{\theta}) = \theta$); this is true, for example, for estimating the means of normal, binomial and Poisson distributions. The quantity $E(\widehat{\theta}) - \theta$ is the *bias*.

Exercise 2.19

The mle of σ^2 in the $N(\mu, \sigma^2)$ model is not exactly unbiased. What is the immediate justification for this statement? What is the bias of $\widehat{\sigma}^2$? How does the bias behave for large n?

Hint   Use the relationship between s^2 and $\widehat{\sigma}^2$ in Solution 2.18.

Maximum likelihood estimators are also generally *consistent*, meaning that, under certain conditions, as well as being asymptotically unbiased, they satisfy

$$V(\widehat{\theta}) \rightarrow 0 \quad \text{as} \quad n \rightarrow \infty.$$

Typically, as n gets large, $V(\widehat{\theta})$ is proportional to $1/n$.

Yet other desirable asymptotic properties, which translate to desirable approximate properties for realistic n, follow, provided certain not-too-restrictive conditions are met. For example, $V(\widehat{\theta})$ is (under certain conditions) smaller than the variance of any alternative estimator. Moreover, very often mles are actually approximately normally distributed. Also, conditions such as all X_is being assumed to arise from the same probability distribution can be relaxed. (You are not expected to know how to derive any of these theoretical properties of mles, nor to understand them in any depth.)

What all of this translates to is that use of maximum likelihood facilitates statistical inference: likelihood-based inference affords point estimators with good properties and affords the ability to assess the variability of estimators. This theme will recur, perhaps not always explicitly, in later chapters of the book.

2.7   The central limit theorem

If $X_1, X_2, \ldots, X_n$ are n independent and identically distributed random observations from *any* population with mean μ and variance σ^2, then for large n the distribution of their mean $\overline{X}$ is approximately normal with mean μ and variance σ^2/n, i.e.

$$\overline{X} \approx N(\mu, \sigma^2/n).$$

This is the famous *central limit theorem*, or CLT.

The central limit theorem drives many useful approximations in statistics and is largely responsible for the ubiquity of the normal distribution in statistical modelling. This is

---

[3] The symbol $\rightarrow$ is read as 'tends to'.

because it can often be conceived that observations are themselves averages of a variety of unobserved contributing effects. It is the CLT that figures prominently in the limiting distributional results for maximum likelihood estimators just mentioned.

For example, for large n, the binomial distribution can be approximated by a normal distribution. In fact, the $B(n, p)$ distribution can be approximated by a $N(np, np(1 - p))$ distribution, a normal distribution with the same mean and variance as the binomial distribution itself. This is so because the binomial distribution can be thought of as the distribution of a sum of n Bernoulli random variables, each with mean p and variance $p(1 - p)$, and so we can apply the sum version of the CLT:

$$T = n\overline{X} \approx N(n\mu, n\sigma^2).$$

Likewise, conceptualizing the Poisson(λ) distribution as a sum of n Poisson(λ/n) distributions leads to a $N(\lambda, \lambda)$ approximation to Poisson(λ). To obtain more accurate approximations to probabilities, there are 'continuity corrections' which will not be discussed here.

A particular use of such approximations is to yield approximate confidence intervals for binomial and Poisson parameters. These were particularly important historically before computer power readily provided exact confidence intervals for these distributions. For instance, for large n an approximate $100(1 - \alpha)\%$ confidence interval for the binomial parameter p is

$$\widehat{p} \pm z_{1-\alpha/2}\sqrt{\frac{\widehat{p}(1 - \widehat{p})}{n}}, \qquad (2.4)$$

where $\widehat{p}$ is the maximum likelihood estimator of p and where $z_{1-\alpha/2}$ is the upper α percentage point of the standard normal distribution itself and not of Student's t distribution. Notice how the unknown variance of $\widehat{p}$ has been estimated in a natural way as $\widehat{p}(1 - \widehat{p})/n$.

Exercise 2.20

(a) What is the approximate distribution, for large n, of the sample mean $\widehat{\lambda} = \overline{X}$ of a random sample from the Poisson distribution? Hence, derive, in analogous manner to (2.4), an approximate formula for a $100(1 - \alpha)\%$ confidence interval for the mean λ of the Poisson distribution.

(b) Obtain such an approximate 90% confidence interval for the mean number of deaths per day in the Montreal hospital using hospdth. What value does the appropriate standard normal percentage point take? How does this approximate confidence interval compare with the exact one obtained in Exercise 2.17?

What remains extremely useful nowadays is that the CLT continues to give an approximate confidence interval for the mean μ of a population even when no standard distributional assumption can be made. Such an approximate $100(1 - \alpha)\%$ confidence interval is

$$\overline{x} \pm z_{1-\alpha/2}\frac{s}{\sqrt{n}}$$

where $\overline{x}$ is the sample mean, s is the sample standard deviation and $z_{1-\alpha/2}$ is the upper α percentage point of the standard normal distribution.

Exercise 2.21

Table 2.7 shows the first few values in the dataset `smith`. Like `hald`, these data are the results of strength testing, this time of 1.5 cm glass fibres. Unlike `hald`, a normal distribution does not fit well, and the quest for an alternative suitable distribution leads us beyond the knowledge assumed here.

Table 2.7

0.55	0.74	0.77	0.81	0.84	...

Source: Smith, R. L. and Naylor, J. C. (1987) 'A comparison of maximum likelihood and Bayesian estimators for the three-parameter Weibull distribution', *Applied Statistics*, **36**, 358–69.
Dataset name: `smith`.

However, approximate confidence intervals for the mean glass fibre breaking strength can be found. The sample size, sample mean and sample variance are 63, 1.507 and 0.1051 respectively. Hence, find an approximate 95% confidence interval for the population mean.

In fact, since $z_{0.975} = 1.96 \simeq 2$, an approximate 95% confidence interval based on the CLT is given by

$$\bar{x} \pm 2 \frac{s}{\sqrt{n}}.$$

This result will be used, usually implicitly, time and time again in this book.

2.8   Categorical and quantitative variables

You should be familiar with the distinction between continuous and discrete random variables. We shall use this distinction when dealing with the response variables in our linear and generalized linear models. However, when considering the explanatory variables in those models, a second distinction will be more important. That distinction is between *quantitative* and *categorical* variables.

Quantitative variables assign meaningful numerical values to observations. Thus, height in metres, weight in kilograms, the number of children in a household, and the dosage of a hypertension treatment in milligrams are all quantitative variables. Notice that quantitative variables can be either continuous or discrete.

Categorical variables, on the other hand, assign observations to categories. The **type** explanatory variable in the gvhd dataset in Chapter 1 was one of these. In this case, patients had one of 'acute myeloid leukaemia', 'acute lymphocytic leukaemia', or 'chronic myeloid leukaemia'. The **type** variable is categorical because all we know about it is that there are different types of leukaemia. Categorical data may well be coded numerically, e.g. **type** was assigned values 1, 2 and 3, but the choices of numbers are arbitrary and are not a meaningful numerical measurement.

Indeed, 'categorical' itself divides up into two further classes, *nominal* and *ordinal*. In the ordinal case, the categories have an ordered relationship one with another, going from 'least something' to 'most something' (or vice versa); nominal variable categories lack any obvious order.

The variable **type** provides an example of nominal categorical data. Other examples might be different drugs being compared in a clinical trial (e.g. different painkillers being assessed for their effectiveness at alleviating headaches), or different types of grape being compared in a wine-tasting experiment, or a range of different hair colours (e.g. grey, black, brown, red, blonde). (In fact, binary variables can be thought of as special cases of nominal categorical data, but in this section we concentrate on situations with three or more categories.)

If, in the leukaemia study, **type** had instead been a measure of severity of one particular kind of leukaemia, e.g. 'mild', 'moderate', 'severe', then it would have had an ordinal form. If, in the clinical trial example, the three treatments were actually all the same painkiller but at 'low dose', 'medium dose' and 'high dose', this too would be an example of an ordinal variable. Other examples might be class of honours degree ('first', 'upper second', 'lower second', etc.) or quality of leisure pursuits (assessed subjectively as 'good', 'bad', or 'intermediate'). Ordinal categorical variables are usually assigned convenient numerical values too, and these values will respect the ordering (e.g. bad = 1, intermediate = 2, good = 3), but the precise values attached remain rather arbitrary. (Remember, however, that if known numerical values can be assigned to the variables – e.g. if, in the clinical trial example, the painkiller treatments had known numerical doses – then the variables would be quantitative.)

Exercise 2.22

Which of the following variables are nominal, which are ordinal, and which are quantitative?

(a) Tonsil size in schoolchildren measured as 'normal', 'large' and 'very large'.

(b) A numerical measurement of tonsil size in, say, millimetres.

(c) In answer to a question on the desirability of divorce compared with staying together in situations of marital difficulty: 'much better to divorce', 'better to divorce', 'don't know', 'worse to divorce', 'much worse to divorce'.

(d) The religion of a respondent to a survey: 'Buddhist', 'Christian', 'Hindu', 'Jewish', 'Muslim', 'other', 'none'.

(e) The number of flies caught in one day on a piece of fly paper.

(f) Snoring frequency measured as 'non-snorers', 'occasional snorers', 'snore nearly every night', 'snore every night'.

(g) In an investigation of salaries of graduates: time since completion of bachelor's degree.

(h) In an opthalmological experiment, colour of eyes: 'brown', 'blue', 'grey', 'green'.

(i) The variable **donmfp** in the example in Chapter 1, in which a male donor was coded '0', a female donor who had never been pregnant was coded '1', and a female donor who had been pregnant was coded '2'.

Summary of methodology

1 You were reminded of the major role of the normal distribution in statistics; of its basic properties; of probability plotting for checking fit and transformations to make data 'more normal'; and of the central limit theorem, which says that the limiting distribution of a sample mean is normal with mean μ and variance σ^2/n whatever the original distribution.

2 You were reminded of other probability distributions: the Bernoulli, binomial and Poisson distributions for discrete data and the t, χ^2 and F distributions for continuous data.

3 You were reminded about confidence intervals. In particular, confidence intervals for the mean of a normal distribution and for the mean of a general distribution (using the CLT) were discussed.

4 You were reminded of hypothesis testing, in particular the approach in which SPs are obtained. The one-sample t test (which can also be applied to matched pairs) and the two-sample t test are specific important examples.

5 You were reminded of maximum likelihood estimation, and the role of likelihood-based inference was stressed.

6 A distinction between quantitative and categorical variables was made. Categorical variables can further be subdivided into nominal and ordinal variables.

3

Introduction to GENSTAT

In this chapter, you will learn the basics of using the GENSTAT statistical software package. Before starting on the chapter, you should have installed GENSTAT on your computer.

The main aim of the chapter is to show you how to do, in GENSTAT, the data analyses that have already been covered in Chapter 2. The chapter should prepare you for the rest of this book, in which GENSTAT will be used to perform all the statistical computing.

3.1 Getting started

The exercises in this chapter are essentially all stand-alone: you can leave GENSTAT at the end of each exercise; but you can also stay in GENSTAT, ready for the next exercise. This allows you to do the exercises one at a time or many in one go, depending on what is convenient for you. Since you have this choice, you will not be told explicitly to start GENSTAT at the beginning of each exercise nor to leave GENSTAT at the end of each exercise, except in Exercise 3.1. If you leave GENSTAT after certain exercises, you may need to re-perform certain tasks from previous exercises before you can continue.

If something goes wrong, a message will appear on the screen saying 'Faults have occurred' or something similar. If this happens, then just press <Return> (or click on the OK button in the **Fault** dialogue box) to get rid of the message, then try to work out what went wrong and have another go. Occasionally, you might get really tangled up; if you do, then stop GENSTAT and start again.

The exercises in this section are intended to be done in order. The instructions in one exercise will assume that you know how to do tasks introduced in earlier exercises. Note, however, that many of the GENSTAT instructions given in the book are not the only ways of achieving certain ends. If you know of alternative ways that you prefer, please do use them.

The first thing to learn with a new computer package is how to start it. Once you can start the package, you can then try to get out of it!

Exercise 3.1
Start Windows, if it is not already running, and start GENSTAT for Windows. A **Genstat 5** window should appear.

(a) What can you see and is there anything that you know, or can guess, how to use? Do not worry if you have no idea what the things that have appeared do. It will not be assumed that you know such things; nevertheless you should always try to work out what something new does.

(b) There should be a list of words at the top of the **Genstat 5** window. This is called the *menu bar*. Most of the commands that you will give GENSTAT will be given using the menu bar. Use the mouse to move the pointer so that it is over the menu bar, and click. What happens? (If nothing happens when you click, then try moving the pointer to somewhere over one of the words on the menu bar, rather than to one of the ends of the menu bar.)

(c) Click with the pointer positioned over the highlighted word on the menu bar. What happens when you click this time?

(d) Now, move the pointer over the menu bar again and click on one of the words. This time, move the pointer along the menu bar. What happens?

Finally, here is a task that you can do using the menu bar. This is one of the ways to shut GENSTAT down.

Move the pointer over the word **File** on the menu bar. If no menu is showing, click to make one appear. One of the items in the menu should be **Exit**. Moving the pointer over **Exit** and then clicking should start the shutdown process. Try clicking on **Exit**.

A dialogue box will appear; this is something that will happen many times in your use of GENSTAT. (If you had used some of the other features of GENSTAT, you would have been asked further similar questions in other dialogue boxes.) You are being asked whether you want to save the text appearing in the **Output** window. Since there isn't anything interesting there, move the pointer over the button labelled **No** and click. Next, the same happens for the **Input Log** and you should respond in the same way.

Having done this, you are asked whether you really want to shut GENSTAT down, which is called *exiting*. To confirm that you want to exit, you have to click on the **Exit** button. You are asked this just in case you clicked on **Exit** by accident. If you did not want to exit, then you would click on the **Cancel** button (either in this box or in one of the boxes asking you about saving earlier on).

If you have not already done so, click on the **Exit** button.

As well as getting in and out of GENSTAT, you also saw the menu bar in Exercise 3.1. Virtually all the tasks that you will do using GENSTAT start off from the menu bar. It would be a bit clumsy to keep instructing you to move the pointer onto the words on the menu bar and then click, and so on. Things such as

Click on the **File** menu; click on **Exit**

or

Click on **Exit** from the **File** menu

will be written instead.

Exercise 3.2

(Start GENSTAT again, if you have not already done so.)

(a) Click on the **Options** menu. Some of the options on the menu will be ticked, i.e. have a ✓ next to them. This means that those options are switched on. Which options are switched on?

(b) Now, click on **Show Toolbar**. What happens?

(c) Click on the **Options** menu again. How has the **Options** menu changed since the last time it was visible?

(d) How do you think the **Options** menu can be changed back to how it was last time? Try the method that you have thought of for changing back the **Options** menu. You can check your answer by looking at Solution 3.2 before you try it, if you like. If you do not check and you do the wrong thing, then no harm will be done: if the method you try does not work, you can try another idea or just click on **Exit** from the **File** menu and start this exercise again.

While you were doing Exercise 3.2, you probably worked out that the row of pictures below the menu bar is called the *toolbar*. The strip at the bottom of the **Genstat 5** window, which has the word Output in it when GENSTAT starts, is called the *statusbar*. We can make the statusbar appear and disappear by using the **Options** menu, in the same way that we made the toolbar appear and disappear in Exercise 3.2.

It is possible to use the statusbar to find out what the buttons on the toolbar do. Exercise 3.3 is about how to use the toolbar, including its connection with the statusbar.

Exercise 3.3

Make sure the statusbar is shown in your **Genstat 5** window.

(a) Move the pointer so that it is over the **A** button on the toolbar. The next bit is difficult to do at the same time as reading these instructions, so read to the end of the next paragraph before attempting to follow the instructions.

Keeping the pointer over the **A** button, press the mouse button and keep it held down. Look at the statusbar, and remember what the words on the statusbar say. Move the pointer away from the **A** button and release the mouse button.

What were the words shown on the statusbar when you held the mouse button down? What does the statusbar show now?

(b) Now, we shall try to do something using the toolbar.

Make sure that the **Output** window is active: if the **Output** window is not active, just move the pointer to the title bar at the top of the **Output** window and click on it. (If the **Output** window is active, then the title bar on the **Output** window will be highlighted and the statusbar will contain the word Output.)

Have a look at the text in the **Output** window and then click on the **A** button. A dialogue box will appear; it will be referred to as the **Font** dialogue box (because that is the title given in its title bar). You are being asked how you would like to change the font used for the **Output** window. A font determines the appearance of the characters in the text: for example **this is a bold font** and *this is an italic font*. You can change the font used in the **Output** window to suit yourself.

There are three characteristics of the font that can be changed: **Family**, **Size** and **Style**. Start by changing the style. What styles are available?

The current font style will be highlighted and is probably **normal**. Change the style to **bold**; to do this you should click on **bold**. At this stage you could try altering the family and size, but for now just click on the **OK** button. How has the appearance of the text in the **Output** window changed?

Try the other styles: how do they change the appearance of the text in the **Output** window?[1]

(c) You can now try changing the font family. In the **Font** dialogue box, under the word **Family**, there should be the name of a font family – perhaps **Courier** on your computer – next to a button with a symbol on it, which is a type of *downarrow*.

What happens when you click on this downarrow button?

Now, click on **Terminal** and then click on the **OK** button. (If you cannot see **Terminal**, then click on the downarrow button again.) How does the appearance of the text in the **Output** window change?

(d) The final characteristic of a font that can be changed is its size. The font size is changed by the same method as the font family; you click on the downarrow button under the word **Size** in the **Font** dialogue box. The number of font sizes available varies between font families.

Try different combinations of font family, font size and font style. Identify the combination that you prefer and make a note of it. Which combination of family, size and style do you prefer? You are advised not to choose font sizes bigger than about 60; larger fonts might take a long time to be displayed (and are no use anyway!).

(e) Most of the buttons on the toolbar are just quick ways to do things that are available through the menu bar. You can use the methods introduced in Exercise 3.1 to work out how to change the font using the menu bar instead of the toolbar.

How do you obtain the **Font** dialogue box using the menu bar, rather than the toolbar?

A menu that is produced when you click on a downarrow button is often referred to as a *pull-down menu*.

Exercise 3.3 introduced the idea that you can change the appearance of GENSTAT to suit yourself. GENSTAT automatically remembers changes you make to its appearance and behaviour when it restarts. You could now arrange the appearance of the **Genstat 5** window to suit you. One of the things you might like to do is to make the **Genstat 5** window and/or the **Output** window larger.

---

[1] You click on the **A** button to get the **Font** dialogue box back again.

From now on instructions will be given using the menu bar not the toolbar. If you know the equivalent button from the toolbar that does what has been described, then feel free to use the toolbar instead.

You should set up GENSTAT's *working directory* correctly as follows. Click on the Options menu; click on **Working Directory**. A dialogue box will appear: change the Working Directory to c:\lmgen; click on the **OK** button.

You have now seen most of the methods of giving GENSTAT instructions. Much of your use of GENSTAT will be with the methods that you have seen already. However, as the final part of this section, we shall look at another way of giving instructions; this method is called *interactive mode*. First, why might you need interactive mode?

The menu system in GENSTAT is just a way of telling GENSTAT what you want to do. Once you have told GENSTAT what you want to do, it converts your instructions into GENSTAT *command language*. The command language instructions are then fed to the GENSTAT *server*,[2] which then carries out your instructions. In interactive mode, you can type your instructions in GENSTAT command language, without using the menu system. We shall occasionally want to use interactive mode because there are some things in this book that can only be done in this way.

Exercise 3.4 uses interactive mode to perform some simple arithmetical and statistical calculations.

Exercise 3.4

(a) Make the Input Log window active (i.e. click on its title bar so that it is highlighted). (You might have to restore the Input Log window if it has been minimized.) Click on the Options menu; click on **Interactive**. The Input Log window itself changes colour when it is in interactive mode.[3]

Now, type the following instruction.

```
PRINT 1+2
```

You can type lower-case rather than capital letters for commands such as PRINT, if you prefer. You are expected to press < Return > at the end of the instruction, otherwise nothing will happen.

What happens in the **Output** window after you type this command?

(b) For a more useful calculation, type

```
PRINT ednormal(0.975)
```

and press <Return>. This command gives what GENSTAT calls the *effective deviate* (ed) of the standard normal distribution. The effective deviate is more usually called a quantile, and ednormal(0.975) gives the 0.975 quantile of the standard normal distribution, i.e. the value of z such that $\Phi(z) = 0.975$ where Φ is the standard normal distribution function. Normal quantiles are mentioned in Section 2.1.2 of this book, and t quantiles in Section 2.2.1. What is the answer?

---

[2] You may see an item labelled **Genstat Server** on your computer screen. If you close it, then most of your instructions will not work. If this happens accidentally, then do the following. Click on the Options menu; click on **Restart Server**. Alternatively, exit from GENSTAT and restart it.

[3] It is also possible to put the Output window into interactive mode, but we shall not be doing so.

You can edit an instruction in the Input Log window using the keyboard in the usual manner, after moving to the desired spot either by using the cursor keys or by clicking at the place you want to move to. (The *cursor keys* are the four keys marked ↑, ↓, ← and →.) Enter the following instruction either by editing ednormal(0.975) or by typing it in from scratch.[4]

```
PRINT ednormal(0.975,0.5,0.95,0.025)
```

What does this command do?

(c) Now try the command

```
PRINT edt(0.975;10)
```

which should give the 0.975 quantile of the $t(10)$ distribution, which is 2.228. Type a semicolon (;) here not a comma. The edt function has two arguments (separated by a semicolon): the first is the quantile you want to find; the second is the degrees of freedom of the t distribution.

Type the following commands.

```
PRINT edt(0.975;10,20)
PRINT edt(0.975,0.95;10)
PRINT edt(0.975,0.95;10,20)
PRINT edt(0.975,0.95,0.975,0.95;10,10,20,20)
```

Which quantiles are calculated for each command? How does GENSTAT know which values correspond to degrees of freedom and which correspond to the quantiles to find?

(d) Now try these commands.

```
PRINT edt(0.975,0.95;10,20,30)
PRINT 1,2+3,4,5
```

What is the rule for deciding which quantile is worked out for each different choice of degrees of freedom, or for choosing which pair of numbers to add together?

Leave interactive mode. You leave interactive mode using the **Options** menu.

The calculations in Exercise 3.4 could also have been performed using the menu system: a **Calculate** dialogue box pops up if you click on **Calculations** on the **Data** menu. You might like to play around with this and decide for yourself whether to use interactive mode or the menu in calculations like this.

A particular convention has been used in GENSTAT commands here, with respect to the use of upper- and lower-case letters. We have chosen to use upper-case letters for command words, such as PRINT. However, this is just for clarity. In most respects, GENSTAT is not case-sensitive and doesn't distinguish between upper and lower case. In some circumstances (see Exercise 3.5), however, the case does matter.

---

[4]Do not forget to press <Return> after completing the editing/typing.

The PRINT commands in the above exercise did not result in anything being printed on your printer; they displayed things in the **Output** window. If you want to print the contents of one of the GENSTAT windows on your printer, make sure it is installed, connected and switched on, and make sure the window you want to print is active.[5] Click on the **File** menu; click on **Print**. The window will be printed in what GENSTAT thinks to be an appropriate way. Note that, if you print a window with a lot of text in it (e.g. the **Output** window) in this way, GENSTAT will print the whole contents of the window, including all the text that has scrolled out of sight. If you wish to print just part of what is in the window, you must first either delete the bits you don't want, or you must copy the material you do want into another window or application and print that.

Occasionally in this book, you will be asked to save output for later reference. Printing is one way you can do this. An alternative is to make sure the **Output** window is active; click on the **File** menu, click on **Save**; and in the resulting dialogue box enter an appropriate filename. The contents of the **Output** window will then be saved as a text file, which you can open again from GENSTAT or from a wordprocessing application.

3.2 Loading, storing, retrieving and manipulating data

The things that we have done in GENSTAT so far have mostly not been especially statistical. The main reason for this is that you do not know how to give data to GENSTAT yet. This section is about:

- how to load, read, input or enter data into GENSTAT;

- how to store or save data for later use;

- how to perform calculations with your data.

Once you can perform these tasks, we can concentrate on doing statistics.

In Chapter 4 we will examine a small set of data on gas consumption in some houses before and after cavity wall insulation. The data are given here as Table 3.1.

Table 3.1

Before insulation	After insulation
12.1	12.0
11.0	10.6
14.1	13.4
13.8	11.2
15.5	15.3
12.2	13.6
12.8	12.6
9.9	8.8
10.8	9.6
12.7	12.4

Exercise 3.5 is about how to load these data into GENSTAT and how to store them for future use. Further details about what these data represent, and their analysis, are deferred until Chapter 4.

---

[5]This method does not work for graphics windows. How to print these is dealt with in Section 3.3.

Exercise 3.5

(a) Click on the **Spread** menu; click on **New**; click on **Blank**.

A dialogue box will appear. The data in Table 3.1 consist of ten rows and two columns, so you need to change the **Number of Columns** to 2. A simple way to do this is to click on the value for **Number of Columns** and edit the value using your keyboard. Change the **Number of Columns** to **2**. Change the **Number of Rows** to **10**, if necessary. Set the **Sheet Type** to **Vector**, if necessary; the simplest way to do this is to click on **Vector**. (If you click on **Scalar** while you are trying out things, then beware that it resets the **Number of Rows** to 1. If this happens, just click on **Vector** and then change the **Number of Rows** to **10**.) Now, click on the **OK** button.

What happens after you click on the **OK** button?

(b) The names C1 and C2 are not very informative, so you should change them. To change the name for the second column from C2 to **after**, do the following.

- Click on any of the cells in the second column – this will make the second column the *current* column. A *cell* is one of the rectangular spaces that contains an asterisk-like symbol at the moment, and will eventually contain a number.

- Click on the **Spread** menu, click on **Column**, and move the pointer over **Edit Attributes** and release the mouse button. A dialogue box will appear: change **Name** from C2 to **after**, using the method that you used to change the **Number of Columns** in part (a).

- Click on the **OK** button.

  Note that the column names are *case-sensitive*; in other words, capital (upper-case) letters are distinct from lower-case letters in column names. For example, the names *after* and *After* are totally different as far as GENSTAT is concerned, because one starts with 'a' and the other starts with 'A'.

Change the name for the first column from C1 to **before**.

Why have we called the columns **before** and **after** rather than something like **consumption_before_insulation** and **consumption_after_insulation**? There are two reasons. The first is that, in its statistical calculations, GENSTAT only takes notice of the first eight characters in a column name. The second is simply to save typing!

(c) Make sure that the **Spreadsheet** window is active. Now, you are ready to type in some data.

Click on the cell in the first row of the **before** column. A rectangle should appear inside the cell and the asterisk should be highlighted. Type

    12.1

which is the first value for 'Before insulation' in Table 3.1.

What happens when you press <Return> after you typed 12.1?

(d) Click on the **Spread** menu; click on **Save As**; click on **Genstat GSH file**. A dialogue box will appear: enter the **File Name** as mydata.gsh; click on the appropriate button to save the file, which may be **OK** or may be **Save**, depending on your computer. These instructions assume that the working directory has been set up as described on page 47. If not then click on the **Cancel** button, follow the instructions on page 47 and then start part (d) again. (If you know enough to make your own choice of working directory, you can do so, but you will have to keep track of the directories yourself.)

Why are you being shown how to save the spreadsheet at this stage?

(e) Turn to Table 3.1 and type in the remaining nineteen values. Remember to type < Return > after each one, including the last. If you are interrupted, you should save the spreadsheet; if, after the interruption, you find that someone has switched the computer off, then you can go to Exercise 3.6, which shows you how to load a spreadsheet. You could then load mydata.gsh and type in the remaining values.

Save the spreadsheet again; you will find that when you get to the **Save Genstat Spreadsheet** dialogue box, the **File Name** is still mydata.gsh; but, when you click on the button for saving the file, you will be asked to confirm that you want to replace the existing file – you do want to, so you should click on the **Yes** button. Then check the numbers that you have typed agree with those in Table 3.1. Correct any mistakes that you have made, then save the spreadsheet again. How do you correct mistakes in the spreadsheet?

Close the **Spreadsheet** window by doing the following. Click on the **File** menu; click on **Close**.

You will not be told when to save your work from now on; the frequency with which you save your work depends on you and varies from person to person. You will be told *how* to save your work when we meet new parts of GENSTAT, but not when to save it. Save your work whenever you want to.

Exercise 3.5 produced a spreadsheet called mydata.gsh. Exercise 3.6 is about how to retrieve the data stored in mydata.gsh.

Exercise 3.6

Click on the File menu; click on **Open**. In the resulting dialogue box, use the pull-down menu for **Files of type** to select **Spreadsheet File** (*.gsh). Enter mydata.gsh as the **File name**. You can do this using the method used in part (a) of Exercise 3.5, or alternatively just click on mydata.gsh and the **File name** will change to mydata.gsh. Now, click on the **Open** button to open the file.

You have now retrieved mydata.gsh, but the GENSTAT server still does not know about these data; so you have to tell the server to read the data from the spreadsheet. This is called *updating the server*. To do this, click somewhere in the **Genstat 5** window other than in the **Spreadsheet** window or the menu bar. What happens in the **Output** window?[6] (If nothing happens in the **Output** window, do the following: make sure that the **Spreadsheet** window is active; click on the **Spread** menu; click on **Update Data**.)

If nothing happened at first in the **Output** window, the reason is probably that the **Spreadsheet Options** had been changed from the default. To remedy this: click on the

---

[6]You can bring the **Output** window (or indeed any window) to the front by clicking on it.

Options menu; click on **Spreadsheet Options**; and in the resulting dialogue box, make sure **Auto Update Server** is checked (i.e. is marked with a tick) by clicking on it if necessary; click on the **OK** button. If you retrieved `mydata.gsh` having saved it and closed it before entering all the data, you may find that it has fewer rows and/or columns than it originally had. To restore the original number of rows/columns you can click on **Spread** and **Insert** and use the resulting options.

It can be confusing sometimes to see a lot of rather complicated GENSTAT commands echoed in the **Output** window. We suggest you turn off this echoing. To do this, click on the **Options** menu; click on **Audit Trail**. In the resulting dialogue box, click on **Echo Commands** to remove the tick from the check box. Then click on **OK**.

It is possible to load `mydata.gsh` directly into the server, without using the spreadsheet at all. Exercise 3.7 is about this.

Exercise 3.7
Use the techniques of part (d) of Exercise 3.1 to find a way to load `mydata.gsh` directly. When you think you have found the right method, try it; if it does not work, then look for a different way and try that. If you have time, keep trying until you work out how to do it. If you are short of time, have a couple of attempts, and then read Solution 3.7. How can `mydata.gsh` be loaded directly into the server?

If you have followed the instructions in the Preface to obtain the datasets used in this book, they will have been stored in `c:\lmgen\data`, rather than in `c:\lmgen`, so that it is harder to alter them accidentally. The datasets are stored as GENSTAT spreadsheets. As you will have already noticed, these have names ending in `.gsh`. We shall often refer to datasets just by the filename without the `.gsh` at the end (e.g. `rubber`); but to use such a file in GENSTAT, you need its full filename (e.g. `rubber.gsh`).

Exercise 3.8
(a) Click on the **File** menu; click on **Open**. In the resulting dialogue box, use the pull-down menu for **Files of type** to select **Spreadsheet File (*.gsh)**. The current directory is displayed: it should be `lmgen` (the full pathname is `c:\lmgen`). The data for the book is in the subdirectory `data`, so you have to change the current directory to this subdirectory. There are several ways to do this. One easy one is to double-click on the `data` item in the directory list. Do this. What happens?

(b) Now, change **File name** to `water1.gsh`. How did you do this? (Your computer might use the filename `Water1.gsh` rather than `water1.gsh`; this makes no difference.)

Exercise 3.9
Load `water1.gsh` into the spreadsheet (you may have already done so in Exercise 3.8).

(a) The data stored in `water1.gsh` are explored in Chapter 4 (Example 4.3), and are given in Table 4.14. There are three columns in `water1.gsh`: mortalty, calcium and north. Each row refers to a town in Britain. Towns in the north of Britain have north = 1, and those in the south have north = 0. The north column thus contains nominal data. Categorical variables are called *group* variables or *factors* in GENSTAT. The only way that GENSTAT knows that a column is a factor is if we tell it. The following instructions tell you how to tell GENSTAT that north is a factor.

Make sure that the **Spreadsheet** window is active. Click on any cell in the north column, so that north becomes the current column. Click on the **Spread** menu; obtain the **Convert** option under **Column**. A dialogue box will appear:[7] click on **Factor**; click on the **OK** button.

What change has this wrought to the spreadsheet?

(b) Now that north is a factor, we can supply text so that south appears instead of 0 and north instead of 1.

To avoid confusion, let us first change the name of the factor from north to region. To do this, click on the **Spread** menu; obtain the **Edit Attributes** option under **Column**; and change the **Name** to region. (If you press <Return>, after typing region, then the dialogue box will disappear. If this happens, get the dialogue box back via the **Spread** menu.)

To supply text labels, click on the **Labels** button at the extreme right-hand side of the box. A dialogue box will appear. The label l corresponds to the lowest value in region, which was 0 for 'south'; you want the first label to be south. If necessary, move to the label l and simply type south and then press <Return>. Change the second label to north and click on the **OK** button. Back in the **Edit Column Attributes** dialogue box, change **Display Factor as** from **Levels** to **Labels**; click on the **OK** button. Factor *levels*, which will play a part later in the book, are simply numerical values assigned to factor labels. In this case the levels 0 and 1 correspond to the labels south and north.

How has the appearance of the spreadsheet changed?

Save the spreadsheet as mywater.gsh. You should change directory to the lmgen default directory. Use **Save As** on the **Spread** menu so as not to overwrite the water1.gsh datafile.

The methods that we have looked at for loading, storing, retrieving and manipulating data are the main ones that you will use in the book.

At some stage, most statisticians need to load data that someone else has stored. It is unlikely that other people will store their data as a GENSTAT spreadsheet. The most portable form of data is a text or *ASCII* file; virtually any package which uses data can read an ASCII file. The file we shall load is one supplied with the data for this book, called anaerob.dat. All the datasets are supplied as text files (.dat) as well as files in GENSTAT spreadsheet format (.gsh). Click on the **Data** menu; click on **Load**; click on **ASCII file**. A dialogue box will appear; fill in the **ASCII Data Filename** field as c:\lmgen\data\anaerob.dat. If you click somewhere in the dialogue box, then the start of the datafile will appear; click on **Read Column Names From File**; click on the **OK** button; click on the **Cancel** button. Make sure that the **Output** window is active, and you should then be able to see a summary saying that GENSTAT has read in two variates called ventil and oxygen. This is as far as we need go with this procedure; the data are now loaded.

---

[7]The little circular buttons such as those in this dialogue box are sometimes referred to as *radio buttons*.

Other formats are used for data as well, including various different types of spreadsheet (such as Excel and Lotus). GENSTAT will read many such types of datafile directly. You can find out the sorts of files that can be read in by clicking on **Open** from the **File** menu. A list of different formats, such as **Excel File**, will appear as a pull-down menu under **Files of type**.

You can also get GENSTAT to save data in forms that can be read by other software. To do this, the data must first be in a GENSTAT spreadsheet. Once the data are in a GENSTAT spreadsheet and its window is active, you can save them in various formats by clicking on the **Spread** menu, clicking on **Save As**, and then clicking on the appropriate file format.

To save instructions becoming too tedious, from now on we shall often use vertical bars to string menu commands together. For example, the set of instructions

Click on the Spread menu; click on New; click on Blank

may be written

Choose the Spread|New|Blank menu item.

3.3   Summaries and graphics

This section is about how to do the sorts of tasks covered in Chapter 2. Section 2.1 included histograms, calculation of the mean and variance, and transformations to normality. These tasks will be covered in the following four exercises, starting with plotting a histogram in Exercise 3.10.

Before you start working with another dataset, you *might* find it helps to clear GENSTAT of the data it already has. This can be done by clicking on the **Data** menu and clicking on **Clear All Data**, i.e. choose the **Data|Clear All Data** menu item. This will give you a dialogue box in which you are asked to confirm your action by clicking on the **Clear** button. You might also want to close any spreadsheet windows that are present. And if you no longer need the information in the **Output** window, you can clear this too by clicking on the button on the toolbar that contains a picture of an **X**. These things are merely ways of tidying up; GENSTAT will generally still work if you don't bother with them. You can do them whenever you feel it would be useful. However, if you are using the Student Version of GENSTAT, and you open lots of datasets at once, you may run out of storage space. If you regularly clear GENSTAT of data you have finished with, this is not likely to happen.

In the next exercise you will be using the magsus dataset which you met in Section 2.1. If you are using the Student Version of GENSTAT, there is a snag. (If you are not, ignore the rest of this paragraph.) This dataset contains 101 values in a single variate, but the default limit on the number of values in a variate in the Student Version is only 100. However, you can change this limit as follows. Choose the **Options|Student Limits** menu item. In the resulting dialogue box, change **Max Length of Structures** to 101 (or something bigger). Click on **OK**. A further dialogue box opens, saying that the new constraints will take effect the next time the server is restarted, and giving you the opportunity to do so. (This will delete any data you already have in the server, so make sure that you have saved anything you want to save.) Click on **Restart Now**. This resets the limits on the Student Version. There remains an overall limit that you cannot alter – the number of items of 'Named Data', which is the product of the 'Number of Structures' and the 'Max Length of Structures', cannot be bigger than 5000. Thus, for example, if you have only one variate, it could contain 5000 values, but if you have 10, they can each contain 500 at most. Changing one of 'Number of Structures' and 'Max Length of Structures' automatically changes the other one appropriately.

Exercise 3.10

Load the data in magsus.gsh into GENSTAT. You can load the dataset either via the spreadsheet, as in Exercise 3.6, or directly, as in Exercise 3.7.

(a) Click on the Graphics menu; click on Histogram: i.e. choose the Graphics|Histogram menu item. A dialogue box will appear: fill in the Data field as magsus. Two ways to fill in the Data field as magsus are:

- click in the space labelled Data so that the cursor appears in the space, then type magsus;

- ensure that the cursor is in the space labelled Data, as above, then double-click on magsus in the list of Available Data.

The relative merits of the two methods are: typing will always work, provided you do not mistype anything; double-clicking avoids typing errors and is faster. Sometimes the Available Data list does not contain what you want, so you have to type it.

Now, click on the OK button. A histogram should appear. Do the data appear to come from a normal distribution?

(b) There are various things that might be changed on this histogram; for example, the boundaries between the bars are at peculiar values – if you had been drawing the histogram by hand, you might have used 1, 2, ... , 14 as the boundaries. (There is a Number of Groups field in the Histogram dialogue box that allows specification of the number of groups but not their precise location. You need not use this.) Also, if you wanted to use the histogram in a report, then you might have wanted a title to appear on the histogram.

Make sure that the Histogram dialogue box is active;[8] fill in the Limits field as !v(1,2...14) and click on the OK button. It is best to type !v(1,2...14) exactly as written here – for instance, the three dots must be one after another, with no spaces between them – but some other variations will work. How has this changed the appearance of the histogram?

How do you think you change the boundaries to 1, 1.5, ... , 13.5 (i.e. in steps of 0.5)?

(c) Now we shall add a title and alter the style of the axes.

Make sure that the Histogram dialogue box is active; fill in the Title field as Magnetic Susceptibility. Now, click on the Axes button. A dialogue box will appear: change Style from X and Y axes to Box; click on the OK button; click on the OK button in the Histogram dialogue box. How has the axis style changed and where does the title appear?

---

[8] You might have to minimize the Genstat Graphics window to see the Histogram dialogue box.

The !v(1,2...14) used in part (b) means a temporary column, or *variate*, containing the values 1, 2, 3, and so on up to 14. An alternative to using !v(1,2...14) would be to create a column containing these values using the spreadsheet. Then, when you came to fill in **Limits**, the column would be available in the list of **Available Data**. Using !v(1,2...14) is just a shortcut. This sort of shortcut will be used later in the book too.

If you want to print your handiwork on your printer, then, *in the* **Genstat Graphics** *window*, choose the File|Print menu item. Do not worry about your ink supplies: GENSTAT prints graphics as black print on a white background, even though it displays them as white (and, as you will see later, as colours) on black. If you want to save your graphical efforts, you can do so using the File|Print menu item in the **Genstat Graphics** window, in three different formats, .gmf (or Genstat Meta File) (which can be opened again in **Genstat Graphics**), and .bmp and .emf, which can be read by certain other Windows software. Whenever you close the **Genstat Graphics** window, you are asked if you want to save the graphics windows, and if you reply **Yes** you can do so in one of these three formats. You are asked the same thing when you exit from GENSTAT if a graphics window is open.

Now that you have finished drawing histograms of these data, click on the **Cancel** button in the **Histogram** dialogue box.

Exercise 3.11

Obtain a histogram of the data in hald.gsh, which:

- has a title;

- has a box around the plotting area;

- uses boundaries at 1.25 up to 3.5, in steps of 0.25.

Compare your histogram with Figure 2.1(a).

There are other types of plots (e.g. boxplots) available through the **Graphics** menu, and we shall use these as necessary. They all work in the same way: you load a dataset and then use the menu system to tell GENSTAT what to draw.

The same basic method is used to obtain summaries such as the mean and variance. This is demonstrated in Exercise 3.12.

Exercise 3.12

Load forearm.gsh into GENSTAT. (This dataset was also used in Exercise 2.1. It contains 140 values in a single variate, so, if you are using the Student Version of GENSTAT, you may need to increase the limit on the maximum length of structures.)

(a) Choose the Stats|Summary Statistics|Summarise Contents of Variates menu item. A dialogue box will appear.

What summary statistics do you think GENSTAT will give if you do not change the **Options**? In particular, since we are looking for the mean and variance, will these be given if you do not change the **Options**?

(b) Alter the Options so that only the mean and variance will be given; fill in the Variate field as forearm; click on the OK button; and click on the Cancel button. Look in the Output window. What are the mean and variance of forearm?

Chapter 2 and the histogram produced in Exercise 3.10 suggest that the magnetic susceptibility data in magsus.gsh are skewed. In Section 2.1.3 the idea of using a transformation to make the normality assumption more appropriate was put forward. Two transformations that seemed to work well for the magnetic susceptibility data were the log and square root transformations.

Exercise 3.13
Load magsus.gsh into GENSTAT (if necessary).

(a) Take the square roots of the magnetic susceptibilities, by obeying the following instructions.

Click on the Data|Transformations menu item. A dialogue box will appear: change Transformation from Linear to Square Root using the pull-down menu; fill in the Data field as magsus; fill in the Save in field as rmagsus (or anything else you like); click on the OK button; click on Cancel.

Produce a histogram of the transformed data. Does this plot suggest that the square-rooted magnetic susceptibility is normally distributed?

(b) Click on Data|Transformations again. Now, transform magsus by taking logs (to base e), saving the transformed data in a column called lmagsus (say). Produce a histogram for the logged magnetic susceptibilities. Do the logged data look normal?

We shall not look at probability plotting in this section, but shall come to it later.

Having considered graphical methods for checking whether data were normally distributed, Chapter 2 moved on to statistical methods based on the normal distribution. These methods included:

- the one- and two-sample t tests;
- confidence intervals for the mean.

The one-sample case is covered in Exercise 3.14. A more complicated two-sample problem is covered in Exercise 3.15.

Exercise 3.14
Load hald.gsh into GENSTAT (if necessary).

(a) First, we shall do a two-sided test.

Choose the Stats|Statistical Tests|One-sample tests menu item. A dialogue box will appear: fill in the Variate field as hald; change Mean Value to 2; click on the OK button; click on the Cancel button. Make sure that the Output window is active.

What hypothesis has been tested, what is the value of the test statistic and what do you conclude from this test?

(b) As you have just seen, the default GENSTAT *t* test is two-sided. We cannot perform a one-sided *t* test using the menu system. Let μ be the mean breaking strength. To perform a one-sided *t* test of $H_0 : \mu = 2$ against $H_1 : \mu > 2$ (as in Exercise 2.6(b)), we have to alter the TTEST command in the Input Log window.

Make sure that the Input Log window is active and enter interactive mode (by clicking on Interactive in the Options menu). Move the cursor so that it is placed immediately after NULL=2 (and before]). Type in ;method=greater so that the command now reads:

TTEST [NULL=2;method=greater] hald

(If the alternative hypothesis had been $\mu < 2$, you would have had to use method =less instead.) Press <Return>.

How does the output from this command differ from that from the previous *t* test?

(c) Suppose instead we wanted a confidence interval for the mean. We also cannot produce a confidence interval using the menu system. Instead, we again have to adapt the TTEST command in the Input Log window.

Make sure that the Input Log window is active and check it is in interactive mode. Change the line that begins TTEST to

TTEST [NULL=2;ciprob=0.95] hald

and press <Return>. One way to remove the words method=greater is to move the cursor to the end of these words and repeatedly press < Backspace >. Leave interactive mode (by using the Options menu).

How does the output from this command differ from that from the previous *t* tests and what is the 95% confidence interval for the mean breaking strength?

The ideas in Exercise 3.14 can be used in many situations. The basic strategy is to use the menu system to generate the appropriate GENSTAT commands; if what is produced is not exactly what you want, then you can sometimes adapt the commands using interactive mode.

Exercise 3.15

Load hcii.gsh into the spreadsheet. Compare the spreadsheet with Table 2.4 and read the description of the data in Exercise 2.10.

The first stage in the analysis of these data was to work out the difference between the plasma measurements before and after haemodialysis. The measurements taken on the first group of people before haemodialysis are in column b1; the measurements taken after haemodialysis are in a1. The corresponding columns for the second group are b2 and a2.

Make sure that the Input Log window is active and enter interactive mode. Enter the command

```
CALC d1=a1-b1
```

(on a new line on its own, and press <Return>). This creates a new variate, called d1, which contains the differences between the plasma measurements before and after haemodialysis in the first group.

Calculate the differences for the second group and store them in a variate called d2. Leave interactive mode.

(a) Produce boxplots of d1 and d2 by obeying the following instructions.

Click on the Graphics|Boxplot menu item. A dialogue box will appear: fill in the Data field as d1,d2; click on the OK button; click on the Cancel button.

Does normality look like a reasonable assumption for these data as far as you can tell from these boxplots? On the same basis, do you think that the variances of the two groups are similar?

(b) Choose Stats|Statistical Tests|Two-sample tests.[9] A dialogue box will appear: fill in the Data set 1 field as d1; fill in the Data set 2 field as d2; click on the OK button; click on the Cancel button.

What are the means and variances of the differences for the two groups? Do you think that the variances are roughly equal? Assuming that a t test is valid, what do the results of this test indicate?

By adding instructions such as method=greater or ciprob=0.95 inside square brackets after TTEST in the command in the Input Log window, one-sided tests and confidence intervals can be obtained.

A quicker, but harder to understand, way to produce the boxplots and the t test would be to proceed as above, but:

- not calculate and store d1 and d2;

- for the boxplots, fill in the Data field as a1,a2−b1,b2;

- for the t test, fill in the Data set 1 field as a1−b1 and fill in the Data set 2 field as a2−b2.

See part (d) of Solution 3.4 for a clue as to why this might work and why filling in the Data field as a1−b1,a2−b2 would give boxplots of a1−b1−b2 and a1−a2−b2.

Exercise 3.16
Read Exercise 2.9. The data in Table 2.3 are stored in carpet.gsh. Use GENSTAT to carry out the analysis required in Exercise 2.9.

---

[9] You might have to minimize the Genstat Graphics window to get back to the menu bar.

In this section, you have seen how to use GENSTAT to summarize data and to perform t tests. In Chapter 4 we shall move on to using GENSTAT to perform simple linear regression.

3.4 Using the help system

Throughout the book, you will be told how to make GENSTAT do what the authors have asked you to do. This section is about additionally using the on-line help system to find out things.

3.4.1 Searching for help on a topic

There is a help system within GENSTAT which works, generally, in the standard Windows way.

Start GENSTAT, if you have not already. Click on the **Help** menu; click on **Search for help on**. A window will appear. At the top of this window there are some words which are referred to as *tabs*. The three tabs, **Contents**, **Index** and **Find**, are different ways of reading the on-line help manual, and work as usual in Windows.

Click on the **Find** tab. As explained in the window that appears, to use the **Find** facility your computer first has to build the list of words used in this manual. Click on **Next** and then on **Finish** to build this list.

(If the **Find** tab is not active after the list building is completed, then your computer probably ran out of memory; just try again until the **Find** tab is active after building the list. If your computer has still not finished the list after three or four attempts, you can temporarily make more memory available by shutting down the **Genstat Server**. There should be a button for **Genstat Server** on the taskbar. Click on it with the right mouse button, and in the resulting menu choose **Close**. In GENSTAT, try building the word list again. After you have done this, you will need to restart the server by choosing the **Options|Restart Server** menu item (or by exiting from GENSTAT and restarting it) before GENSTAT will work properly.)

Once the list is built and the Find tab is active, you can make use of the **Find** facility. Suppose you want to find out about facilities for graphical display in GENSTAT. You can fill in the **Type the word(s) you want to find** field as graphics. This produces a list of over 100 topics. This is because the word 'graphics' will occur many times in any document about fitting models statistically. This is a case where the index would be better, as the index will lead us to sections about graphics rather than to sections that happen to contain the word 'graphics'.

Click on the **Index** tab; fill in the **Type the first few letters of the word you're looking for** field as graphi, so that the graphics index only is highlighted; then click on the **Display** button. The resulting dialogue box contains a list of topics about graphics. To select **Histogram (Genstat 5 Help)**, say, press the H key (<H>) and/or scroll down till you come to this entry and select it; click on the **Display** button. (Pressing the H key finds the first of the topics beginning with the letter H; in this case the first of these is **HISTOGRAM**. This entry gives information only about the HISTOGRAM command, which you could use in interactive mode, and not the menu, so use **Histogram (Genstat 5 Help)** instead.) This produces a window containing text. The text is about how to complete the dialogue box for a histogram, which is used in Section 3.3.

The word 'variate' is highlighted. This means that if you click on that word, or press <Tab> and then <Return>, a definition of that word will appear. Try it. Either click or press <Return> to make the definition go away again.

Suppose you wanted other information on statistical graphics. There are several ways to get this information. One is to click on the **Help Topics** button, click on the **Display** button and choose a different topic. A second way is to click once or more on the << button or once or more on the >> button: the << button takes you to the previous page in the manual and the >> button to the next page. Clicking on the **Help Topics** button enables you to search for help on any topic. Clicking on the << and >> buttons keeps you within one particular area of the help manual (in this case, within help on 'graphics').

Suppose you want further information on the histogram menu item. The first page of the 'Histogram' section of the help manual is a contents page. This contents page has highlighted words. Some of the highlighted words have dashed lines beneath them, others have solid lines. The dashed lines mean that a small definition will appear if you click on that word. The solid lines mean that you will go to a new page if you click. Now, click on **Axes**, and the information needed for modifying the axes of the histogram will appear. Having read this information, you can click on the **Back** button to return to the page about the histogram, and then click on **Exit** to exit from the help system.

The **Find** and **Index** methods of searching for information are what you use when you do not know where the information is, or even whether it is in the manual or not. However, you might find it more convenient to start looking for information in a different way. Click on the **Help** menu; click on **Contents**. The resulting window is a contents page that leads to information about many aspects of GENSTAT, including each menu on the menu bar. For example, click on the **Graphics Menu** button and, from the new page, click on **Histogram**. The page that you then reach is one that you saw above. Click on the **Back** button twice, to get back to the contents page. The contents page is just a different way of getting to information in the manual. You might well find that the contents page is a good place to begin in the help system, because you will often be looking for help on something that is in the menu system and the contents page lists each menu.

When you start learning GENSTAT, you might well find it useful to look at the **Visual Guide** and the **Spreadsheet Guide** sections several times. The **Getting Started** and **Introduction** sections are worth a look as well. Later you will probably use the **Genstat Language Reference** section more than all the other sections on the contents page of the help system. This section is about the commands used in interactive mode, and an example of how you might use it is contained in the next subsection.

3.4.2 Genstat Language Reference

In preparing Exercise 4.1 (to follow), the authors decided that it might be better to use dots instead of crosses as the plotting symbols on a scatterplot. Initially they could not see a way to change the plotting symbol using the menu system. What they could see, though, was that the commands for drawing the graph in interactive mode were in the **Input Log** window. The commands were similar to the following.

```
PEN 1;SYMBOL=1;METHOD=point
DGRAPH [SCREEN=clear] loss;hardness
```

This meant that they had to find out about two commands, PEN and DGRAPH. In fact, the PEN command looked the more promising, because the word SYMBOL appears in that command. This is how to find out about PEN, using the contents page.

Start the help system and go to its contents page (see Section 3.4.1). First, click on the **List of directives** button under **Genstat Language Reference**. Since we know that we are looking for PEN, we can use the < PageDown > key to move through a list of GENSTAT commands (referred to as 'directives'); the list is in alphabetical order and so you can keep using <PageDown> to get to PEN. Alternatively you could use the scroll bar. If you did not know the name of the command you were looking for, then you could click on **Graphics** under the **Overview** button to get a list of graphics commands. If you know the name of the command but cannot find it in the list, then you can click on the **List of procedures** button (also under **Genstat Language Reference**) and look there. There are two types of GENSTAT command. A *directive* is a command that is built into GENSTAT. A *procedure* is a command that has been defined in terms of GENSTAT directives. From our point of view directives and procedures are used in the same way and it does not matter whether a command is a directive or a procedure. The only reason you need to know that there are directives and procedures is that they are in different parts of the manual.

Anyway, PEN is a directive, so you should be able to find it: click on **PEN**. Looking at the new page, you can see that you are allowed to use text for a plotting symbol. Therefore, to change from crosses to dots, you could change the PEN line in the Input Log window to

```
PEN 1;SYMBOL='.';METHOD=point
```

and then repeat the commands to draw the scatterplot using interactive mode.

Eventually, the authors decided to use the crosses after all, because they wanted to keep things simple at this stage of the book!

So this example shows again how you can use the menu system to produce something close to what you would like, and then adapt the commands generated by the menu system to get exactly what you want. The original scatterplot was close to what was wanted, but you can get precisely what you want by reading the help manual and adapting the commands used.

The only other part of the **Genstat Language Reference** section that you might use regularly is the list of functions in GENSTAT. Go back to the contents page for the **Genstat Language Reference** section and click on **Overview**. The new page has (among other things) a list of different types of function. If you click on **Probability functions**, you can find information on the functions ednormal and edt used in Exercise 3.4. You can also see that there are equivalent functions for other distributions, such as the F distribution.

3.4.3 The help system

The authors hope that you will not need to use the help system to any great extent in conjunction with this book. Usually you will be given instructions on how to carry out a new task in GENSTAT. One reason why the help system might be less useful to you than it was to the book authors is that it is written assuming that the reader is familiar with the statistical techniques being used. The authors are familiar with the techniques, but you will be learning them in this book. Thus, things that helped the authors might be meaningless to you until you have completed the book.

3.5   Some useful hints about GENSTAT

Here are a couple of points about GENSTAT that may not be obvious when you first start to use it, but which might save you a lot of time when you are regularly using it. Basically, these are things that the authors wished they had realized when they started!

GENSTAT works like other Windows applications

In particular, you can cut and paste between GENSTAT windows and between text fields in GENSTAT dialogue boxes, and you can cut text out of GENSTAT windows and paste it into other Windows applications such as wordprocessors. (You can do the same with graphics from the **Genstat Graphics** window, but only if you save them as a bitmap (.bmp) file or Enhanced Windows Metafile (.emf) first and open them with some other Windows program that can deal with such files – and you may well find that the quality of the resulting diagram is worse than if you printed the diagram out from **Genstat Graphics** directly.) In addition, if you have to type in a long command in interactive mode, or perhaps a long list of variables in a dialogue box field, it often happens that some or all of the text you have to type appears somewhere in the **Input Log**, and it may be quicker to cut and paste it than to retype it.

GENSTAT's interactive mode has pros and cons

If you are a speedy typist, then you might find when you get used to GENSTAT that it is quicker to type short commands in interactive mode than to use the menu system. For example, if you want to calculate a log transformation of a variable called **response**, rather than go via the menus you might find it quicker to type CALC lresp=log(response). If this way of working suits you, do use it. But be warned that you do have to be careful with the spelling and punctuation of interactive commands. If you get semicolons and commas mixed up, for instance, you will not get the result you want, and GENSTAT may even completely fail to understand what you mean and give less than helpful (or worse than useless) error messages.

Finally, do remember that GENSTAT is not perfect! Like any large and complicated piece of software, GENSTAT has some bugs, and in some (reasonably rare) circumstances things can go wrong even though you have not done anything wrong. For instance, the server may occasionally not update automatically when you load a new dataset into the spreadsheet, even though GENSTAT's spreadsheet options have been set so that it should update. Or **Genstat Graphics** may display a graph on a peculiar grey background or at completely the wrong scale. Usually, if this sort of thing happens, the best thing to do is to exit from GENSTAT completely and start again. It will almost always work properly after you have done this. None of the exercises in the book involves you in a great deal of manipulation of data, so that if you do have to start again in the middle of an exercise you will normally not have lost much time or effort. In some cases, because of the fact that GENSTAT is really three applications (the main **Genstat 5** application, the **Genstat Server** that does the number-crunching, and **Genstat Graphics** that draws the pictures), you may be able to get things going again without stopping GENSTAT entirely. For example, closing the **Genstat**

Graphics window and drawing the graph again might fix a graphics problem. But, often, if something goes wrong with one part of GENSTAT then things are wrong with the other parts too, and it is safer to start the whole thing again.

Summary of methodology

1 Most tasks in GENSTAT start with using the menu bar. Exercise 3.1 starts GENSTAT, introduces the menu bar and shows how to use it to exit from GENSTAT.

2 The behaviour of GENSTAT can be altered using the **Options** menu. These tasks are covered in Exercises 3.2 and 3.3.

3 Interactive mode is introduced in Exercise 3.4. Interactive mode allows you to perform tasks that are not offered by the menu system.

4 GENSTAT has several methods of loading and storing data. We shall use GENSTAT spreadsheet files most of the time. Entering data, changing data in spreadsheets and saving spreadsheets is covered in Exercise 3.5. Loading spreadsheets is covered in Exercises 3.6, 3.7 and 3.8. Exercise 3.9 is about how to change a column so that GENSTAT knows that it contains categorical data. Dealing with ASCII files is treated at the end of Section 3.2.

5 Histograms are presented in Exercise 3.10 and boxplots are in Exercise 3.15.

6 Transformations from the ladder of powers are used in Exercise 3.13.

7 The hypothesis testing and confidence intervals of Chapter 2 are carried out using GENSTAT in Exercises 3.14 and 3.15.

8 The GENSTAT on-line help system is described in Section 3.4.

4

Linear regression with one explanatory variable

The object of this chapter is to review the ideas and techniques of simple linear regression and to carry out such analyses in GENSTAT. You should be familiar with much (but probably not all) of this material from your previous studies, though the emphasis in this chapter may be on rather different aspects from those you have concentrated on before. The terminology 'simple linear regression' will be used in this book because it is standard rather than because it is unambiguously descriptive.

In simple linear regression, there is only *one* explanatory variable, plus a response variable. The response variable is continuous. In many applications the explanatory variable is continuous as well, but this need not be the case, as we shall see. The relationship between the explanatory variable and the response variable is assumed to be linear, and the distribution of the random variables that measure the differences between the observed values of the response variable and the values predicted by the linear regression equation is taken to be normal. This sounds like rather a limited model, with rather a lot of assumptions. It is. But the model is appropriate for a surprisingly wide range of data situations, and (more importantly for this book) by generalizing the model a bit at a time – adding more explanatory variables, using different distributions, and so on – we can derive all the other data models that will be introduced in the book. It is thus very important that you understand this basic linear regression model and that you can use it effectively.

Regression models are the focus of the book, but regression relationships are not the only way in which variables can be related. You should also be familiar with the idea of association or correlation, which is briefly reviewed in Section 4.8: it also has a role, albeit a comparatively minor one, to play in the book.

4.1 The simple linear regression model

In this section we begin by looking briefly at an example of data that can be analysed using simple linear regression, to remind you of what is involved.

Exercise 4.1
The data in Table 4.1 come from an experiment to investigate how the resistance of rubber to abrasion is affected by various factors. Each of 30 samples of rubber was tested for

hardness (measured in degrees Shore: the larger the number, the harder the rubber), and was then subjected to steady abrasion for a fixed time. The weight loss due to abrasion was measured in grams per hour.

Table 4.1

Abrasion loss (g/h)	Hardness (degrees Shore)	Abrasion loss (g/h)	Hardness (degrees Shore)
372	45	196	68
206	55	128	75
175	61	97	83
154	66	64	88
136	71	249	59
112	71	219	71
55	81	186	80
45	86	155	82
221	53	114	89
166	60	341	51
164	64	340	59
113	68	283	65
82	79	267	74
32	81	215	81
228	56	148	86

Source: Davies, O. L. and Goldsmith, P. L. (eds) (1972) *Statistical Methods in Research and Production*, 4th edition, Edinburgh, Oliver and Boyd.
Dataset name: `rubber`.

Load `rubber.gsh` into the spreadsheet, and confirm that the first two columns of data (loss and hardness) correspond to those in Table 4.1. (The third column, **strength**, contains another variable which we will come back to later.)

Plot **loss** against **hardness**. To do this: click on the **Graphics** menu; click on **Point Plot**. A dialogue box will appear. Fill in the **Y Coordinates** field with **loss** and the **X Coordinates** field with **hardness**. Click on the **OK** button; click on the **Cancel** button.

On the basis of the plot of these data, how are abrasion loss and hardness related?

You will be aware from your previous studies that the basic model for linear regression involves several different ideas, which will be reviewed here.

Suppose the data consist of a set of points (x_i, y_i), $i = 1, 2, \ldots, n$. We wish to find a model that will allow us to say something useful about what the response will be for any given value of the explanatory variable. Thus the explanatory and response variables have very different roles in the model. In setting up the regression model, these different roles are taken account of by treating the values of the explanatory variable as fixed and by treating the response as a random variable. To be precise, denote by x_i the value of the explanatory variable for the ith datapoint, and denote by Y_i the random variable representing the response for the same datapoint.

The first assumption made is that the mean of the response variable Y_i depends on the value x_i of the explanatory variable in a linear fashion, where the linearity is in x as well as in the parameters. This can be written as

$$E(Y_i) = \alpha + \beta x_i \tag{4.1}$$

where α and β are (unknown) constants – the intercept and slope of the regression line, respectively. By way of interpretation, the intercept α provides the mean value of Y when $x = 0$. The slope parameter β is such that an increase of one unit in x results in an increase of β units in $E(Y)$. α and β are known collectively as the *regression coefficients* of the regression line.

The second assumption is that the variation of the value of the response variable Y_i about this mean is represented by a random variable ϵ_i which has a normal distribution. The ϵ_is are often called random 'errors'. The use of the word 'errors' does not imply that random errors are incorrect in any way; the terminology is historical and derives from (actual) errors in measurement. The model becomes

$$Y_i = \alpha + \beta x_i + \epsilon_i. \tag{4.2}$$

For the mean of Y_i to be as given in (4.1), $E(\epsilon_i)$ has to be zero; but what about its variance? The third assumption is that the variance of ϵ_i is the same for all values of the explanatory variable. It is called the *error variance* (or, sometimes, the 'residual variance') and is usually denoted by σ^2. Therefore $\epsilon_i \sim N(0, \sigma^2)$.

There is a fourth assumption. It is not always stated explicitly, but it remains important. It is that the different ϵ_i, $i = 1, 2, \ldots, n$, are *independent* of each other. In other words, the deviations of the response from the mean are taken to be purely random; they do not affect each other.

There is an alternative, but equivalent, way of writing model (4.2). Considering the value of each x_i to be fixed, Y_i is made up of a fixed quantity, $\alpha + \beta x_i$, plus a $N(0, \sigma^2)$ random variable. But from property (2.1) in Section 2.1.4, this means that Y_i is itself normally distributed with mean $\alpha + \beta x_i$ and variance σ^2. (To see this, set $X = Y_i, a = 1$ and $b = \alpha + \beta x_i$ in property (2.1).) The simple linear regression model can therefore be written as in Box 4.1.

Box 4.1

The **simple linear regression** model can be written

$$Y_i \sim N(\alpha + \beta x_i, \sigma^2)$$

where the Y_is are independent random variables and the x_is are values of the explanatory variable.

This formulation specifies the *conditional distribution* of each response random variable Y_i given (or, conditional on) the corresponding value x_i of the explanatory variable. Each Y_i has a normal distribution, but a different normal distribution, its parameters – in fact, just its mean – depending on x_i. This conditional distribution specification will prove particularly amenable to generalization later in the book.

Exercise 4.2

For the data in Exercise 4.1, do you think the simple linear regression model is appropriate, given the scatterplot in Solution 4.1?

For some data, there are reasons to use a slightly simpler model, in which the regression line is constrained to pass through the origin; that is, the (mean) response must be zero if the value of the explanatory variable is zero. In this case, the model is that

$$Y_i = \gamma x_i + \epsilon_i$$

where again the error terms ϵ_i are normally distributed with mean 0 and variance σ^2 and are independent of each other. Equivalently,

$$Y_i \sim N(\gamma x_i, \sigma^2)$$

with the Y_is independent.

Exercise 4.3

The data in Table 4.2 (which you met first, briefly, in Table 3.1 in Section 3.2) come from a study of the effectiveness of cavity wall insulation. The total energy consumption (in MWh) of each of ten houses of a particular type in the Fishponds area of Bristol was recorded over one winter. Then cavity wall insulation was installed, and the total energy consumption was recorded again over the next winter.

Table 4.2

Before insulation	After insulation
12.1	12.0
11.0	10.6
14.1	13.4
13.8	11.2
15.5	15.3
12.2	13.6
12.8	12.6
9.9	8.8
10.8	9.6
12.7	12.4

Source: data from the Electricity Council, via The Open University (1983) MDST242 *Statistics in Society*, Unit A0, *Introduction*, 1st edition, Milton Keynes, The Open University.
Dataset name: bristol.

A possible model for these data is one in which the energy use after insulation, y, is (on average) a certain percentage of the energy use before insulation, x; that is

$$E(Y) = \gamma x$$

where γ is an unknown constant. If this model is appropriate, a straight line constrained to pass through the origin should be fitted to these data.

Load the dataset into GENSTAT and produce a scatterplot, using the method of Exercise 4.1.

It is not easy, from GENSTAT's default scatterplot, to see whether a straight line through the origin is appropriate, because the origin $(0, 0)$ does not appear on the plot. To produce a more helpful plot, return to the **Point Plot** dialogue box. (If you have closed it, choose Graphics|Point Plot again.) With after as the Y Coordinates and before as the X Coordinates, click on the Axes button. A new dialogue box opens; enter 0 for the Origin for *both* the Y Axis Options and the X Axis Options. This alters where the axes cross on the resulting plot. Click on OK; then click on OK in the Point Plot dialogue box.

Use the resulting diagram to see whether, on the face of it, a straight line constrained to pass through the origin is appropriate for these data.

Of course, the assumptions of simple linear regression do not hold for all data consisting of pairs of x and y values.

Example 4.1

The scatterplot in Figure 4.1 is based on data collected in an experiment in kinesiology. A person performed a standard exercise task at a gradually increasing level. The two variables are oxygen uptake and expired ventilation, which is related to the rate of exchange of gases in the lungs. How does expired ventilation depend on oxygen uptake?

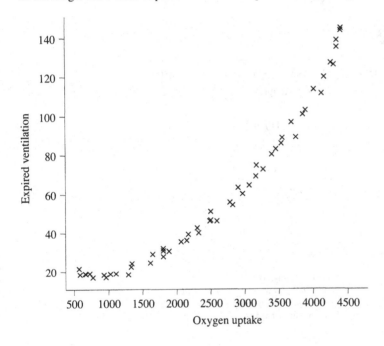

Fig. 4.1

There is clearly no point in fitting a straight line through these data! It is, however, possible to fit more complicated models, using methods described in later chapters. For instance, Figure 4.2 shows the results of fitting a quadratic function of oxygen uptake, i.e. one of the form $\alpha + \beta x + \gamma x^2$. The quadratic model is linear in its parameters but not in x. Fitting such models is discussed in Chapter 6. ∎

Exercise 4.4

The data whose first five pairs of values are given in Table 4.3 were collected as part of a study of extinctions in past geological ages. Geologists divide time since the start of the

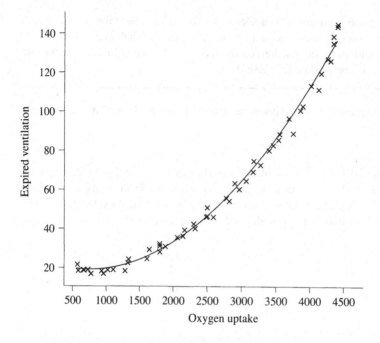

Fig. 4.2

Table 4.3

Percentage extinct	Time (mybp)
22	265
23	258
61	253
60	248
45	243
⋮	⋮

Source: Manly, B. F. J. (1991) *Randomization and Monte Carlo Methods in Biology*, London, Chapman & Hall. Dataset name: marine.

Permian period, some 290 million years ago, into a series of 48 stages (on what is known as the Harland timescale). For each of these stages, the table records when it ended (measured in millions of years before the present time, or mybp) and also the percentage of genera of marine animals that became extinct during the period. The aim of the study was to investigate how extinction rates had changed over time, and also to identify the stages at which 'mass extinctions' (abnormally high extinction rates) occurred.

Produce a scatterplot of the data, with time (variate mybp) as the explanatory variable. Comment on the appropriateness of the simple linear regression model for these data.

In the solution to Exercise 4.4, you were reminded of the phenomenon of *outliers* – that is, points which do not conform to the general pattern of the majority of the dataset.

4.2   Fitting lines and making inferences

We have a linear regression model, but it contains several unknown parameters. In the case where the regression line is not constrained to pass through the origin, these are the intercept α, the slope β and the error variance σ^2. In this section, we shall discuss how to estimate these parameters and how to make inferences about their values.

The principle of estimation that is generally used in simple linear regression is that of *least squares*. The idea is to look at the discrepancies, or *residuals*, between the values of the response variable that were actually observed and the values that would be predicted by the estimated regression line. Suppose the data are (x_i, y_i), $i = 1, 2, \ldots, n$, and the parameters α and β are estimated by $\widehat{\alpha}$ and $\widehat{\beta}$. Then, the values of the response variable predicted by the regression line are

$$\widehat{\alpha} + \widehat{\beta}x_i, \qquad i = 1, 2, \ldots, n,$$

and the residuals are given by[1]

$$r_i = y_i - (\widehat{\alpha} + \widehat{\beta}x_i), \qquad i = 1, 2, \ldots, n.$$

If the estimated line is to fit the data as closely as possible, then we want the residuals to be small. In this context, 'small' is taken to mean that we want to choose $\widehat{\alpha}$ and $\widehat{\beta}$ to make the sum of the squared residuals, that is the quantity

$$\sum_{i=1}^{n} r_i^2 = \sum_{i=1}^{n} \left(y_i - (\widehat{\alpha} + \widehat{\beta}x_i) \right)^2,$$

as small as possible. (This explains the term 'least squares'.) Then algebra can be used to show that the least squares estimate of the slope is

$$\widehat{\beta} = \frac{n \sum x_i y_i - \sum x_i \sum y_i}{n \sum x_i^2 - \left(\sum x_i \right)^2}$$

and the least squares estimate of the intercept is

$$\widehat{\alpha} = \bar{y} - \widehat{\beta}\bar{x}$$

where $\bar{x}$ and $\bar{y}$ are the sample means of the x and y values. (The regression line thus goes through the point with coordinates $(\bar{x}, \bar{y})$.) For regression through the origin, where the model is $Y_i = \gamma x_i + \epsilon_i$, the least squares estimate of the slope γ is $\widehat{\gamma} = \sum x_i y_i / \sum x_i^2$.

These estimates are not difficult to calculate by hand, though the process is pretty tedious except for small datasets. However, in the rest of this chapter and generally in this book, you will be calculating the estimates using GENSTAT rather than applying the formulae directly.

In most of this book, the method of maximum likelihood is used to find estimators of unknown quantities. So why is another method being used here? In fact, with the assumptions we have made about the random errors ϵ_i (i.e. normality, constant variance and independence), the maximum likelihood estimators of α and β are exactly the same

---

[1] Some other books use other symbols (often e_i or w_i) for residuals.

as the least squares estimators. If other assumptions are made about the distribution of the ϵ_is, however, the maximum likelihood estimators and the least squares estimators no longer coincide, and things can get much more complicated.

The least squares procedure for estimating α and β does not provide an estimate of the other parameter in the model, the error variance σ^2. In the case of the unconstrained line (i.e. not necessarily through the origin), the estimator of σ^2 which is used is

$$S^2 = \frac{\sum (Y_i - \widehat{Y}_i)^2}{n - 2} \tag{4.3}$$

where $\widehat{Y}_i = \widehat{\alpha} + \widehat{\beta} x_i$ is the (estimated) fitted value of Y at the ith datapoint. The numerator of the expression for S^2 is the value of the quantity (the sum of the squared residuals) that is minimized in finding the least squares estimates of the intercept and slope. The usual term for this is the *residual sum of squares* or *RSS*. The denominator is $n - 2$ essentially because two parameters, α and β, have already been estimated from the n datapoints.

For regression through the origin, the estimator of σ^2 is $S^2 = \sum (Y_i - \widehat{Y}_i)^2/(n - 1)$ where $\widehat{Y}_i = \widehat{\gamma} x_i$ is again the fitted value of Y at the ith datapoint. The denominator is $n - 1$ because only one parameter has been estimated.

The residuals may appear, at first glance, to be the observed values of the normally distributed random errors ϵ_i that appear in the simple linear regression model. In fact, they are not. If the true values α and β of the intercept and slope were known, then they would be. But in practice the intercept and slope have to be estimated, and it is $\widehat{\alpha}$ and $\widehat{\beta}$ that appear in the definition of the residuals. However, it is generally true that the residuals are reasonable estimates of the ϵ_is, so they can be used to investigate whether the modelling assumptions about the ϵ_is are satisfied. (There is more on this later.)

Exercise 4.5

In this exercise you will use GENSTAT to fit a straight line to the data on abrasion loss and hardness in Table 4.1, using the method of least squares.

Load `rubber.gsh` into GENSTAT's spreadsheet again if necessary. Choose the Stats|Regression Analysis|Linear menu item. A dialogue box will appear. The Regression field contains Simple Linear Regression, which is what we want. Fill in the Response Variate field as loss; fill in the Explanatory Variate field as hardness; click on the OK button. Do not click on Cancel in the Linear Regression dialogue box yet.

(a) Look in the Output window at the output produced by fitting the line. There will probably be parts that you do not recognize, but some parts should be familiar. Which part of the output do you recognize and what does it tell you?

(b) Now produce a scatterplot of the data with the fitted line plotted on it. To do this, go back to the Linear Regression dialogue box (which you should have left open – if you did not, you will have to go through the process of fitting the line again before you can produce the plot). Click on the Further Output button. Another dialogue box will appear; click on the Fitted Model button to produce the plot. Comment on whether the line seems to fit the data well.

Exercise 4.6

The data in peanuts comprise, for 34 batches of peanuts, the average level of the fungal contaminant aflatoxin in a sample of 120 pounds of peanuts and the percentage of non-contaminated peanuts in the whole batch. The data, five values of which are given in Table 4.4, were collected with the aim of being able to predict the percentage of non-contaminated peanuts (variate **percent**) from the aflatoxin level (variate **toxin**) in a sample.

Table 4.4

Percentage non-contaminated	Aflatoxin level (parts per billion)
99.971	3.0
99.979	4.7
99.982	8.3
99.971	9.3
99.957	9.9
⋮	⋮

Source: Draper, N. R. and Smith, H. (1981)
Applied Regression Analysis, 2nd edition,
New York, John Wiley.
Dataset name: peanuts.

(a) Make a scatterplot of the data in peanuts. Describe the relationship between the two variables. Would you say that the percentage of non-contaminated peanuts in a batch could be predicted accurately from the level of aflatoxin in a sample?

(b) Use GENSTAT to fit a straight line to these data.

Exercise 4.7

For the data in Table 4.2 on house insulation, you saw that a straight line through the origin (treating the 'before' data as the explanatory variable) looked reasonable, on the basis of a scatterplot at least. To fit such a line using GENSTAT, load the data (bristol.gsh), and obtain the Linear Regression dialogue box as in Exercise 4.5. Enter after as the Response Variate and before as the Explanatory Variate. Click on the Options button. In the resulting dialogue box, click on Estimate Constant Term to remove its checkmark. (This causes GENSTAT to leave out the constant term from the regression equation, so the line is forced to go through the origin.) Click on OK, and then on OK in the Linear Regression dialogue box. What is the equation of the fitted line?

Plot the line on a scatterplot of the data, using the same method as in Exercise 4.5. Comment on the fit.

Exercise 4.8

An experiment was carried out to investigate the effect of the stimulant caffeine on performance on a simple physical task. Thirty male college students were trained in finger tapping. They were then divided at random into three groups of ten, and the groups received different doses of caffeine (0, 100 and 200 mg). Two hours after this treatment, each man was required to do finger tapping and the number of taps per minute was recorded. The data are given in Table 4.5.

Table 4.5

0 mg caffeine	242	245	244	248	247	248	242	244	246	242
100 mg caffeine	248	246	245	247	248	250	247	246	243	244
200 mg caffeine	246	248	250	252	248	250	246	248	245	250

Source: Draper, N. R. and Smith, H. (1981) *Applied Regression Analysis*, 2nd edition, New York, John Wiley.
Dataset name: taps.

The aim was to investigate whether caffeine affected performance on this task and, if it did, to describe how the effect was related to the dose.

At first, these data do not look like typical regression data. Table 4.5 does not consist of a list of paired measurements. However, the question of interest is about the relationship between an explanatory variable x (dose of caffeine, variate **dose**) and a response variable y (number of taps, variate **taps**); and one can think of the data in the form of (x, y) pairs: $(0, 242), (0, 245), \ldots , (0, 242), (100, 248), (100, 246), \ldots , (100, 244), (200, 246), (200, 248), \ldots , (200, 250)$.

(a) Produce a scatterplot of the data in Table 4.5. Do you think it is appropriate to fit a straight line?

(b) Fit a straight line to the data. If you have not closed the **Linear Regression** dialogue box since you fitted the line through the origin in Exercise 4.7, GENSTAT will again fit a line through the origin, which is not appropriate here. To avoid this, click on the **Reset** button in the **Linear Regression** dialogue box before starting to fit this line. (This resets everything to its default value.)

The least squares method provides estimates of the slope and intercept of the regression line; but they are only estimates, and – like all estimates from sample data – are subject to error. In order to use statistical methods to make inferences from the sample estimates of the parameters of the regression line to the population from which the sample was drawn, we must use some results on the sampling distributions of the estimators involved. The key results are as follows.

For the unconstrained regression model, assuming that the random errors ϵ_i are normally distributed, the intercept and slope estimators $\widehat{\alpha}$ and $\widehat{\beta}$ are both normally distributed:

$$\widehat{\alpha} \sim N\left(\alpha, \frac{\sigma^2}{n}\left(1 + \frac{n\bar{x}^2}{S_{xx}}\right)\right), \qquad \widehat{\beta} \sim N\left(\beta, \frac{\sigma^2}{S_{xx}}\right),$$

where

$$S_{xx} = \sum_{i=1}^{n} (x_i - \bar{x})^2.$$

However, $\widehat{\alpha}$ and $\widehat{\beta}$ are not independent (except when $\bar{x}$ happens to be zero). In practice these results are not very useful for making inferences because the variances of both estimators involve the error variance σ^2, which is unknown. The estimator S^2 of the error variance (as defined in equation (4.3)) has the following distribution:

$$\frac{(n-2)S^2}{\sigma^2} \sim \chi^2(n-2).$$

This can be used to obtain the following very useful result:

$$\frac{\widehat{\beta} - \beta}{S/\sqrt{s_{xx}}} \sim t(n-2). \tag{4.4}$$

This result can be used, for instance, to test the null hypothesis that the true value β of the slope of a regression line is zero. With $\beta = 0$, the left-hand side of (4.4) can be calculated from the data as the t statistic for this t test, and then compared with the t distribution with $n - 2$ degrees of freedom to calculate a significance probability. You will see how to use the same result to calculate a confidence interval for β in Section 4.3.

Closely related results hold for fitting a line constrained to go through the origin. The details are not given here.

Exercise 4.9

For the rubber data of Table 4.1, carry out the regression analysis again with loss as the response variable and hardness as the explanatory variable, as in Exercise 4.5. Look at the output and see if you can work out how the information in the Estimates of parameters section relates to (4.4).

What of the rest of the output from the regression of abrasion loss on hardness? It is, sensibly enough, headed Regression Analysis. You are then reminded that you took loss to be what GENSTAT calls the Response variate and you are also reminded of the explanatory variable used, hardness, in what GENSTAT calls Fitted terms. In fact, GENSTAT gives the useful reminder that you are fitting both a constant (i.e. a non-zero intercept) and one explanatory variable (hardness).

Next comes a table headed Summary of analysis. You will not have seen such a table before, but in ensuing chapters tables like this will become very familiar objects. Later on, they will often be called 'analysis of variance' tables, for reasons that can remain obscure for now. Here is the one associated with the analysis of Exercise 4.9.

	d.f.	s.s.	m.s.	v.r.	F pr.
Regression	1	122455.	122455.	33.43	<.001
Residual	28	102556.	3663.		
Total	29	225011.	7759.		

The table consists of three rows and five columns, not counting row and column labels (and not all row/column combinations have an entry in them).

The rather abstract explanation that follows will become clearer in the specific contexts of later chapters. Do not worry if you do not grasp every detail of it at this stage.

Explanation of the table is facilitated by starting with the meaning of the column label
s.s.. This stands for 'sum of squares'. As you already know, the sum of squares of
numbers $z_1, z_2, \ldots, z_N$ is simply

$$z_1^2 + z_2^2 + \cdots + z_N^2 = \sum_{i=1}^{N} z_i^2.$$

Let us look down the s.s. column and identify the corresponding row labels. All that
is done to the data to obtain the values in the row labelled Total is that the responses
$y_1, y_2, \ldots, y_n$ have their overall mean $\bar{y}$ subtracted off. Thus s.s.(Total) is just the sum
of squares of what is left after this operation:

$$\sum_{i=1}^{n} (y_i - \bar{y})^2.$$

The word Total refers to this being a measure of the total variability (about the mean) in
the data.

The s.s.(Residual) entry is the residual sum of squares. Recall that each residual is
the difference between y_i and its value $\hat{y}_i = \hat{\alpha} + \hat{\beta} x_i$ predicted by the regression model.
So s.s.(Residual) is the amount of variability remaining in the data once the regression
model is fitted.

The first two entries in the s.s. column add up to the last; by definition,

s.s.(Regression) + s.s.(Residual) = s.s.(Total),

or equivalently,

s.s.(Regression) = s.s.(Total) − s.s.(Residual).

We can therefore say that s.s.(Regression) is the amount of variation in the (response)
data that is taken account of by the linear relationship between the mean response and the
explanatory variable.

The d.f. heading stands for 'degrees of freedom'. The d.f.(Total) entry is $n - 1$
where n is the sample size, $30 - 1 = 29$ in the abrasion loss case. If there were no
explanatory variable, and we had only the set of response data, we would estimate the
variance of the responses by dividing s.s.(Total) by precisely this quantity $n - 1$. This,
in fact, is what is done in the m.s. column:

$$\text{m.s.(Total)} = \frac{\text{s.s.(Total)}}{\text{d.f.(Total)}}.$$

The abbreviation m.s. stands for 'mean square'. The $n - 1$ is the degrees of freedom of
the χ^2 distribution associated with the variance estimator m.s.(Total) under normality
(Section 2.4.1).

Also, d.f.(Residual) $= n - 2$ (which is 28 in this case). This is precisely the divisor
of s.s.(Residual) used at (4.3) to form the estimate s^2 of the error variance (or residual
variance) σ^2. In fact,

$$\text{m.s.(Residual)} = \frac{\text{s.s.(Residual)}}{\text{d.f.(Residual)}} = s^2.$$

So, the estimate of the error variance in this case can be read from the table to be 3663. Again, the d.f. is the degrees of freedom of the associated χ^2 distribution.

As sums of squares of regression and residual add up to that of the total, so degrees of freedom add up in the same way. (Note that *mean* squares do not add up in this way.) Thus,

$$\texttt{d.f.(Regression)} = \texttt{d.f.(Total)} - \texttt{d.f.(Residual)}$$
$$= (n-1) - (n-2) = 1.$$

In fact, the 1 corresponds to the one extra parameter that has been fitted by the regression, the slope parameter β. The intercept, α, is not an extra parameter in this sense, because in the Total row of the table the sample mean has already been subtracted, which essentially corresponds to fitting the model $E(Y) = \alpha$. And

$$\texttt{m.s.(Regression)} = \frac{\texttt{s.s.(Regression)}}{1} = \texttt{s.s.(Regression)}.$$

The table directly addresses one particular null hypothesis, and that is whether $\beta = 0$. The test statistic it uses to test this hypothesis is given under v.r., which stands for 'variance ratio'. In fact,

$$\texttt{v.r.} = \frac{\texttt{m.s.(Regression)}}{\texttt{m.s.(Residual)}}.$$

Now, under the null hypothesis that $\beta = 0$, it turns out that m.s.(Regression) and m.s.(Residual) are both estimates of σ^2 (hence the term 'variance ratio'). It turns out also (and you are not expected to see why) that the two estimates are independent, that

$$\frac{\texttt{m.s.(Residual)}}{\sigma^2} \sim \chi^2(\texttt{d.f.(Residual)})$$

and that, under H_0,

$$\frac{\texttt{m.s.(Regression)}}{\sigma^2} \sim \chi^2(\texttt{d.f.(Regression)}).$$

This is just the setting of Section 2.4.2, implying that the ratio v.r. has, under the null hypothesis, an F distribution on d.f.(Regression) and d.f.(Residual) degrees of freedom. GENSTAT works out the SP for this test for you and gives it under F pr., which stands for 'F probability': explicitly, in this case, it is $P(F \geq 33.43)$ where $F \sim F(1, 28)$. The important thing to grasp from this paragraph is that, under $H_0 : \beta = 0$, v.r. $\sim F(\texttt{d.f.(Regression)}, \texttt{d.f.(Residual)})$.

Exercise 4.10
The GENSTAT output from Exercise 4.9 has given you two tests of $H_0 : \beta = 0$. What are they? Are the two SPs the same? Is there some very close connection between the two tests, particularly the two test statistics?

Hint For the last question, you *might* like to compare the logs of the (absolute values of the) test statistics.

The line starting Change can be ignored.

Beneath the Summary of analysis table, GENSTAT prints out some further information before coming to the Estimates of parameters. The first of these is to give you the Percentage variance accounted for. This statistic is interpreted as a measure of the strength of the linear association of the response variable with the explanatory variable. (Its formula is not important just now.) A perfect linear relationship would produce 100%; a value close to 100 is generally interpreted as a 'strong' linear relationship, whereas a value close to zero is interpreted as 'weak'. However, precisely what is meant by 'strong' or 'weak' in this context is difficult to enunciate. The abrasion loss's 52.8% variance accounted for is in a grey area, which to many people is not very strong, but to some is.

The value in the Standard error of observations line is simply the estimated error standard deviation (or residual standard deviation), i.e. the square root, s, of the m.s. (Residual). In this case, $s = \sqrt{3663} = 60.5$.

Underneath these lines is the space GENSTAT reserves for warning MESSAGEs. In this case, it mentions 'units' which seem to have 'high leverage'. This is a concept to do with model checking that you should ignore for now, and indeed not worry about until it is considered properly in Chapter 11. Often, but not this time, GENSTAT also uses this area to warn of particularly large residuals.

Exercise 4.11

For the data in Table 4.5 (dataset taps), investigate whether caffeine does indeed have an effect on performance of the tapping task by testing the hypothesis that the true slope of the regression line is zero. Report your conclusions.

4.3 Confidence intervals and prediction

This section investigates other statistical inferences that can be made on the basis of regression models, namely confidence intervals for regression parameters and inferences to do with prediction.

First, let us look at confidence intervals. From (4.4), a $100(1 - \delta)\%$ confidence interval for β can be found as

$$\widehat{\beta} \pm [t_{1-\delta/2}(n - 2)]\frac{s}{\sqrt{s_{xx}}}.$$

The notation δ, rather than the previously used α, is introduced here to avoid confusion with the parameter α in the equation of the regression line. The quantity $t_{1-\delta/2}(n - 2)$ is the $1 - \delta/2$ quantile of the $t(n - 2)$ distribution. Just like some of the confidence intervals in Chapter 2, this interval is of the form $\widehat{\beta} \pm$ multiple of s.e.$(\widehat{\beta})$, where s.e.$(\widehat{\beta})$ is the estimated standard error of $\widehat{\beta}$. A similar (but rather more complicated) expression exists for a confidence interval for α – we return to the details later in this section. But, to calculate the confidence intervals in GENSTAT, the detailed formulae are not required. The interval for α is also of the form $\widehat{\alpha} \pm$ multiple of s.e.$(\widehat{\alpha})$, and in fact the multiplier is exactly the same t percentage point as for the confidence interval for β. GENSTAT gives us the estimates $\widehat{\alpha}$ and $\widehat{\beta}$, together with their estimated standard errors. Thus forming the confidence intervals is reasonably simple, if a little indirect. How does this work out for the abrasion loss data?

Exercise 4.12

Load rubber.gsh into GENSTAT again, and fit a straight line to these data, using loss as the response variable and hardness as the explanatory variable.

From the Linear Regression dialogue box, click on the Save button. A dialogue box will appear: click on Estimates so that it becomes checked. This will cause a new field to become available: fill in the In field as est. Now, click on Standard Errors and fill in the In field as se. (Do not press <Return> after each insertion.) Click on the OK button.

Make sure the Input Log window is visible and active. Enter interactive mode. Enter the following command

```
PRINT est,se
```

and compare the contents of est and se with the output from fitting the regression.

You should be able to see that est contains $\widehat{\alpha}$ and $\widehat{\beta}$, and se contains their standard errors. Don't worry if values are given to different accuracies in different places. Internally, GENSTAT stores these values to greater accuracy than is displayed. Given the estimates and their standard errors, all we need to work out confidence intervals for α and β is the relevant quantile of the $t(28)$ distribution. The degrees of freedom, 28 in this case, are given in the Summary of analysis table. Finding quantiles of the t distribution was done in Exercise 3.4.

Enter the commands

```
CALC lower=est-se*edt(0.975;28)

CALC upper=est+se*edt(0.975;28)

PRINT lower,upper
```

to produce 95% confidence intervals for α and β. What confidence intervals do these give?

An alternative is to use the command PRINT est+-1,1*se*edt(0.975;28). This is quicker, but harder to understand.

Exercise 4.13

The data whose first five values are in Table 4.6 were collected by the Open University's Energy Research Group in the early 1980s. For each of 15 houses of similar design in the Neath Hill district of Milton Keynes, they recorded over a period of time the average temperature difference (in °C) between the inside and the outside of the house, and the average daily gas consumption (in kWh). The aim was to investigate how these quantities were related in real lived-in houses.

According to physical theory, the heat energy needed to sustain a particular temperature difference is proportional to the temperature difference. Thus, if the only energy input into the houses came from the gas, the theory predicts that gas consumption (y, variate gascons) against temperature difference (x, variate tempdiff) should be well fitted by a regression line through the origin. That the temperature difference is the explanatory variable is justified by the fact that the occupants had (partial) control over temperature by setting the thermostat; interest centres on how this control affects gas consumption.

Table 4.6

Daily gas consumption (kWh)	Temperature difference (°C)
69.81	10.3
82.75	11.4
81.75	11.5
80.38	12.5
85.89	13.1
⋮	⋮

Source: The Open University (1984) MDST242 *Statistics in Society*, Unit A5, *Review*, 2nd edition. Milton Keynes, The Open University.

Dataset name: `temperat`.

The data were collected in winter, and the houses had gas central heating, so gas is certainly the predominant energy input. However, gas is not the only energy input; for instance, there was energy from electricity and from the body heat generated by the inhabitants. Also, the data give the energy content of the gas supplied to the house, and not all this energy is converted into heat within the house, mainly because the efficiency of the central heating boiler in turning chemical energy in gas to heat energy is less than 100% and because the gas also provides hot water.

(a) Produce a scatterplot of the data. Does it indicate that a regression line constrained to pass through the origin would provide a reasonable model?

(b) Fit an unconstrained straight line to the data. Calculate a 95% confidence interval for the intercept of this line. Comment on the appropriateness of calculating such a confidence interval for these data. What does the 95% confidence interval tell you about the wisdom of trying to fit a straight line through the origin to these data?

Very often the main aim of fitting a regression line to data is to produce an equation that can be used to predict new responses from given values of the explanatory variable. If all that is required is a point estimate of the response, things are very easy; one merely substitutes the given value of the explanatory variable into the fitted regression equation. That is, if x_0 is the value of the explanatory variable for an individual whose response Y_0 is not known, the obvious predictor of Y_0 is

$$\widehat{Y}_0 = \widehat{\alpha} + \widehat{\beta} x_0.$$

But this does not take account of two further issues. First, the true regression line is not known; we merely have an estimate for it. Second, even if the true line were indeed somehow known, then all we would have is a value for the mean of the response variable, as in (4.1). We do not have a value for the random quantity that defines how far the actual value of Y will be from its mean, so it would still not be possible to predict the response exactly.

One way of taking into account these uncertainties is to calculate interval estimates. The first interval estimate we shall consider is a *confidence interval for the mean* of the response variable Y_0 at a new value x_0 of the explanatory variable. Such a $100(1 - \delta)\%$ interval is given by

$$\widehat{\alpha} + \widehat{\beta} x_0 \pm [t_{1-\delta/2}(n-2)]s \sqrt{\frac{(x_0 - \bar{x})^2}{s_{xx}} + \frac{1}{n}}. \qquad (4.5)$$

This is in fact again of the form estimate ± multiple of s.e.(estimate), and the multiplier is the same t percentage point as in the confidence intervals for α and β. Note that $\bar{x}$ and s_{xx} depend on $x_1, x_2, ..., x_n$ only, and not on x_0. Also, for values of x_0 close to the mean $\bar{x}$, the value of $(x_0 - \bar{x})^2$ is small and the interval will be relatively narrow. For x_0 a long way from $\bar{x}$, the interval will be relatively wide.

The other type of interval we shall consider is a *prediction interval* for the response Y_0 at a new value x_0 of the explanatory variable. The confidence interval for the mean of Y_0 takes into account the sampling variability inherent in the estimation of the regression line, so that it gives a range of plausible values for the mean of Y_0, but does not take into account the fact that Y_0 varies about its mean value. A prediction interval takes into account both types of variability, so that it provides a range of plausible values for the value of Y_0 that will actually be observed. The prediction interval thus has to be wider than the confidence interval for the mean of Y_0. The expression for the prediction interval is

$$\widehat{\alpha} + \widehat{\beta} x_0 \pm [t_{1-\delta/2}(n-2)]s \sqrt{\frac{(x_0 - \bar{x})^2}{s_{xx}} + \frac{1}{n} + 1}. \qquad (4.6)$$

Again, prediction intervals are (relatively) narrower for values of x_0 near the mean $\bar{x}$.

There are corresponding expressions for the confidence interval for the mean and for the prediction interval in the case of regression lines constrained to pass through the origin, but they are not given here.

Note that the intercept is the mean value of the response variable when the explanatory variable takes the value 0. Thus (4.6), with $x_0 = 0$, gives the confidence interval for $\widehat{\alpha}$, that you calculated in Exercises 4.12 and 4.13.

Formulae (4.5) and (4.6) are intuitively sensible, reflecting the facts that estimation and prediction close to the bulk of the data are reasonable things to do, but that estimation and prediction beyond the range of the data are much less reliable. In fact, the latter type of estimation and prediction, commonly referred to as extrapolation, additionally puts great trust in the continued suitability of the model away from the data.

Formulae (4.5) and (4.6) look rather complicated, but in practice your computer will help with calculating them.

Exercise 4.14

(a) In this part of the exercise you will use the regression line for the data on contaminated peanuts in Table 4.4 (Exercise 4.6) to predict the mean percentage of non-contaminated peanuts in batches for which the aflatoxin level in a sample is 13.2 parts per billion, in the form of a point estimate and also a 90% confidence interval. Forming a confidence interval for the mean response is very similar to forming confidence intervals for the slope and intercept. We have to obtain the estimated means and their standard errors and then proceed as in Exercise 4.12.

First load peanuts.gsh into GENSTAT again, and fit a straight line to these data, using percent as the response variable and toxin as the explanatory variable. In interactive mode, enter the commands

```
DELETE [redefine=yes] se
PREDICT [pred=yhat;se=se] toxin;13.2
```

which have the following effects. The DELETE command is simply to allow us to use the name se in the next line. GENSTAT is rather fussy about the circumstances in which it allows you to overwrite data. Here, if you do not explicitly delete se, then GENSTAT may refuse to save the standard errors in se. This is because se was set up in Exercise 4.12 (and used in the same way in Exercise 4.13) to contain two values, and only one is needed here. If you did not stop GENSTAT after Exercise 4.12 and 4.13, it remembers this and complains. This fussiness can be very frustrating, but is for your own protection.

The PREDICT command tells GENSTAT to calculate the estimated mean response at the value of toxin given at the end of the line, i.e. 13.2, and store the estimate in yhat and the corresponding standard error in se. Notice that the GENSTAT command is called PREDICT even though we are not currently predicting individual responses.

Now, work out the 90% confidence interval for the mean response, using the same basic method as in Exercise 4.12. To work out the confidence interval, you might have to delete lower and upper; alternatively, you could use different names to store the upper and lower bounds.

(b) Now calculate a 90% prediction interval for the percentage of non-contaminated peanuts in batches for which the aflatoxin level in a sample is 13.2 parts per billion, as follows. Prediction intervals are slightly more complicated than confidence intervals. We still have to calculate the equivalent of the standard error,

$$s\sqrt{\frac{(x_0 - \bar{x})^2}{\sum (x_i - \bar{x})^2} + \frac{1}{n} + 1},$$

given in (4.6). We already have the values of

$$s\sqrt{\frac{(x_0 - \bar{x})^2}{\sum (x_i - \bar{x})^2} + \frac{1}{n}}$$

stored in se, from part (a). (See formula (4.5).) So if we knew s^2, then we could get the equivalent of the standard error by squaring se, adding s^2 and taking the square root of the result.

So that is the plan. Enter the following commands to obtain the equivalent of the standard error, which will be stored in pse.

```
RKEEP deviance=rss;df=df
CALC s2=rss/df
CALC pse=sqrt(se**2+s2)
```

The se**2 means 'raise se to the power 2'; similarly se**3 would cube se. Also, sqrt means 'take the square root of the quantity in brackets'. The RKEEP line stores the residual sum of squares from the regression (i.e. s.s.(Residual)) in rss, and the residual degrees of freedom (d.f.(Residual)) in df. For now, just accept *deviance* as another name for the residual sum of squares; the general meaning of deviance will be met later. RKEEP works with the most recently fitted regression model. Then s^2 is m.s.(Residual), which has to be calculated by dividing s.s.(Residual) by d.f.(Residual). Finally, we do the calculation explained above.[2]

To get a 90% prediction interval for an aflatoxin level of 13.2, enter the following commands.

```
CALC plower=yhat-pse*edt(0.95;32)
CALC pupper=yhat+pse*edt(0.95;32)
PRINT plower,pupper
```

What is the 90% prediction interval you were asked for?

GENSTAT happens to make rather a big meal of producing prediction intervals; other programs may well produce them more readily.

4.4 Checking the assumptions

This section is concerned with only one class of methods for checking regression assumptions, but it is a very useful class. You will probably be familiar from your previous studies with the idea of using residuals to investigate how well a regression model fits. Recall that residuals are the (observed) differences between the observed values of the response variable, y_i, and the corresponding fitted values, $\widehat{\alpha} + \widehat{\beta} x_i$, predicted by the regression line, that is

$$r_i = y_i - (\widehat{\alpha} + \widehat{\beta} x_i), \qquad i = 1, 2, \dots, n. \tag{4.7}$$

(This is the basic idea, which will be developed somewhat later in the section.) As mentioned in Section 4.2, the residuals can be thought of as estimates of the random errors ϵ_i in the regression model. Plotting the residuals in various ways can provide useful checks on the assumptions about the random errors. Two types of residual plot that are especially useful in most simple linear regression situations are the following.

First, plotting the residuals against the explanatory variable or against the fitted values can indicate whether there is some aspect of the relationship between the explanatory variable and the response variable that has not been taken account of by the straight-line model. This sort of plot can also show up problems with the assumption of constant variance. Since the fitted values are a linear function of the values of the explanatory variable, a plot of residuals against fitted values looks the same as a plot of residuals against the explanatory variable, apart from a change of scale on the horizontal axis. (The horizontal scale will be reversed if $\widehat{\beta}$ is negative.) In this book we will plot residuals against fitted values rather than against the explanatory variable; this is because the plot against fitted values is rather easier to do in GENSTAT, and because this sort of plot is easier to extend to situations where there are several explanatory variables.

Second, producing a normal probability plot of the residuals can give a guide as to whether the assumption of normality of the ϵ_is is appropriate.

---

[2] Alternatively you can calculate pse directly, but in a rather more obscure way, using the command
PREDICT[pred=yhat;se=pse;scope=new]toxin;13.2

Each of these plots can also highlight abnormally large residuals, which can draw attention to potential outliers in the data.

Other types of residual plot can also be useful in some contexts. A third such plot, useful when the data have been collected in time order, is to plot the residuals in order of time. Such a plot can provide evidence of trends in time in the residuals. Such trends can be problematic, because they indicate that the ϵ_is are not independent.

A fourth type of residual plot is to plot the residuals against the values of some other potential explanatory variable apart from the one you have used in fitting the line. This can indicate that you need a more complicated model that takes into account the new explanatory variable as well.

You should make it a habit to look routinely at the first two types of residual plot whenever you fit a regression line. The third and fourth types are not always appropriate, but can be useful when they are. In later parts of the book, notably Chapter 11, you will meet other ways of checking the assumptions behind regression data, some of them also based on residuals.

Exercise 4.15

Load the rubber.gsh data into GENSTAT again, and once more carry out a simple linear regression with loss as the response variate and hardness as the explanatory variate. Make sure the **Linear Regression** dialogue box is active.

(a) To produce a plot of residuals against fitted values, click on the **Further Output** button. A dialogue box will appear: click on the **Model Checking** button. Another dialogue box will appear. Change **Type of Graph** from **Composite** to **Fitted Values**; change **Type of Residual** from **Pearson** to **Simple**; click on the **OK** button. 'Simple' residuals are the type defined in equation (4.7). In the resulting plot, the straight red line across the middle shows where zero comes on the vertical axis, and the curved blue line gives an indication of the general pattern of the residuals. What does the plot tell you about the appropriateness of the linear regression model?

(b) Repeat the procedure that gets you to the **Model Checking** dialogue box and select simple residuals again. Now, change **Type of Graph** from **Composite** to **Normal**; click on the **OK** button. What sort of diagram is produced, and what does it tell you about the appropriateness of the model?

The default **Type of Residual** in the **Model Checking** dialogue box (for linear regression models) is **Pearson**. These are the simple residuals divided by their estimated standard errors. (More details are given in Chapter 11 and, in a slightly different context, in Chapter 5.) The idea is to 'calibrate' the vertical scale on residual plots by producing standardized residuals which, although not independent, individually have approximately a standard normal distribution. (The residuals themselves are normally distributed, and have mean 0. But division by the estimated standard error spoils the exact (standard) normality of the ratios.) It may seem counter-intuitive that the residuals are not independent of each other, since they are estimates of the random terms ϵ_i in model (4.2) and the ϵ_is are independent. But the residuals have to sum to zero (as can be shown from the formulae for the parameter estimates in Section 4.2). Therefore, if you know all but one of them, you can work out the missing one. Hence they cannot be independent. In future, we shall generally look at Pearson residuals, and you should assume from here onwards that all linear regression residuals are Pearson residuals, unless stated otherwise.

The default Type of Graph in the Model Checking dialogue box is Composite. This produces a window with four different graphs in it; the two types you produced in parts (a) and (b), a histogram of the residuals, and a half-normal plot (which will be used later in the book). A half-normal plot just consists of the negative half of the normal probability plot superimposed on the positive half; it thus assesses absolute values of residuals rather than residuals themselves.

(c) Obtain the Model Checking dialogue box again, and this time leave all the items at their default values to produce a composite graph of Pearson residuals. Does it tell you anything about the model you did not already know?

Note that if you fit two or more different regression models during a GENSTAT session, then the residual plots will correspond to the most recent regression. If the residuals you want to plot are not from the most recent regression, then you have to repeat the appropriate regression, or you have to have had the foresight to have saved them; to save the residuals, use the same method as for saving estimates and standard errors (see Exercise 4.12).

Exercise 4.16
Produce a composite plot of the (Pearson) residuals for the data on peanut contamination in Table 4.4. Comment on the appropriateness of the model.

Exercise 4.17
The iron content of crushed blast-furnace slag can be determined by an accurate chemical test at a laboratory or estimated by a cheaper, quicker magnetic test. Data were collected to investigate the extent to which the results of the chemical test of iron content could be predicted from the magnetic test. The observations, the first five of which are shown in Table 4.7, are given in order of the time they were made.

Table 4.7

Chemical	Magnetic
24	25
16	22
24	17
18	21
18	20
⋮	⋮

Source: Roberts, H. V. and Ling, R. F. (1982) *Conversational Statistics with IDA*, New York, Scientific Press/McGraw-Hill.
Dataset name: iron.

(a) Produce a scatterplot of these data, treating the magnetic test value as the explanatory variable and the chemical test value as the response variable. Describe how the two variables appear to be related. Fit a regression line. Produce the 'default' composite residual plot.

(b) Investigate whether there is a change in the relationship between the two measurements over time by plotting the residuals in time order. This can be done in GENSTAT by obtaining the Model Checking dialogue box and choosing Index as the Type of Graph. The resulting *index plot* plots each residual against the corresponding row number in the dataset. It is thus a plot of r_i (or to be precise the Pearson version of r_i) against i. In this case, since the data are given in time order, this plots the residuals against time.

(c) Comment on the appropriateness of the simple linear regression model for these data.

Exercise 4.18

Ice cream consumption (in pints per capita, variate pints) was measured over 30 successive four-week periods from 18 March 1951 to 11 July 1953. One of the variables thought to influence consumption was the mean temperature (°F, variate temp). (This dataset contains values for other variables, which you should ignore.) The first five of the data values, which are given in time order, are shown in Table 4.8.

Table 4.8

Consumption (pints per capita)	Temperature (°F)
0.386	41
0.374	56
0.393	63
0.425	68
0.406	69
⋮	⋮

Source: Kadiyala, K. R. (1970) 'Testing for the independence of regression disturbances', *Econometrica*, **38**, 97–117.
Dataset name: cream.

Produce a scatterplot of these data and fit a regression line. Produce the default composite residual plot, and produce a plot of the residuals in time order. Comment on the appropriateness of the simple linear regression model for these data.

In the above exercise, GENSTAT flagged a residual as corresponding to a possible outlier. It uses a rather complicated procedure, based on the Pearson (standardized) residuals to decide which points to flag. Now each of these residuals follows, approximately, a standard normal distribution. So, for each residual, GENSTAT checks whether it is smaller than -2 or greater than 2. This is based on the fact that the central 95% of the probability mass of the standard normal distribution lies in $(-1.96, 1.96)$, and 1.96 is approximately 2. Therefore, under normality one can flag any points appearing outside this interval as being in the most extreme 5% of points and hence candidates for consideration as possible outliers. However, this calculation assumes independence of points: the residuals are not independent of each other (though their dependence is generally small). Partly because of this, GENSTAT only uses the '2' rule if d.f.(Residual) $\leq$ 20, and uses a higher threshold, details of which need not concern us here, otherwise.

Exercise 4.19

In the experiment on abrasion loss in rubber, more data were collected than were shown in Table 4.1. The tensile strength of the rubber (in kg/cm^2) was also recorded, and the overall aim was to investigate how abrasion loss was determined by both hardness and tensile strength. The first five values of the full rubber dataset are shown in Table 4.9.

Table 4.9

Abrasion loss (g/h)	Hardness (degrees Shore)	Tensile strength (kg/cm^2)
372	45	162
206	55	233
175	61	232
154	66	231
136	71	231
⋮	⋮	⋮

Fit (yet again) the regression line of abrasion loss against hardness. In the Linear Regression dialogue box, click on the Save button. In the resulting dialogue box, click on Residuals so that it becomes checked. Fill in the corresponding In field as resids. Click on OK. This saves the residuals in a variate called resids. Plot these residuals against tensile strength (variate strength). What do you find?

4.5 Transformations

In the previous section you saw several examples of data for which the simple linear regression model is not appropriate. This raises the question of how else one might model the data. Answers to this question will be covered in several places in the rest of this book. This section discusses an approach that you have already met – transforming the data. The idea here is that, if the simple linear regression model is not appropriate to model the relationship between two variables y and x, it might be possible to find a transformation of one or both of y and x such that the relationship between the transformed variables *is* well fitted by the simple linear regression model.

To summarize very crudely, there are broadly two approaches to finding appropriate transformations. One approach is to use insights from known scientific (or other) theory or from background knowledge that describe how the quantities involved might be expected to be related. Often such theory or knowledge can suggest appropriate transformations that can lead to a linear relationship. Another approach is simply to look at plots of the data and choose transformations that appear to lead to straight-line relationships with the right properties. This empirical approach often involves a certain amount of trial and error before an appropriate transformation is found.

In this chapter we confine ourselves to transformations on the ladder of powers (including the logarithmic transformation; see Section 2.1.3). (Of course, these are not the only transformations that can be used.) One consequence of restricting ourselves to power transformations is that they will not help us to analyse data like those shown in Figure 4.1 (which seems to indicate a slight upturn in the mean towards the left-hand end). The reason is that all the power transformations (including log) are *monotonic* (when applied to positive data) i.e. they are either increasing or decreasing. For instance, the graph of x^a against x, where a is a positive constant, is increasing. (It slopes upwards from left to right for all positive x.) The same is true of $\log x$. If a is negative, the graph of x^a against x is decreasing. In no case does the graph increase and then decrease (or vice versa).

Example 4.2

Table 4.10 contains the first few points of a famous dataset on memory retention, collected in an experiment by a psychologist named Strong. Subjects memorized a list of disconnected items, and then average percentage memory retention (p) was measured against passing time (t, measured in minutes). The measurements were taken five times during the first hour after the items were memorized, and then at various times up to a week later.

Table 4.10

p	t
0.84	1
0.71	5
0.61	15
0.56	30
0.54	60
$\vdots$	$\vdots$

Source: Mosteller, F., Rourke, R. E. K. and Thomas, G. B. (1970) *Probability with Statistical Applications*, 2nd edition, Reading (MA), Addison-Wesley. Dataset name: strong.

A scatterplot of these data is shown in Figure 4.3. The relationship between the variables is obviously not linear. However, it is monotonic, in that p generally decreases as t increases. Thus it may be possible to 'straighten out' the plot by transforming one or both of the variables, using a transformation from the ladder of powers. In this case, psychological theory suggested that a model of the form $p = C \exp(-\beta t)$ might be appropriate, where C and β are unknown positive constants. Taking logs of both sides of this equation leads to

$$\log p = \log C - \beta t,$$

which is a linear equation relating $\log p$ to t. If this model fits, then a plot of $\log p$ against t should be linear. Figure 4.4 shows the resulting scatterplot. So much for the theory in this case! The plot is clearly still nowhere near straight.

But what about the empirical approach of finding an appropriate power (i.e. trying out some transformations and seeing what happens)? You are asked to try this approach in the next exercise. However, before you attempt it, note that, when trying to straighten out a relationship, it is often good practice to concentrate on transforming the explanatory variable rather than the response variable.

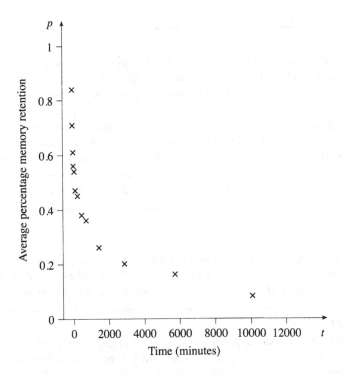

Fig. 4.3

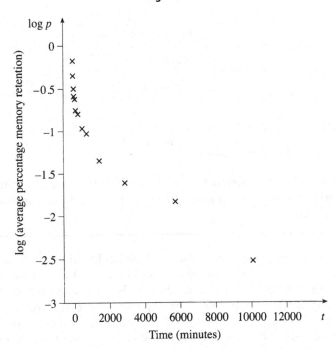

Fig. 4.4

Exercise 4.20

Look back at the original scatterplot of p against t (Figure 4.3). A transformation of t can move points to the right or left relative to one another in the scatterplot. In this case, we need to bring the points at the lower right of the scatterplot back towards the others. In other words, we need to bring the high values of t closer together compared with the low values. This can be achieved by moving *down* the ladder of powers (see Section 2.1.3). Load the file strong.gsh into GENSTAT, and find an appropriate transformation for these data, using the Transformation dialogue box you met in Exercise 3.13. (Power and log transformations appear as options in this dialogue box.) Fit a straight line to the transformed data, and check the appropriateness of the simple linear regression model by producing residual plots.

■

Finding an appropriate transformation for the memory retention data was reasonably straightforward because there was only one major problem in the original scatterplot: the curvature. For the data in the next exercise, there are more things to take account of.

Exercise 4.21

The tensile strength of cement depends on (among other things) the length of time for which the cement is dried or 'cured'. In an experiment, different batches of cement were tested for tensile strength after different curing times. The data are given in Table 4.11.

Table 4.11

Curing time (days)	Tensile strength (kg/cm^2)				
1	13.0	13.3	11.8		
2	21.9	24.5	24.7		
3	29.8	28.0	24.1	24.2	26.2
7	32.4	30.4	34.5	33.1	35.7
28	41.8	42.6	40.3	35.7	37.3

Source: Hald, A. (1952) *Statistical Theory with Engineering Applications*, New York, John Wiley.
Dataset name: cemstren.

Produce a scatterplot of these data, treating curing time (variate **curetime**) as the explanatory variable. Comment on how appropriate the simple linear regression model would be for these data.

Can problems with the model like those you found in Exercise 4.21 be dealt with by transformation? It is often a good strategy in cases like this to deal with the non-constant variance first. The problem is with the variance in the y direction, the direction of the response variable, so a transformation of the response variable may correct things. The aim in this case is to reduce the spread of the higher values of the tensile strength variable compared with the lower values. Such a transformation is one towards the left on the ladder of powers – something like a square root, a logarithm or a reciprocal (i.e. power of -1). (By the way, transformation is not the only way of dealing with non-constant variance. A method called weighted least squares provides one alternative. Another alternative, using a non-normal error distribution, is dealt with in Chapters 9 and 10.)

Exercise 4.22

For the data in Table 4.11, try, by trial and error, to find an appropriate transformation of the tensile strength that produces a constant variance. (Such a transformation will, typically, also help with the normality of the responses.)

The problem after transforming the response variable is that the relationship is even less linear than it was originally (compare the scatterplots in Solutions 4.21 and 4.22). A transformation of the explanatory variable, curing time, may, however, straighten it out. Again the requirement is to reduce the spread of the higher values of curing time compared with the lower values, so again we want to move down the ladder of powers.

Exercise 4.23

(a) Find an appropriate transformation of curing time that makes a plot of log(tensile strength) against transformed curing time reasonably linear.

(b) Fit a regression line with log(tensile strength) as the response variable and curing time transformed as in part (a) as the explanatory variable. Produce residual plots to investigate the fit of the model.

Exercise 4.24

The data underlying Table 4.12 were obtained in a study of a wind generator; they record the direct current (DC) output at different wind speeds.

Table 4.12

DC output	Wind speed (miles per hour)
0.123	2.45
0.500	2.70
0.653	2.90
0.558	3.05
1.057	3.40
⋮	⋮

Source: Joglekar, G., Schuene-meyer, J. H. and LaRiccia, V. (1989) 'Lack-of-fit testing when replicates are not available', *The American Statistician*, **43**, 135–43. Dataset name: wind.

(a) Produce a scatterplot of these data with DC output, y, as the response variable and wind speed, x, as the explanatory variable.

(b) Decide the appropriate direction on the ladder of powers in which you might transform y to straighten the curve in the plot. Try to find an appropriate transformation, and plot the transformed data. Are the assumptions of simple linear regression appropriate for the transformed data?

(c) Repeat part (b), but this time leave y untransformed and transform x.

(d) Choose the most appropriate transformation from those you investigated in parts (b) and (c). Fit a regression line to the transformed data, and check the model using appropriate residual plots.

Exercise 4.25

Janka hardness is an important structural property of Australian timbers, but is difficult to measure directly. However, it is related to the density of the timber, which is comparatively easy to measure, and therefore it would be useful to fit a model allowing Janka hardness to be predicted from the density. The Janka hardness and density of 36 Australian eucalypt hardwoods form the dataset underlying Table 4.13.

Table 4.13

Janka hardness	Density
484	24.7
427	24.8
413	27.3
517	28.4
549	28.4
⋮	⋮

Source: Williams,   E.   J.
(1959) *Regression Analysis*,
New York, John Wiley.
Dataset name: hardness.

Produce a scatterplot of these data with density as the explanatory variable. Explain why simple linear regression might be inappropriate for these data without transformation. Carry out appropriate transformations so that a simple linear regression model does fit the transformed data reasonably well. Produce residual plots to investigate how well the model fits.

Notice that, as in Solution 4.25, a conclusion that there is nothing amiss with (some of) the flagged residuals is often to be expected, since we should expect something like 5% of the datapoints ($0.05 \times 36 \simeq 2$ points in this case) to be flagged anyway.

We have dwelt on transformations at some length, because the ability to transform data appropriately is a very important tool in the statistical modeller's toolbox. Two aspects have been emphasized: the desire to make the variance constant and the need to straighten out the mean. Transformations in the y and x directions, respectively, are useful ways of coping with each of these in turn (if indeed transformations are powerful enough to produce a linear regression model for the data). Note that transformations of y and/or x retain the linearity in the parameters of the regression equation.

4.6   Comparing slopes

The next topic covered in this chapter is a method of carrying out an approximate hypothesis test to compare the slopes of two different regression lines. This topic is useful for two main reasons. In some examples, there are important questions whose answers depend on whether different regression lines have the same slope, and as you will see these questions can be answered using simple extensions of statistical ideas you have already met. Furthermore, developing these ideas will begin to illustrate how the basic regression models covered in this chapter can be extended to the more complicated models covered in the rest of this book.

Example 4.3

The data indicated in Table 4.14 were collected in an investigation of environmental causes of disease. They comprise the annual mortality rate per 100 000 for males, averaged over the years 1958–1964, and the calcium concentration (in parts per million) in the drinking water supply for 61 large towns in England and Wales. (The higher the calcium concentration, the harder the water.) The data also identify which towns are at least as far north as Derby.

Table 4.14

Mortality rate	Calcium concentration (parts per million)	North? (1 is yes)
1247	105	0
1668	17	1
1466	5	0
1800	14	1
1609	18	1
⋮	⋮	⋮

Source: Professor M. J. Gardner, MRC Environmental Epidemiology Unit, Southampton, via The Open University (1986) M345 *Statistical Methods*, Unit 3, *Examining Straight-line Data*, Milton Keynes, The Open University.
Dataset name: water.

A scatterplot of mortality rate against calcium concentration is shown in Figure 4.5. There is a clear relationship between mortality rate and calcium concentration, with lower mortality rates in areas with harder water, on average. But many of the areas with soft water in England and Wales are in the north, and so one possible explanation for these data is that the mortality rate is higher in the north for reasons that have nothing to do with the water supply itself.

We can investigate this further by identifying on the plot those points that correspond to towns in the north, as in Figure 4.6, where the towns in the north are shown by crosses. This shows that in fact there is a range of calcium concentrations in the north and in the south. The mortality rate is generally rather higher in northern towns; but, within each group of towns, the mortality rate is lower in towns with harder water.

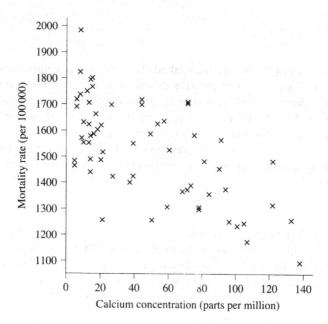

Fig. 4.5

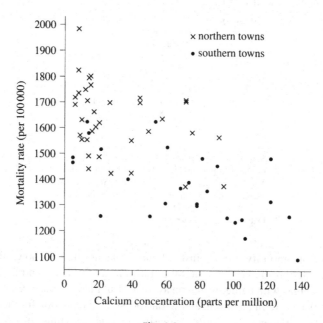

Fig. 4.6

Exercise 4.26

Load the water.gsh spreadsheet into GENSTAT. Fit a regression line to the data on mortality rate and calcium concentration for the northern towns only. One way to do this is as follows.

Make sure the Spreadsheet window is active. Choose the Spread|Restrictions|By Column Values menu item. In the resulting dialogue box, choose north in the top left field (to work with the north column). Set the Restriction Type to Include, and set Use Factor to Levels (which tells GENSTAT to work with the 0 and 1 values given in the north column). Make sure Add Restriction to Existing Units is unchecked by clicking on it (if necessary) to remove the check mark. Set the Choose Units with Values Equal To field to 1. All this tells GENSTAT to include in its analyses only those rows of the spreadsheet where north = 1, i.e. just the northern towns. Click on OK. The rows corresponding to southern towns disappear from the spreadsheet. (You can get them back, if you need to, using the Spread|Restrictions|Remove All menu item.) Because this changes the information in the spreadsheet, when you eventually close this Spreadsheet window, GENSTAT will ask you if you want to save the spreadsheet. You are advised to answer No to this question, to avoid overwriting the existing data file. Click somewhere in the Genstat 5 window outside the spreadsheet to update the server. Now GENSTAT will act as if the southern towns were not in the dataset, and you can fit the regression line in the usual way. Note the estimated standard error of the slope of the regression line.

Now go through the same procedure, but this time using the data for the southern towns only.

These regression lines are shown on the scatterplot in Figure 4.7. Perhaps the most obvious feature of these lines is that they are close to being parallel. In other words, they appear to have almost the same slope. The question then arises of whether the slopes differ significantly, or whether the small difference between them could be attributed to sampling error.

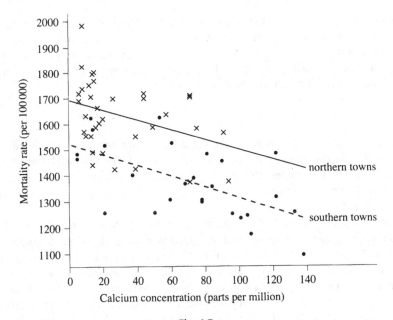

Fig. 4.7

A hypothesis test can be constructed as follows. An obvious estimate of the true difference between the slopes is the difference between the slopes of the fitted lines, which is $(-1.931) - (-2.093) = 0.162$. Since the two slope estimators are calculated from independent datasets, they are independent. Each of the slope estimators has a normal distribution, so their difference has a normal distribution too. Using results in Section 2.1.4, under the null hypothesis of equal slopes, the mean of the difference between the slopes of the fitted lines is zero. The variance of the difference is the sum of the variances of the two slopes; and, though these are not known exactly, we have sample estimates for them, namely the squares of their estimated standard errors. Thus the estimated variance of the difference in slopes is $0.848^2 + 0.566^2$, or 1.0395. The estimated standard deviation of the difference between slopes is then $\sqrt{1.0395}$, or 1.0195.

Exercise 4.27
Without doing any calculations, what do you think the outcome of this test is likely to be? In what way is the information given insufficient for you to calculate the *SP* precisely?

■

This hypothesis test, as presented here, is indeed a little crude and approximate; however, to improve it we need to make a further assumption about the data, namely that the error variances are the same in both groups. The technique for performing the test in the light of this further assumption will be covered in later chapters.

4.7   A look forward

Much of this book extends simple linear regression by adding more explanatory variables. To model many datasets fully, one would need a model with more than one explanatory variable. For instance, in Exercise 4.19 there was a single response variable, abrasion loss (Y), and two explanatory variables, hardness (x) and tensile strength (z). The most straightforward model for data like these is

$$Y_i = \alpha + \beta_1 x_i + \beta_2 z_i + \epsilon_i, \qquad i = 1, 2, \ldots, n,$$

where α, β_1 and β_2 are unknown constants and the ϵ_is are normally distributed random variables with the same variance σ^2, the ϵ_is being assumed independent. If there are more than two explanatory variables, the model is extended by adding on more terms in the obvious way. The process of fitting models like this is called *multiple regression* and is the topic of Chapter 6.

The other main extension of basic linear regression in the earlier part of the book is to categorical explanatory variables. All the explanatory variables in the examples in this chapter have been quantitative. But there is nothing in the regression model to say that this has to be the case. Regression with categorical explanatory variables is a way of solving a surprisingly large range of problems, as you will see. One example is the following.

In Example 4.3, you saw that a plausible model was that, in both the south and the north, the mortality rate, Y, is linearly related to calcium concentration, x, and that both these lines have the same slope, but their intercepts are different. We dealt with these data in the previous section by using two separate, independent models:

$$E(Y) = \alpha_S + \beta x \qquad \text{for the south;}$$
$$E(Y) = \alpha_N + \beta x \qquad \text{for the north;}$$

where α_S, α_N and β are three parameters to be estimated from the data. But suppose we are particularly interested in the difference, δ say, between the intercepts. Since the regression lines are parallel, this quantity represents the vertical distance between them; in other words, it is the excess of the mortality rate in the north compared with the south, when comparing two towns with the *same* calcium level in the drinking water supply. We could write the models as:

$$E(Y) = \alpha_S + \beta x \qquad \text{for the south;}$$
$$E(Y) = \alpha_S + \delta + \beta x \qquad \text{for the north;}$$

where this time the three parameters to be estimated are α_S, δ and β.

Now define a new variable z which takes two possible values, 0 for towns in the south and 1 for towns in the north. We can now combine the two models into one:

$$E(Y) = \alpha_S + \delta z + \beta x.$$

This single model incorporates both of the separate models above; and it has exactly the same form as the multiple regression model that was described for the abrasion loss data. Thus multiple regression methods with categorical explanatory variables can be used to answer questions about parameters such as differences between regression intercepts. You will see several examples of this general approach in the rest of the book.

The standard multiple regression models, like those for simple linear regression, involve an assumption that the random terms in the model are normally distributed. In later chapters of the book, you will see how this assumed distribution can be replaced by different ones, such as Poisson or Bernoulli, in order to accommodate responses that are in the form of counts or are binary.

4.8 Correlation

Regression models are concerned with explanatory relationships in which values of one variable, y, are considered to be dependent on the values of one, or more, explanatory variables, x. That is, there is a directionality in the relationship, 'from x to y'. However, there need be no such directionality in a relationship. In many cases, it is more appropriate to consider x and y to be on an equal footing: certainly, there is some kind of relationship between x and y (for instance, high values of x may tend to go with high values of y), but it is not sensible to think of y as being 'driven' by x or indeed of x as being driven by y.

As well as relationships of this sort between observed data values, similar relationships often exist between quantities calculated from data. For instance, in Section 4.2 it was pointed out that $\widehat{\alpha}$ and $\widehat{\beta}$ are (in general) dependent, and in later chapters you will see that residuals are related to each other.

Example 4.4

Digoxin is a drug that is eliminated in the urine largely unchanged. Table 4.15 gives the first five data pairs from a dataset of measurements, for each of 35 heart patients, of the 'renal clearance' of both digoxin and the naturally occurring substance creatinine. Each variable has been logged, the original units being ml/min/1.73m^2.

Table 4.15

Log of creatinine clearance	Log of digoxin clearance
2.970	2.862
3.207	3.550
3.277	2.434
3.437	3.378
3.444	2.632
⋮	⋮

Source: Halkin. H., Sheiner, L. B., Pech. C. C. and Melmon. K. L. (1985) 'Determinants of the renal clearance of digoxin', *Clinical Pharmacology and Therapeutics*, **17**, 385–94.
Dataset name: `digoxin`.

There does not seem to be any obvious directionality to any relationship there might be between digoxin clearance and creatinine clearance; it seems best to treat them on an equal footing. Figure 4.8 is a scatterplot of the logs of digoxin clearance and creatinine

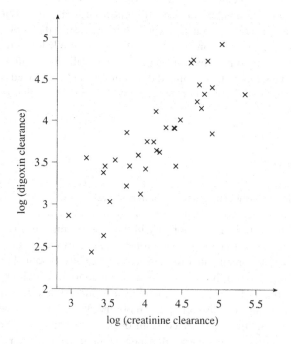

Fig. 4.8

clearance. Note that since we are just considering the *association* between the variables and not a regression relationship, the choice of which variable to put on which axis is arbitrary; it would have been equally justifiable to plot the variables the other way round. The scatterplot suggests quite a strong relationship between digoxin and creatinine clearances. The relationship is positive in the sense that large values for digoxin tend to go with large values for creatinine (and small with small). The strength of this association can be measured by the *Pearson product-moment correlation coefficient*, or *correlation coefficient* (or even just *correlation*) for short. In this case, the value of the correlation coefficient turns out to be 0.836.   ∎

The formula for the sample version r of the correlation coefficient is given by

$$r = \frac{\frac{1}{n-1} \sum_{i=1}^{n} (x_i - \bar{x})(y_i - \bar{y})}{s_x s_y}$$

where s_x and s_y are the sample standard deviations of the x_is and the y_is respectively. The correlation coefficient takes values between -1 and $+1$. The sign of r reflects the type of relationship: positive if high values of one variable go with high values of the other, negative if high values of one go with low values of the other. The stronger the relationship, the closer the value of r to 1 (if positive) or -1 (if negative). No association means $r = 0$. The value of 0.836 found for the digoxin example above reflects the strong positive association between the variables seen in Figure 4.8.

There are, however, caveats about the use and interpretation of r. First, the Pearson correlation coefficient is a measure of *linear* association between variables; that is, it is high (in absolute value) when the relationship between the variables is strong and roughly linear (in the straight-line sense). However, it is distinctly possible to have a very strong relationship between variables which is not well reflected by a straight line, in which case r could well be small. Second, if X and Y are independent, then the population correlation coefficient, the population version of r, is zero. However, because of the first caveat, the converse does not hold: it is perfectly possible to have a zero correlation coefficient but a strong dependence between variables. Nonetheless, r is still a useful measure.

Exercise 4.28
Let us consider the two variables hardness and strength from the rubber dataset; these refer to the hardness, in degrees Shore, and to the tensile strength, in kg/cm², respectively, of samples of rubber. In Exercise 4.19, these variables were considered as explanatory variables for the response variable abrasion loss. But hardness and tensile strength are not necessarily unrelated themselves. However, there does not seem to be any obvious directionality to any relationship they may have; it seems best to treat them on an equal footing.

Load the rubber data into GENSTAT again. Make a scatterplot of hardness against tensile strength. Use the following method to calculate the correlation coefficient. Choose the Stats|Summary Statistics|Correlations menu item. Enter the two variables whose correlation is required, hardness and strength, in the Data field. Make sure the Print Correlations item is checked, and click on OK. Look in the Output window. You will see what GENSTAT calls the `Correlation matrix` of the variables. Two of the entries in this matrix are not very interesting; they are the correlation of **hardness** with itself, which has to be 1, and the same for **strength**. The remaining entry is the one required. What do the scatterplot and correlation coefficient tell you about the relationship between hardness and tensile strength?

One way of assessing the smallness of a correlation is to make a hypothesis test of whether the (population) correlation is zero, but this will not be recounted here. We shall make only subjective assessments of the size of the correlation in this book.

The main role of correlation in this book will be in respect of relationships between explanatory variables. In particular, as already suggested in Chapter 1, if explanatory variables exhibit strong relationships with one another, it may well be that one need only include a small subset of these explanatory variables in the regression model; the reason is that there is little extra information to be gained by including further variables that are closely related to those already included, since they behave much like repeats of those already there.

Finally, be aware that correlation between variables is not necessarily indicative of a *causative* effect. (Nor indeed are regression relationships necessarily causative.) For example, in the USA, there is a high positive correlation between teachers' pay and alcoholism in teachers, yet the alcoholism is not caused directly by increasing pay nor are teachers paid more for being alcoholic! Such an observed relationship probably arises (perhaps in a complicated way) from other variables measuring stress levels and/or social activity. High correlation between smoking and lung cancer is now generally accepted to reflect a causative effect (i.e. smoking causes lung cancer) but one argument against this was that perhaps both variables are closely linked to a third variable reflecting lung types that had a propensity both to make people smoke and to develop lung cancer.

Causation, which is not an entirely well-defined concept, has to be established by routes other than correlation and regression. Typically, the aim is to carry out a study such that any posited causal relationship can be the only plausible explanation for the effects of interest.

Summary of methodology

1 You were reminded of the basic model for linear regression with a normally distributed response variable and a single explanatory variable, given in Box 4.1. This is called the simple linear regression model.

2 The least squares method for fitting regression lines was reviewed, and the way to employ it in GENSTAT was described. Sampling distributions of parameter estimators were given and used to make inferences (hypothesis tests, confidence intervals) about the parameters.

3 Confidence intervals for the mean of the response variable were considered and contrasted with the wider prediction intervals for values of the response variable itself. The way to calculate such intervals in GENSTAT was described.

4 Several residual plots were explored for checking the assumptions of the regression model. The most generally useful ones are plots of residuals against fitted values,

together with histograms and normal probability plots of residuals. Others, useful on some occasions, plot residuals against time order or against the values of another explanatory variable. By default, GENSTAT standardizes regression residuals by dividing by their standard errors (to give 'Pearson' residuals).

5 Transformations using the ladder of powers were explored in a regression context. A strategy of transforming the response variable first if variance is not constant and then transforming the explanatory variable to straighten out the mean was proposed. Bringing large values closer together relative to small values requires a transformation down the ladder of powers; bringing small values closer together requires a higher power.

6 You were reminded of relationships in which variables are on an equal footing, and of one way of quantifying such relationships using correlation. Neither correlation nor regression necessarily implies causation.

5

One-way analysis of variance

The linear regression model discussed in Chapter 4 concerned the situation of a single continuous response variable Y and its relationship to a single quantitative explanatory variable x: the model in fact prescribed a straight-line relationship between the explanatory variable and the mean of the response, and a normal distribution for deviations about that mean. There are many variations on this model needed to cope with the rich variety of real-world situations, a number of which are covered in the rest of this book.

This chapter presents the first important variation on the basic model. All that will be changed is the type of explanatory variable. Instead of the single quantitative x dealt with in Chapter 4, we shall now consider the case of a single *categorical* explanatory variable: properties of the model concerning the response variable Y, in particular its normality, remain the same. The new model is often called the *one-way analysis of variance* model, for reasons that will become apparent later.

Though, in terms of the model, this is a relatively small change in what you have already seen, it turns out that the new model has many uses. One is in extending the two-sample t test so that more than two population means can be compared at once.

5.1 Regression with a continuous response variable and a categorical explanatory variable

Example 5.1

The dataset etruscan contains data from an anthropometric study in which the maximum head breadths of skulls (in mm) of 84 Etruscan males are given along with the same measurements on 70 modern Italian males. Comparison of such skull measurements was of interest to the archaeological researchers. Table 5.1 gives the first five values in each of the groups.

Table 5.1

Etruscan skulls	141	148	132	138	154 ...
Modern Italian skulls	133	138	130	138	134 ...

Source: Barnicot, N. A. and Brothwell, D. R. (1959) 'The evaluation of metrical data in the comparison of ancient and modern bones' in Wolstenholme, G. E. W. and O'Connor, C. M. (eds) *Medical Biology and Etruscan Origins*, Boston, Little, Brown and Co.
Dataset name: etruscan.

Exercise 5.1

(a) Start up GENSTAT and open the datafile `etruscan.gsh` as a GENSTAT spreadsheet. Investigate how the data are arranged in the file. (This dataset has 154 values in each of the two columns. Thus, if you are using the student version of GENSTAT, you may need to alter the student limits on structure size, as explained in Section 3.3.)

(b) Update the GENSTAT server. Use GENSTAT to calculate summary statistics for the Etruscan and the modern skulls separately, by doing the following. Select the Stats|Summary Statistics|Summaries of Groups (Tabulation) menu item. In the resulting dialogue box enter breadth as the Variate and origin as the Groups. Note that the Available Data changes to include only factors when you click on the Groups field. Select the check boxes for No. of Observations, Means and Variances; click on OK. Look at the results in the Output window. Briefly summarize what you see.

(c) Use GENSTAT to produce boxplots of the two samples as follows. Select the Graphics|Boxplot menu item. Choose breadth as the Data and origin as the Groups. Ensure that the Box and Whisker Display and Fixed Boxwidth buttons are selected, and click on OK. Briefly summarize what you see in the resulting Graphics window. Do the results of parts (a) and (b) lead you to believe that the usual assumptions for a t test of the hypothesis that the population mean skull breadths are the same for both groups are justified?

(d) Perform the t test mentioned in part (c), as follows. Choose the Stats|Statistical Tests|Two-sample tests menu item. In the resulting dialogue box, choose t-test (unpaired) as the Test. The default when the dialogue box opens is for Two Sets to be marked as the Data Arrangement. This is what was needed for data laid out in the way you met in Chapter 3; but for data laid out as in `etruscan`, you should choose the other option, One Set with Groups. Choose breadth as the Data set and origin as Groups: this tells GENSTAT to treat the data in the breadth variate as two samples, one for cases where origin is Etruscan and the other for cases where origin is modern. Click on OK. Summarize the results of the t test that appear in the Output window.

Exercise 5.1 provided an adequate analysis of these data, but (as will become clearer later) not one that is easy to generalize. Let us look at it another way. Let x take the value 0 if a particular skull is Etruscan and 1 if it is modern Italian. A different way of thinking about the question of whether Etruscan and modern Italian skulls are the same is to consider how the response variable Y, the maximum head breadth in millimetres, depends on the explanatory variable x, which indicates 'group membership', i.e. whether the person was modern Italian or Etruscan. With these definitions of x and Y we get the scatterplot shown in Figure 5.1.

Let us now plough ahead and apply the linear regression techniques of Chapter 4 to these data, especially since normality of the response for each separate value of x seems to be a reasonable assumption.

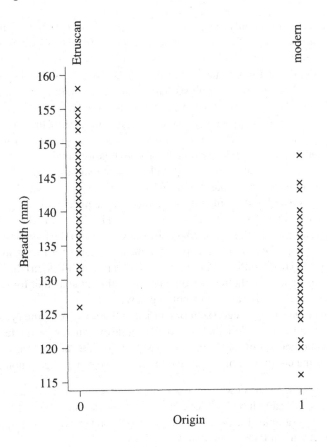

Fig. 5.1

Exercise 5.2

The regressions you have done in GENSTAT have all used a quantitative explanatory variable. In the dataset etruscan, origin is a factor. In GENSTAT, calculate a variate orvar which takes the values defined above for x by selecting the Input Log window, entering interactive mode, and typing CALC orvar=origin. This sets up orvar correctly because of the way the factor origin was defined, with its levels being 0 for Etruscan and 1 for modern; you need not worry about the details. Perform a simple linear regression of the response variable breadth on the explanatory variable orvar. It will be useful to keep a copy of the output from the analysis of Exercise 5.2 for reference later.

The output from the regression in Exercise 5.2 consists mainly of a Summary of analysis table and a table of estimates of regression coefficients.

The estimates of the regression coefficients tell us that the estimated mean of Y is given by

$$E(Y) = 143.774 - 11.331x.$$

Now, for these data, there are only two possible values of x, namely 0 for the Etruscan skulls and 1 for the modern skulls. That is, $E(Y) = 143.774 - 0 = 143.774$ for the Etruscan skulls and $E(Y) = 143.774 - 11.331 = 132.443$ for the modern skulls. In fact, hardly surprisingly, these are the sample means for the two samples of skulls, as you calculated in Exercise 5.1(b). The regression analysis has so far told us nothing we did not already know from the t test. The reason is that the assumptions behind the regression and the t test are identical. The t test assumptions are that the populations from which the two samples are drawn have normal distributions with the same variance (but possibly different means), and that all the observations are independent. The regression analysis also assumes that all the values of Y are normally distributed with the same variance, and that they are all independent. It further assumes that the mean of Y is $E(Y) = \alpha + \beta x$; but, since x can take only the values 0 and 1, this amounts to saying that the mean of Y for Etruscan skulls ($x = 0$) is α and the mean of Y for modern skulls ($x = 1$) is $\alpha + \beta$. In other words, the two means might differ or, if $\beta = 0$, they might be the same. A test of the hypothesis that $\beta = 0$ is just a test of the hypothesis that the population mean breadths for Etruscan and modern skulls are the same. This is reflected in the fact that (apart from the sign) the t value given for **orvar** in the output from the regression analysis is the same as that for the t test. So, again, the regression analysis has told us nothing new.

The other main part of the regression output is the Summary of analysis table. As with those you saw in Chapter 4, in the context of a single explanatory variable this provides little more than another test of the hypothesis that $\beta = 0$. The variance ratio is the square of the t value for the coefficient of **orvar**, and the SP is the same. Again nothing new. ■

You are probably wondering why anyone would bother with an analysis such as that in Example 5.1, as we already have an adequate way to analyse such data. But what if there were more than two samples for the t test, or (in regression terms) what if the explanatory variable is categorical but takes more than two values?

Example 5.2

Table 5.2 gives data collected by R. Lowe of the Iowa Agricultural Experiment Station on the amount of fat absorbed by doughnuts when cooking. For each of four different fats, which will arbitrarily be labelled Fat A, Fat B, Fat C and Fat D, six batches each of 24 doughnuts were cooked. The response variable was the total amount, in grams, of fat absorbed by each batch. Interest lay in whether, on average, the amount of fat absorbed was different for different fats. To put this more precisely: if we can treat the six batches of doughnuts cooked with a particular fat as a sample from a population of batches of doughnuts cooked in that fat, is the population mean response the same for all the fats? Or does it differ and, if so, how?

Exercise 5.3

The data in Table 5.2 are stored in the file **fats** in a similar way to the data in **etruscan** – there is one variate called **absorb** containing the responses, in grams of fat absorbed, for all $4 \times 6 = 24$ batches of doughnuts, and one factor, **fat**, with four levels (one for each fat), indicating the fat type corresponding to each response. The four levels have corresponding labels A, B, C, D.

Load the data into GENSTAT, and produce tables of summary statistics and comparative boxplots to compare the four different groups of doughnut batches, as you did for the two

Table 5.2

Fat A	Fat B	Fat C	Fat D
164	178	175	155
172	191	193	166
168	197	178	149
177	182	171	164
156	185	163	170
195	177	176	168

Source: Snedecor, G. W. and Cochran, W. G. (1967) *Statistical Methods*, 6th edition, Ames (IA), Iowa State University Press. Dataset name: `fats`.

groups of head breadths in parts (b) and (c) of Exercise 5.1. Does it look plausible that the population means are all the same?

For the t test and regression analysis of the `etruscan` data, it was necessary to assume that the populations involved had normal distributions with equal variances. On the basis of your tables and plots, does it look reasonable to assume that the four populations of fats have normal absorption distributions with equal variances?

Given Example 5.1, one way of proceeding with the analysis of the `fats` data may well have occurred to you: make comparisons in pairs, compare Fat A with Fat B, Fat A with Fat C, and so on. There are in fact six such pairwise comparisons for four fats, so (assuming normality) the data could be analysed with a series of six t tests. This may be manageable, but what if there had been five fats, or six, or 10, or 100? Well, a computer could deal with that; but there are two reasons why this approach is incorrect anyway. First, you would be making many t tests; remember that if testing at, say, a 5% fixed significance level, even if all pairs of means are equal, roughly 5% of tests will give erroneous 'significant' results. Second, these t tests are not independent of each other (for example, comparing Fat A with Fat B is based on half the same data as comparing Fat A with Fat C, and so on), so that, for instance, all tests associated with one particular fat might be erroneously significant. The carrying out of multiple t tests is very prone to produce misleading results.

What about the regression approach? Perhaps the most obvious thing to try here would be to construct an explanatory variate x that takes the value 1 for batches fried in Fat A, 2 for batches fried in Fat B, and so on. Labelling the response variable as Y, a plot of Y against x is given in Figure 5.2. This again indicates that the means for the different fats may well differ; also it shows that the relationship between Y and x does not look linear.

Figure 5.2, however, suggests another approach. Suppose we redefine x so that Fat D has $x = 1$, Fat A has $x = 2$, Fat B has $x = 4$, and Fat C still has $x = 3$. Then, plotting Y against the new version of x, we obtain Figure 5.3. It now looks as though a linear regression model might fit the data reasonably well. (Perhaps if we made the fats correspond to non-integer values, we could get an even better straight-line relationship.) However, making the fats correspond to particular numerical values is quite arbitrary, so the numerical values really have no meaning in terms of the original data. What would values between those assigned to the fats mean? Couldn't we obtain almost any kind of relationship by relabelling fats as we feel fit?

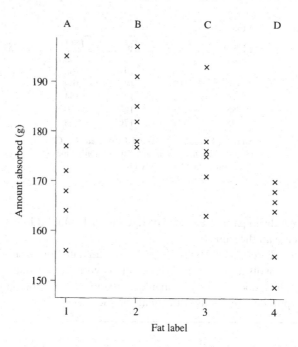

Fig. 5.2 Scatterplot of amount of fat absorbed against fat label.

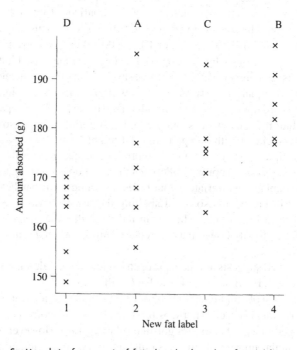

Fig. 5.3 Scatterplot of amount of fat absorbed against fats with new labels.

It should be clear that the approaches you know about already will not (quite) cope with the current data structure, and so a new methodology for such data is needed. This is the topic of the remainder of the chapter. The new method will be developed from scratch in the rest of this section and in Section 5.2, but we shall see later that ideas of t testing and regression remain relevant and closely related to the new developments.

The essential idea behind the new approach, the *analysis of variance*, is to develop another version of the Summary of analysis table you already know about from regression, but which can be used in other circumstances. To introduce this idea, let us return to the skulls.

Exercise 5.4

Load the etruscan dataset into GENSTAT again (if necessary). Choose the Stats|Analysis of Variance menu item. When the resulting dialogue box opens, the Design field at the top says General Analysis of Variance. Click on this field to change it to One-way ANOVA (no Blocking). The change of the Design field results in the appearance of two more fields, called Y-Variate and Treatments (and another field called Covariates which we shall ignore in this chapter). In the Y-variate field insert breadth, and insert origin as Treatments. Click on OK. Compare the output in the Output window with the output you obtained in the regression analysis of these data (Exercise 5.2).

ANOVA is an abbreviation for ANalysis Of VAriance, and the analyses of variance you will meet in this chapter are called *one-way ANOVA* in contrast to more complicated analyses that you will meet first in Chapter 7. You will learn about 'blocking' in Chapter 8, but all the analyses of variance in this chapter have no blocking. Notice too that, while the Linear Regression dialogue box produces a table headed Summary of analysis, the analogous table produced by the Analysis of Variance dialogue box is headed Analysis of variance.

Essentially, for data like those in etruscan, which can be modelled as the regression of a continuous response variable on a categorical explanatory variable *with only two values*, all the Analysis of Variance command in GENSTAT does is to provide another way of calculating the same Summary of analysis table that you would get from the Regression Analysis command. Again, nothing new here. But the Analysis of Variance command can be used easily with categorical explanatory variables that take *more than two values*, and it provides the basis for a test of the null hypothesis that the mean response is the same for every value of the explanatory variable. That is, it provides a correct alternative to the very dubious procedure of carrying out multiple t tests, in many situations where (in GENSTAT terminology) the response variable is a (continuous) variate and the explanatory variable is a factor. (It is also possible to analyse data of this sort using a more complicated version of regression analysis; you will learn about this in Chapter 6.)

More details on the kind of data that can be analysed using one-way analysis of variance, and on the statistical model behind it, are given in Section 5.2. But, without worrying about the details, let us see how GENSTAT deals with the doughnuts.

Exercise 5.5

Load the `fats` data into GENSTAT once more (if necessary). Choose the Stats|Analysis of Variance menu item in the resulting dialogue box; choose One-way ANOVA (no Blocking) in the Design field, absorb for the Y-Variate and fat as the Treatments. Click on OK, and look at the resulting output. Do not worry about whether any specific conditions for the test to be applicable actually hold; we shall consider that, along with more details of how the ANOVA table is constructed and what it means, later in this chapter.

In the `Analysis of variance` table in the output, there is a row labelled `fat`, which contains a variance ratio and a corresponding SP for a test of the null hypothesis that the mean responses for all of the fats are the same. What do you conclude about this hypothesis?

5.2 One-way ANOVA: data and model

In this section you will be introduced to some more of the kinds of data that can be analysed using one-way ANOVA, and you will learn about the model used in the analysis. There are no computing exercises in this section.

5.2.1 The completely randomized experiment

Before building a statistical model suitable for analysing data consisting of a continuous response variable Y and a single explanatory factor x, let us consider a little more the kinds of situation in which such data typically arise.

Sometimes, one is interested in an *observational* study in which individuals, chosen randomly from the population of interest, have measurements made which one wishes to relate to another, categorical, variable. An example is a study of sickle cell disease in which the response of interest is the haemoglobin level for each patient, and it is desired to relate this to the particular one (of three) types of sickle cell disease that the patient may have. Another example is from a study of longevity (Y) comparing workers in each of several different industries (x).

Alternatively, such data often arise from a situation in which the investigator has *control* over values of the explanatory variable. For example, the investigator, whom we might now call an *experimenter*, might be interested in the effects of three different drugs on the amount of morning stiffness in people with rheumatoid arthritis. The explanatory factor x will be the drug that a patient receives. The patients do not come with their x values already defined, but the experimenter can assign patients to different treatments. Another example is an agricultural experiment to assess the effects of different types of grass on fattening lambs, in which lambs can be assigned to different fields. (Think for a moment about why the investigator cannot *assign* individuals to values of the explanatory factor in the situations in the previous paragraph.)

Exercise 5.6

Which of the following are necessarily observational studies and which could be controlled experiments?

(a) A study of the effects of four different exercise regimes allocated to infants with a view to understanding the effects of the regimes on the time at which the infants first learn to walk by themselves.

(b) A study of salinity values for samples from three separate water masses in the Bimini Lagoon, Bahamas.

(c) In a study of the effect of Vitamin C on the risk of catching cold, each of a number of families consisting of a mother, father and three children is asked to follow, for one year, a diet which is one of high, medium or low in Vitamin C. The number of colds caught in the family in the year is measured.

Chapter 8 deals in some depth with many aspects of experimentation; here is a very brief introduction to some of the jargon and a few of the key ideas. First, some jargon.

In drug or agricultural trials and other similar experimental situations, 'treatment' is often a very natural word for the categorical explanatory variable x. We shall also sometimes use the word 'group', as GENSTAT does, to denote the same type of controlled explanatory variable.

The term 'experimental unit' is often used for the individual person, or animal, or field, or plot of ground, to which 'treatments' are applied. The term is fairly self-explanatory, but note that it need not refer to an individual person, animal or object, but possibly to a group of individuals treated, and measured, as a single entity (e.g. a family, a batch of seeds).

Exercise 5.7

Identify the experimental units, treatments and responses in the *experimental* situations identified in Exercise 5.6.

Where the experimenter has control over how experimental units are assigned to particular values of the explanatory factor (i.e. to particular treatments), the question of how that control should be exercised becomes important. The assignment should be done in a way that allows as much information as possible to be obtained from the experiment. These ideas are dealt with in more detail in Chapter 8.

As an example, consider the arthritis drug trial mentioned above. Different arthritis sufferers are not all going to have exactly the same characteristics. Some people might simply have a greater propensity to react to, say, Drug A than others. Responses for male and female patients may be different. Responses might be related to age: Drug B might not be effective in elderly patients, but very effective in younger ones. So, in this last case, if Drug B happened only to be given to younger patients, and other drugs only to older patients, Drug B might appear to be more beneficial overall than it really is.

Of course, knowledge such as 'Drug B is very beneficial for younger arthritis patients' could be very important information indeed. Ideally, then, complicated models incorporating all kinds of extra explanatory variables beyond just 'which treatment?' would be very valuable. But, at least as an initial evaluation, the question of interest here is more likely to involve the overall effects of the drugs on patients of all ages and types. In that case, the above assignment of drugs to patients, in which Drug B was only given to younger patients, would have *biased* the results and given a misleading answer. Moreover, in practice it will not often be possible to obtain all the information that one might like about each experimental unit in order to build a comprehensive statistical model that takes into account all the important variables that might affect the response. Indeed, the experimenter may not even be aware of some of the important variables.

Exercise 5.8

What possibly biasing aspects can you think of in the lamb fattening experiment mentioned just before Exercise 5.6?

The way around introducing biases of this type is to allocate experimental units *randomly* to treatments. For example, if there were just two treatments, then each patient in a clinical trial, say, could be assigned to either Drug A or Drug B according to the value of a uniform random number, u (i.e. a random number uniformly distributed between 0 and 1): $u \leq \frac{1}{2} \rightarrow$ Drug A, $u > \frac{1}{2} \rightarrow$ Drug B (equivalently, toss a fair coin). In this way, it is intended that older patients/younger patients, male patients/female patients, patients with a greater/lesser propensity to react, etc., are 'spread out' in an 'even-handed' manner among all the treatments. Note that this solution is only available in controlled experiments and not in observational studies, where it remains important to cope with possibly biased 'allocations' in some other way.

Box 5.1

A **completely randomized experiment** is one in which experimental units are allocated uniformly at random to the treatments.

Example 5.3

Individuals with high serum cholesterol levels (the experimental units) were randomized into three groups receiving different levels of advice (the treatments) on how to reduce their cholesterol. This experiment was conducted by M. Crouch, and reported in Brown, B. W. and Hollander, M. (1978) *Statistics: a biomedical introduction*, New York, John Wiley. Quoting from the source, 'After initial advice, one group received no follow-up information and help, one group received further contact by phone and mail, and a third group received more costly and time-consuming personal attention.' The response is the change in cholesterol level after one year.

The only point to which your attention is drawn is that random allocation to groups does not necessarily imply equal numbers of experimental units per treatment. It turned out that ten subjects were allocated to the 'Personal Follow-up' and 'No Follow-up' groups, and eight to the 'Mail/Phone Follow-up' group. This should be no surprise to you if the patients were being allocated to treatments by (in principle) rolling a fair three-sided die; the population distribution of the results of the allocation is uniform, but a sample from a uniform distribution is rarely exactly uniform itself. Moreover, in this case there were 28 patients to allocate to three groups! ∎

The data from completely randomized allocations conducted like that in Example 5.3 are said to have *unequal replications*. The term 'replication' is used frequently in discussing experiments, usually to denote the number of experimental units to which a particular treatment is applied. However, it is often preferred to end up with equal numbers in each group, but with the allocation completely random except for this constraint (the mechanics of such an allocation need not concern us here). In such a case, we have *equal replications* within a completely randomized experiment. An example is the fat absorption in doughnuts experiment described in Example 5.2. One reason for preferring equal replications is mentioned in Section 5.5. Also, in more complicated experiments, where there is more than one explanatory factor, it is sometimes important for several reasons to obtain equal replications; this matter is dealt with further in Chapters 7 and 8. You will not have to worry much about the distinction between equal and unequal replications in one-way analysis of variance, because GENSTAT will cope with both.

5.2.2   The basic one-way analysis of variance model

You have now met several data situations where there is a continuous response variable and a single categorical explanatory variable, and you have seen that such data can be analysed using one-way analysis of variance. In order to put this procedure on a firm footing, let us consider the model that lies behind the analysis.

Consider for a moment just the data in the continuous response variable/categorical explanatory variable situation that corresponds to a single value of the explanatory variable, value j, say. An example might be the collection of fat absorption responses in Example 5.2 associated with just Fat A, say. Then if Y_{ij} denotes the response (e.g. fat absorption) of the ith experimental unit (e.g. batch of doughnuts) in this jth group (e.g. fat), it may well be reasonable to do as we did for the t test and model $Y_{ij}, i = 1, 2, \ldots, n_j$, as a random sample from a normal distribution with some mean μ_j, say, and some variance σ_j^2, say. Notice that the mean and variance have subscript j and not i because they depend on the group the experimental unit is in but not on the individual experimental unit within the group. Also, n_j is the number of experimental units in group j and, if there are k groups all together, $n_1 + n_2 + \cdots + n_k = n$, the total sample size. For equal replications, $n_1 = n_2 = \ldots = n_k = n/k$. Often it will be reasonable to make this assumption of normality for all the different possible values of the explanatory variable; and, provided each data value does indeed correspond to a different experimental unit, it will also be appropriate to assume independence of responses across groups. The resulting model can then be written

$$Y_{ij} = \mu_j + \epsilon_{ij}$$

where each $\epsilon_{ij} \sim N(0, \sigma_j^2)$ and all the ϵ_{ij}s are independent of one another.

Looking back to Example 5.2, and in particular at Solution 5.3, and remembering the rather small sample sizes within each group, we can see that, as a first approximation at least, this model seems to be appropriate for this dataset. (The question of appropriateness of this model will be looked into more closely in Section 5.4.)

Another thing that we concluded from the boxplots in Solution 5.3, allowing for the small sample sizes, is that it *may* not be unreasonable to assume that the variance associated with each treatment group might be taken to be the same, i.e. $\sigma_1^2 = \sigma_2^2 = \ldots = \sigma_k^2 = \sigma^2$, say.

Example 5.4

Table 5.3 gives sets of salinity values (response, parts per thousand) for the three separate water masses (groups) in the Bimini Lagoon, Bahamas, as mentioned in Exercise 5.6(b).[1] Boxplots are given as Figure 5.4. The sample variances for the three groups in this dataset are, respectively, 0.3281, 0.2824 and 0.2609. On the basis of these values and Figure 5.4, the equal variance assumption seems reasonable for these data.

Table 5.3 The first five salinity values for each water mass are given: there were, all together, 12 measurements for Water Mass I, 8 for Water Mass II and 10 for Water Mass III.

I	II	III
37.54	40.17	39.04
37.01	40.80	39.21
36.71	39.76	39.05
37.03	39.70	38.24
37.32	40.79	38.53
⋮	⋮	⋮

Source: Till, R. (1974) *Statistical Methods for the Earth Scientist*, London, Macmillan.
Dataset name: salinity.

In this case, the assumption was clearly reasonable; but, as you saw in Exercise 5.3, it is much less clearly reasonable for the fats data. ∎

The equal variance assumption will be made from here on (for suitable datasets), and not just because it can often be justified on empirical grounds as above. There is also a theoretical reason for making this assumption in completely randomized experiments, provided that the effect of each treatment is simply to add a constant quantity to the response which would otherwise have been obtained for that unit. If this is the case, then the variance of the responses in each group will not be affected by the treatment, and the only other thing which could cause the variance to differ between treatment groups is the allocation of experimental units to the treatment groups. In a completely randomized experiment this is done by random choice, so it follows that the observed variances in the different treatment groups are all estimates of the same variance, σ^2.

The crucial assumption in this argument is that the effect of a treatment on an experimental unit is to *add* a constant quantity, positive, negative or zero, to the response which would otherwise have been obtained. This is called the assumption of *unit-treatment additivity*. Where the treatment effects are small enough to be partially obscured by random variation, which is where statistical analysis is important, the assumption of unit-treatment additivity is usually valid.

So, here is what will be called the *one-way analysis of variance model*.

---

[1] You need not load this into GENSTAT.

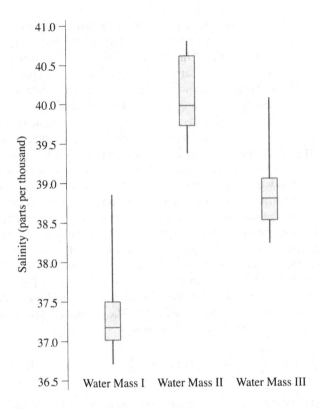

Fig. 5.4 Boxplots of salinity values for different water masses.

Box 5.2

The **one-way analysis of variance (ANOVA) model** applies to a single categorical explanatory variable and a continuous response variable, and postulates that the response of the ith experimental unit for the jth value of the categorical explanatory variable is

$$Y_{ij} = \mu_j + \epsilon_{ij} \tag{5.1}$$

where the $\epsilon_{ij} \sim N(0, \sigma^2)$ independently of one another.

A small change that is sometimes made to model (5.1) is to rewrite each μ_j as $\mu_j = \mu + \tau_j$, say. This change to the basic model is not needed at all for the data analyses in the present chapter. It is introduced here because it forms the basis for useful ways of writing the somewhat more complicated models addressed in some of the following chapters. Another reason for doing this is to make the analysis of variance model look more like linear regression models, which include a constant term like μ which does not vary with the explanatory variable. But a snag arises. Instead of the k parameters $\mu_1, \mu_2, \ldots, \mu_k$ that we had in (5.1) to describe the k group means, we would now have $k + 1$ parameters $\mu, \tau_1, \tau_2, \ldots, \tau_k$. And this would give us a major problem, as follows. All we are trying to do here is to re-express model (5.1), which had k parameters, in a useful way. If $\widehat{\mu}$ and $\widehat{\tau}_1$, say, were estimates of μ and τ_1 derived from the data in some way, then a natural estimate of μ_1 would be $\widehat{\mu} + \widehat{\tau}_1$. However, if the same two parameters were estimated instead by, say, $\widehat{\mu} - 3$ and $\widehat{\tau}_1 + 3$, then the corresponding estimate of μ_1 would still be $(\widehat{\mu} - 3) + (\widehat{\tau}_1 + 3) = \widehat{\mu} + \widehat{\tau}_1$ as before! In other words, there are different sets of estimates of the new parameters that lead to the same estimates of the original parameters μ_j. But the values of μ_j describe everything we want in the model, so the different estimates of the new parameters are not telling us anything relevant about the data.

Basically, this problem arises because there are too many parameters describing the treatment means ($k + 1$ instead of k). An obvious possibility is simply to leave out one of the parameters. If we leave out μ we are back to model (5.1). What is often done instead is to leave out τ_1, so that the mean of the first group is μ, the mean of the second group is $\mu + \tau_2$, and so on. You will see this done by certain GENSTAT regression commands later.

However, the GENSTAT analysis of variance commands take a different approach. Instead of leaving out a parameter, they impose a *constraint* on the parameters as follows. Suppose, for a moment, that we have equal replications. Then, without any loss of generality, we add the single constraint $\tau_1 + \tau_2 + \cdots + \tau_k = 0$ to the model – that is, we insist that the parameters satisfy this equation. There would remain just k new parameters, since any τ_j is given in terms of the other $k - 1$ parameters as $-(\tau_1 + \tau_2 + \cdots + \tau_{j-1} + \tau_{j+1} + \cdots + \tau_k)$. In other words, once you know $k - 1$ of the τ parameters, the final one is defined by the constraint, so it is not really a new parameter at all. The parameters now have meaningful interpretations as follows. First, the overall mean is

$$\frac{1}{k} \sum_j \mu_j = \frac{1}{k} \sum_j (\mu + \tau_j) = \mu + \frac{1}{k} \sum_j \tau_j = \mu + \frac{1}{k} \times 0 = \mu.$$

The other new parameters, $\tau_1, \tau_2, \ldots, \tau_k$, are *treatment effects*: they describe what effect (if any) treatment k has in terms of increased or decreased mean, *relative to the overall mean*. That is, $\mu_j - \mu = \tau_j$. $\tau_j < 0$ means treatment j has a smaller mean than overall; $\tau_j > 0$ means treatment j has a larger mean than overall. The τ_js are particularly useful because no effect for treatment j (relative to the overall mean) corresponds to $\tau_j = 0$.

For unequal replications, the constraint used by GENSTAT is rather more complicated and it is not given here, since you will not need to use it for any of the data analyses. (Indeed, GENSTAT will not provide parameter estimates in such an ANOVA model unless you request them specifically.)

Now that the model has been set up, let us return to analysing data.

5.3 Testing for equality of means

The first question that can be addressed with the one-way analysis of variance model is whether the mean response in each treatment group is the same: i.e. the null hypothesis $H_0 : \mu_1 = \mu_2 = \ldots = \mu_k$ (or equivalently, $H_0 : \tau_1 = \tau_2 = \ldots = \tau_k = 0$) can be tested. You saw in Section 5.1 that this can be done with the help of the analysis of variance (ANOVA) table.

Example 5.5
The file plywood.gsh contains data on the mean dry shear strength (in pounds per square inch) of batches (experimental units) of five test pieces of birch plywood. Ten batches were

tested per treatment. The data were obtained from an experiment to compare the strengths of plywood produced with six different glues, labelled Glue A to Glue F. The first five responses for each treatment group are given in Table 5.4.

Table 5.4

		Glue			
A	B	C	D	E	F
502	470	500	520	551	620
458	483	502	510	556	643
445	478	480	582	592	589
479	493	519	530	562	576
468	498	459	495	566	576
:	:	:	:	:	:

Source: Black, J. M. and Olson, W. Z. (1947) 'Durability of room-temperature-setting and intermediate-temperature-setting resin glues cured to different degrees in yellow birch plywood', US Department of Agriculture Forest Product Laboratory Report 1537. Dataset name: plywood.

Exercise 5.9

(a) Load the data into the GENSTAT spreadsheet and check how they are laid out. Responses are in the variate **strength** and treatment labels in the factor **glue**. Since there are six treatment groups and ten experimental units per treatment, each of these columns in the spreadsheet contains $6 \times 10 = 60$ rows.

Update the GENSTAT server and produce tables of summary statistics and boxplots for the six treatment groups, in the same way that you did for the **fats** data in Exercise 5.3. Does it seem plausible that the assumptions of the one-way analysis of variance model (normal distributions within each treatment group, and the same variance for all groups) are satisfied?

(b) Obtain the ANOVA table using the same procedure as in Exercise 5.5. Choose **strength** as the Y-Variate and **glue** as the Treatments. Look at the resulting output. (Keep a copy of the (complete) output from this analysis for use in the next section.)

Let us look only at the first part of this output for the moment, the ANOVA table. It is reproduced below.

```
Source of variation   d.f.       s.s.       m.s.     v.r.   F pr.
glue                     5    115280.4    23056.1    35.08   <.001
Residual                54     35486.9      657.2
Total                   59    150767.2
```

The table contains, among other things, the result of a test of the hypothesis that all treatment means are equal. In particular, the value of the test statistic is given under the heading v.r.; here, it happens to have value 35.08. This alone means nothing until it is compared with the appropriate null distribution and the significance probability is evaluated. This too has been done for us in the ANOVA table: under the heading F pr., the SP is given as less than 0.001, which is very small.

The conclusion we come to is that there is strong evidence that the null hypothesis is not true and hence that the treatment means are not all equal. This is not surprising if we look at boxplots of these data, as given in Solution 5.9. Plywoods produced with Glues E and F seem to be considerably stronger, on average, than, say, plywoods produced with Glues A and B, and there may well be other important differences too.

Let us not attempt any further conjectures on exactly what the treatment differences might be for now. We turn instead to consider in more detail where the different parts of the ANOVA table come from, and what they mean. ∎

Recall the notation and model described in Section 5.2.[2] Each treatment mean μ_j, $j = 1, 2, \ldots, k$, is most easily estimated by the corresponding sample treatment mean

$$\widehat{\mu}_j = \overline{Y}_j = \frac{1}{n_j} \sum_{i=1}^{n_j} Y_{ij}.$$

Also, recall the sample estimator of the overall mean

$$\overline{Y} = \frac{1}{n} \sum_{j=1}^{k} \sum_{i=1}^{n_j} Y_{ij}.$$

The right-hand summation here involves adding up the responses within the jth treatment group, and the left-hand summation involves adding up the resulting sums for all the treatment groups.

The ANOVA table starts from the idea that there are two reasons why the responses for different experimental units vary. First, different experimental units receive different treatments. The treatment means may not all be the same, and, if they are not, that is a reason for different responses. Second, whether or not the treatment means are the same, there is random variability (modelled by normal distributions) in the data, and that is another reason why the responses differ. Analysis of variance is basically a method for dividing up the variability in the responses according to these two different sources of variability in the data.

The overall variability of a dataset is often measured by its variance, which can be calculated by squaring all the differences between observations and their mean, adding them up, and then dividing by $n - 1$. In ANOVA we proceed in a similar way, by squaring differences and adding them; however, the last step of dividing by an appropriate number occurs rather later in the process. We begin with the overall or *total sum of squares* (*TSS*) of the entire dataset, which is defined to be

$$\sum_{j=1}^{k} \sum_{i=1}^{n_j} (Y_{ij} - \overline{Y})^2.$$

This quantity reflects the total variability in the data. For the plywood data, the value of *TSS* is 150 767.2 (see the ANOVA table in Example 5.5).

---

[2] You have seen some of the ideas here in Section 4.2.

Now consider the first source of variability mentioned above – variability caused by differences between the treatment means. To measure this, we must take away the random variability *within* treatment groups, and this is done by imagining that all the responses in any particular treatment group are equal. In fact, suppose that all the responses in the jth treatment group were equal to the mean of that treatment group, namely $\overline{Y}_j$. If this were the case, the sum of squared differences between the responses and their overall mean (across the whole dataset) for the quantities in the jth treatment group would be $n_j(\overline{Y}_j - \overline{Y})^2$ (because there are n_j experimental units in the group). The sum of squared differences across the whole dataset would thus be

$$\sum_{j=1}^{k} n_j(\overline{Y}_j - \overline{Y})^2.$$

This quantity is called the *ESS*, for *explained sum of squares*, because (in a particular sense) it 'explains' how much of the overall variability in the data can be attributed to differences in the treatment means.

For the plywood data, the treatment means for the six glues are (respectively) 478.8, 492.4, 503.4, 528.8, 578.6 and 596.5. The overall mean response $\overline{Y}$ is 529.75,[3] and the n_j are all equal to 10. Thus the *ESS* is calculated by taking all ten responses in the first treatment group as 478.8, all ten values in the second treatment group as 492.4, and so on, and then adding the squares of the differences between these values and the overall mean, 529.75. Thus $ESS = 10(478.8-529.75)^2 + 10(492.4-529.75)^2 + \cdots + 10(596.5-529.75)^2$, which indeed comes to 115280.4 as given in the glue row of the ANOVA table in Example 5.5.

Now what about the other source of differences between responses – random variability? By this is meant the variability left over after we have taken into account any differences in treatment means. This can be measured as follows. Within the jth treatment group, subtract the treatment mean $\overline{Y}_j$ from each response. This removes the effect of the differences between treatment means. The resulting values are known as *residuals*; we can write the residual corresponding to the response Y_{ij} as

$$r_{ij} = Y_{ij} - \overline{Y}_j.$$

(There is more on residuals in Section 5.4.) If there were no random variability, the residuals would all be zero. We proceed as before, and calculate the sum (over the whole dataset) of the squared residuals. This is *RSS*, the *residual sum of squares*:

$$RSS = \sum_{j=1}^{k} \sum_{i=1}^{n_j} (Y_{ij} - \overline{Y}_j)^2.$$

For the plywood data, the first two responses for Glue A are 502 and 458, and the corresponding treatment mean is 478.8, so the residuals corresponding to these responses are $502 - 478.8 = 23.2$ and $458 - 478.8 = -20.8$. The residual sum of squares is thus $23.3^2 + (-20.8)^2 + $ [another 58 squared residuals]. This is not difficult, though very tedious, to work out with a calculator; it does indeed come to 35486.9, as given in the Residual row of the ANOVA table in Example 5.5.

---

[3] The GENSTAT output gives the overall mean as 529.7. The exact value is 529.75. (GENSTAT does use the exact value in its calculations.)

The key to the analysis of variance is the following rather remarkable fact: *TSS* can be split up (or decomposed or analysed) as $TSS = ESS + RSS$, or, in terms of the algebraic definitions,

$$\sum_{j=1}^{k}\sum_{i=1}^{n_j}(Y_{ij} - \overline{Y})^2 = \sum_{j=1}^{k} n_j (\overline{Y}_j - \overline{Y})^2 + \sum_{j=1}^{k}\sum_{i=1}^{n_j}(Y_{ij} - \overline{Y}_j)^2.$$

This can be proved fairly easily by algebraic manipulation, but rather than doing so here let us concentrate on interpretation. If you have an algebraic bent and want to try the proof, write $(Y_{ij} - \overline{Y})^2$ as $[(Y_{ij} - \overline{Y}_j) + (\overline{Y}_j - \overline{Y})]^2$ and expand the square into three terms: the two square terms become the right-hand side while the 'cross' term can be shown to equal zero. For the `plywood` data, this equation says that $150767.2 = 115280.4 + 35486.9$, which is indeed true (allowing for rounding).

If all the (sample) treatment means were equal, and hence equal to the overall (sample) mean, then *ESS* would be zero. Conversely, if there is great disparity between treatment means, *ESS* would be large. It is clear, then, that *ESS* is very relevant to testing $H_0 : \mu_1 = \mu_2 = \cdots = \mu_k$. In fact, a good test statistic should be based on the size of *ESS* relative to *TSS* or, equivalently, to *RSS*: the test statistic that we shall use is based on the ratio *ESS/RSS*. This works because, if there is a lot of variability due to differences in treatment means (*ESS*) relative to the residual variation reflected in the *RSS*, the null hypothesis is in doubt, while if *ESS* is small relative to *RSS*, the null hypothesis seems tenable. The technical statistical justification for looking at *ESS/RSS* rather than (say) *ESS/TSS* is that *ESS* and *RSS* are independent quantities. This is related to the fact that they describe two independent sources of variability in the responses.

The column in the ANOVA table headed s.s., for 'sum of squares', therefore contains *ESS*, *RSS* and *TSS* in the first, second and third rows, corresponding to treatment, residual and total, respectively. Table 5.5 shows the first three columns of the ANOVA table in generic form.

Table 5.5

Source of variation	d.f.	s.s.
Treatment	$k - 1$	ESS
Residual	$n - k$	RSS
Total	$n - 1$	TSS

The column heading d.f. stands, as usual, for 'degrees of freedom'. Concentrate for a moment on the *RSS*. The number of degrees of freedom, v, for this sum of squares is defined to be

$$v = n - \text{(the number of unknown parameters estimated in the model)}.$$

Two important properties of *RSS* and its degrees of freedom are:

(a) $E\left(\dfrac{RSS}{v}\right) = \sigma^2$,

(b) $\dfrac{RSS}{\sigma^2} \sim \chi^2(v)$.

The first property says that RSS/v is unbiased as an estimator of σ^2. The second explains why the number of degrees of freedom for RSS is defined in this particular way: the degrees of freedom for RSS match the degrees of freedom for the χ^2 distribution involved. The 'number of unknown parameters' here is the number of unknown parameters other than σ^2, since we are currently estimating σ^2 via the RSS. In the current ANOVA model, k parameters are being estimated, namely $\mu_1, \mu_2, \ldots, \mu_k$. Hence the d.f. associated with the RSS in Table 5.5 is $n - k$.

Likewise, TSS can be thought of as a residual sum of squares because it is the residual sum of squares for the very simple model

$$Y_{ij} = \mu + \epsilon_{ij}.$$

This is the model of Box 5.2 with μ_j taking some constant value μ for $j = 1, 2, \ldots, k$ and is thus the model associated with the null hypothesis of equal treatment means. In fact, this model is simply $Y_{ij} \sim N(\mu, \sigma^2)$ and $TSS/(n - 1)$ is then the usual estimator of σ^2, as in Chapter 2. Since this model has only one (overall) mean parameter, the degrees of freedom associated with TSS is $n - 1$. Degrees of freedom is therefore a generalization of the idea of the 'number of pieces of information' about the variance in the data.

The number of degrees of freedom for ESS is defined to be the difference between the number for TSS and the number for RSS: $(n - 1) - (n - k) = k - 1$. This is the number of treatment means, minus one for the overall mean.

Not only is $RSS/\sigma^2 \sim \chi^2(n - k)$ but it turns out that, under the null hypothesis H_0 of equal means,

$$\frac{ESS}{\sigma^2} \sim \chi^2(k - 1)$$

independently of RSS. So, under H_0, the random variable $B = ESS/RSS$ is distributed as a ratio of independent χ^2 random variables. It is more usual, however, to consider the random variable

$$F = \frac{ESS/(k - 1)}{RSS/(n - k)} = \frac{n - k}{k - 1} B.$$

The distribution of F, under H_0, is the F *distribution on $k - 1$ and $n - k$ degrees of freedom*, written $F(k - 1, n - k)$, as introduced in Section 2.4.2. And F is the corresponding F test statistic, sometimes called the F *statistic* for short.

Now we can complete the ANOVA table. After the s.s. column comes a column headed m.s., for 'mean square'. This consists of $ESS/(k - 1)$ and $RSS/(n - k)$ in the Treatment and Residual rows, respectively, i.e. the entry in the s.s. column divided by the d.f. Under the null hypothesis of equal treatment means, both quantities are estimates of σ^2; however, if the null hypothesis is false, $ESS/(k - 1)$ is no longer an estimate of σ^2, but $\hat{\sigma}^2 = RSS/(n - k)$ still is, since it directly estimates σ^2 in the most natural way (using residuals from separately estimated treatment means) from the one-way ANOVA model of Box 5.2.

The next column of the ANOVA table, headed v.r. (for 'variance ratio'), contains the ratio of the two mean squares: this is the F statistic.

Finally, GENSTAT compares the F statistic with the $F(k - 1, n - k)$ distribution and prints out, under the heading F pr. (for 'F probability'), the significance probability for the test for equal treatment means. This is $P(F \geq \text{v.r.})$ where $F \sim F(k - 1, n - k)$.

Box 5.3

The one-way analysis of variance (ANOVA) table for the model of Box 5.2 has the following entries.

Source of variation	d.f.	s.s.	m.s.	v.r.	F pr.
Treatment	$k-1$	ESS	$\dfrac{ESS}{k-1}$	$\dfrac{ESS/(k-1)}{RSS/(n-k)}$	$P(F \geq \text{v.r.})$
Residual	$n-k$	RSS	$\dfrac{RSS}{n-k}$		
Total	$n-1$	TSS			

Here $F \sim F(k-1, n-k)$.

Exercise 5.10

Measurements of the heights (Y, in inches) of male singers in the New York Choral Society were made. The singers were classified according to four voice parts, with vocal range decreasing in pitch going from Tenor 1 to Bass 2. These data comprise 107 responses (heights) of individuals grouped into four groups (voice parts). The data are plotted in Figure 5.5. (Source: Chambers, J. M., Cleveland, W. S., Kleiner, B. and Tukey, P. A. (1983) *Graphical Models for Data Analysis*, Boston, Duxbury Press. Dataset name: `singers`.) The data could be analysed entirely using GENSTAT; but, instead, you are invited to check your understanding of how the ANOVA table is constructed by doing the following.

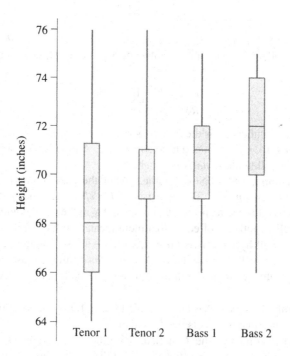

Fig. 5.5   Boxplots of heights of male singers for different voice parts.

It turns out that, for these data, $TSS = 786.785$ and $RSS = 705.670$. With this information, complete the ANOVA table by hand for this dataset, all bar the F pr. entry. To obtain the latter, you need to do a calculation in GENSTAT. Go into the Input Log window in interactive mode and enter the appropriate form of

PRINT cuf (*v.r.* ; *d.f.(Treatment)* ; *d.f.(Residual)*).

(Here *v.r.*, *d.f.(Treatment)* and *d.f.(Residual)* represent numbers; you should enter the appropriate numbers rather than the characters printed here.) Notice that cuf has three arguments (separated by semicolons), namely the value of the F statistic and the two d.f. parameters of the F distribution (in that order). This command works out the probability that a random variable with the F distribution on *d.f.(Treatment)* and *d.f.(Residual)* degrees of freedom is greater than *v.r.*

What do you conclude in relation to the null hypothesis that the mean heights for all the voice parts are the same?

To complete this section, the mystery of the name 'analysis of variance' can, at last, be solved. Although the analysis you have learned to carry out is to do with the comparison of *means*, the comparison is done by comparing quantities (*ESS*, *RSS*) which arise from a decomposition (the 'analysis') of *TSS*. And *TSS* (or $TSS/(n-1)$) is a combination of $ESS/(k-1)$ and $RSS/(n-k)$, all these mean squares being, under appropriate versions of the model, estimates 'of variance'.

5.4 Checking the model

Before doing anything further using the one-way ANOVA model, let us briefly consider how we can investigate whether the model is a reasonable one for the data at hand. There are several useful checks that GENSTAT provides, one in the output that has already been obtained, others available with one or two further clicks of your mouse. The checks generally resemble quite closely those used in Chapter 4 in investigating the appropriateness of regression models.

As usual in any data analysis, a good place to start checking the assumptions is by plotting the data. In our GENSTAT analyses so far, we have generally begun by producing boxplots of the responses in each treatment group. These plots, together with summary tables including the group variances, give a good basis for checking whether the assumption of equal variance is valid.

There exist formal tests for equal variance. (One is implemented in GENSTAT in the VHOMOGENEITY command.) We do not recommend their routine use.

Example 5.5 continued
Let us consider the plywood data once more. You saw in Exercise 5.9 that the assumption of equal variances looked reasonable for these data. Let us see whether GENSTAT indicates any departures from the model.

The first extra item beneath the ANOVA table in the GENSTAT output for the plywood data reads as follows.

```
* MESSAGE: the following units have large residuals.

*units* 50            59.4    s.e.  24.3
```

Here, GENSTAT is trying to alert us to a possible outlier in the model. To do this it looks for large residuals from the fitted model. In fact, you can obtain, in your output, all 60 residuals from fitting the one-way ANOVA model, by clicking on the **Options** button in the **Analysis of Variance** dialogue box and clicking on the **Residuals** check box when the **ANOVA Options** dialogue box appears; but we do not recommend this overload of information.

The unstandardized residuals are $r_{ij} = Y_{ij} - \overline{Y}_j$, i.e. differences between observed responses and estimated treatment means. To assess 'largeness' of residuals, as in Chapter 4, we need to take account of the standard error associated with the residuals. Since

$$r_{ij} = Y_{ij} - \frac{1}{n_j} \sum_{l=1}^{n_j} Y_{lj}$$

$$= Y_{ij} - \frac{1}{n_j} \left[Y_{ij} + \sum_{l=1, l \neq j}^{n_j} Y_{lj} \right]$$

$$= \left(1 - \frac{1}{n_j}\right) Y_{ij} - \frac{1}{n_j} \sum_{l=1, l \neq j}^{n_j} Y_{lj}$$

and the Y_{ij} are assumed to be independent with constant variance σ^2, the formulae for the variance of a linear function of a random variable and for a sum of independent random variables (Chapter 2, Section 2.1.4) yield

$$V(r_{ij}) = \left(1 - \frac{1}{n_j}\right)^2 \sigma^2 + (n_j - 1) \left(\frac{1}{n_j}\right)^2 \sigma^2 = \left(1 - \frac{1}{n_j}\right) \sigma^2.$$

The standard error of r_{ij} is thus $\sqrt{(1 - (1/n_j))\sigma^2}$, and its estimated standard error is obtained by replacing σ^2 by its estimator which, as pointed out in Section 5.3, is given by the m.s.(Residual) from the ANOVA table. For the plywood data, $n_j = 10$ for all treatment groups and m.s.(Residual) $= 657.2$, so the estimated standard error of each residual is $\sqrt{591.5} = 24.3$, as stated in the GENSTAT output.

GENSTAT uses these standard errors in much the same way as in simple linear regression (see Section 4.4). It looks at the ratios residual/s.e.(residual), where the denominator is estimated as above. These are the (standardized) Pearson residuals, and are again approximately independent and normally distributed. GENSTAT uses the same rule as for simple linear regression to decide which residuals to flag.

A first thing to notice is that the flagged residual, 59.4, is 2.44 times its standard error, and hence is not *very* extreme. We can also check the flagged datapoint against boxplots of the data which, for the plywood data, were given in Solution 5.9. All datasets in this chapter are arranged in order of treatment: all the responses in the first treatment group appear first, all the responses in the second treatment group next, and so on. So, 'units' 50 (i.e. datapoint 50) comes from the fifth treatment group, Glue E, and must be the largest value (because of the sign of the residual) in that group. So we need to look at the boxplot for Glue E, which does not indicate that anything is amiss. In this case, the potential outlier does not seem to provide cause for concern about the model.

As in simple regression, we can check the assumptions further by producing residual plots. After fitting the basic model, click on the **Further Output** button in the **Analysis of Variance** dialogue box, and then click on the **Residual Plots** button in the **ANOVA Further Output** dialogue box. Click on **OK** (i.e. do not change any default options) and obtain something like Figure 5.6. The plots are essentially the same as those you obtained for regression models in Chapter 4; but for analysis of variance, simple rather than Pearson residuals are plotted, and the four plots appear in a different order and in different colours. (Also, the extra curve is not included in the plot of residuals against fitted values.)

All these plots help us assess whether normality is indeed a reasonable assumption (as well as giving further information on the possible outliers). Figures 5.6(a) and (b) are normal and half-normal plots. Two or three points show up (most clearly in the half-normal plot) as giving a slightly curved end to the line of points. Figure 5.6(c) is a histogram of these same residuals, and more clearly suggests a skewness in the distribution of residuals. However,

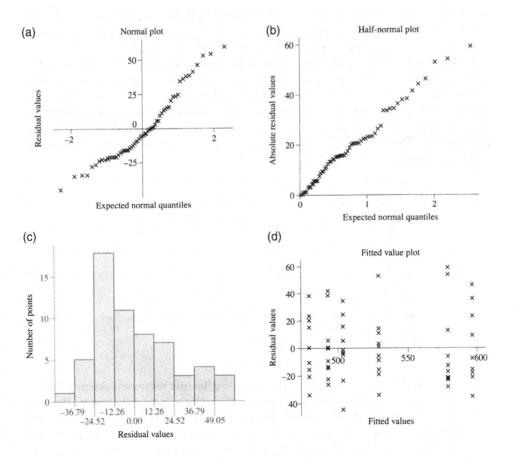

Fig. 5.6   Residual plots for the plywood data.

any departures from normality that might be indicated by these plots seem not to be serious enough to worry about. Remember that we are dealing with relatively small samples, and natural random variation means that mild departures from apparent 'exact normality' happen all the time. Moreover the results of the ANOVA method are not especially sensitive to small departures from normality: in the technical jargon, one-way ANOVA is fairly *robust* with regard to deviations from normality. It is therefore only severe departures from normality indicated by such plots that need seriously worry us.

Figure 5.6(d) plots residuals against fitted values – which in this case are treatment sample means. If there were marked differences in variance between the treatment groups, it would show up in this plot, but here there is no such problem. There is perhaps a slight hint of curvature in this plot, indicating that the model has not taken full account of the way the data vary. If this were a linear regression analysis, such curvature might indicate that a straight-line model is inappropriate. Here, it is more likely to be a consequence of the skewness of the residuals, but, in any case, it is not very marked and does not throw serious doubt on the model assumptions.   ■

Exercise 5.11

Data from an experiment in plant physiology conducted by W. Purves are contained in the file `pea.gsh`. Purves recorded the lengths (responses) of pea sections grown in an appropriate tissue culture in the presence of each of five different growth mediums (treatments) with a view to testing the effects of different sugars on pea growth. Lengths were given 'in coded units', implying that, while meaning something to the experimenter, the units do not mean anything to us. The five mediums differed only in the presence or absence of different sugars; one medium ('control') contained no sugars, while the others contained different quantities of different sugars as given in Table 5.6. Ten pea section lengths were recorded per treatment. The responses for the first five experimental units in each treatment group are also given in Table 5.6.

Table 5.6

Control	2% glucose	2% fructose	1% glucose + 1% fructose	2% sucrose
75	57	58	58	62
67	58	61	59	66
70	60	56	58	65
75	59	58	61	63
65	62	57	57	64
⋮	⋮	⋮	⋮	⋮

Source: Sokal, R. R. and Rohlf, F. J. (1981) *Biometry*, 2nd edition, San Francisco, W. H. Freeman.
Dataset name: pea.

(a) Load the data into GENSTAT. The section lengths (the responses) are in a variate called **length** and the treatment groups are defined by a factor called **sugar**. Produce tables of summary statistics and boxplots for the five treatment groups. Does it seem plausible that the assumptions of the one-way analysis of variance model (normal distributions within each treatment group, and the same variance for all groups) are satisfied?

(b) You probably concluded in part (a) that the assumption of equal variance was not appropriate for these data. In some situations, it is possible to deal with this difficulty by transforming the response variable. With these data, no simple transformation seems to have the required effect (though a reciprocal transformation improves things to some extent). Ignoring this problem for now, obtain the ANOVA table (for the untransformed response variable), and describe the results of the test of the hypothesis that all mean lengths for different growth mediums are equal.

(c) Observe what GENSTAT flags in the way of possible outliers, and identify the positions of the flagged value(s) on the boxplots you produced in (a). Comment on what you find.

(d) Look at the residual plots provided by the **Further Output** button, and comment.

(e) What are your conclusions about the suitability of the one-way ANOVA model for these data?

Exercise 5.11 raised the issue of whether we should call a particular point an outlier and, more importantly, even if we do, whether its inclusion is actually making any difference to the inferences we draw. One rather crude but (sometimes) effective way of investigating this is simply to remove the suspect datapoint from the dataset, and to perform the analysis again on the reduced dataset. If the conclusions are unchanged, then one can usually be confident that the breakdown in the one-way ANOVA model assumptions indicated by the presence of the suspect point is not, in fact, leading you to incorrect conclusions. In such circumstances, one says that the analysis is *robust* to the presence of the outliers. But sometimes the analysis is not robust to the presence of outliers, as the following exercise illustrates.

Exercise 5.12

Does 'active exercise' for infants in the first few weeks of life speed up the time at which the infants learn to walk alone? This question was addressed in a completely randomized experiment from which the data in Table 5.7 are taken. The response variable is the age of first walking alone (in months) and there are four treatments: 'active exercise' and three other regimes not expected to have such an effect and thus used as 'controls' against which to measure any effect of active exercise, namely 'passive exercise', 'no exercise', and an '8-week control group'. There were 6, 6, 6 and 5 infants respectively under these exercise regimes, a total of $n = 23$ subjects in $k = 4$ groups. The data are in babies.gsh, the responses being in the variate **walktime** and the treatments in the factor **exercise**.

(a) Load the data into the GENSTAT spreadsheet and update the server. Rather than producing boxplots for these very small groups, produce a scatterplot with the response variable on the vertical axis and the treatment factor on the horizontal axis. You can do this by choosing the **Graphics|Point Plot** menu item, and, in the resulting dialogue box, entering **walktime** for the Y Coordinates and **exercise** for the X Coordinates. On the horizontal axis of the resulting plot, the treatment groups are identified by numbers rather than the factor labels: they are in the order given in Table 5.7. The advantage of this plot over boxplots for such small datasets is that

Table 5.7

Active exercise	Passive exercise	No exercise	8-week controls
9.00	11.00	11.50	13.25
9.50	10.00	12.00	11.50
9.75	10.00	9.00	12.00
10.00	11.75	11.50	13.50
13.00	10.50	13.25	11.50
9.50	15.00	13.00	

Source: Zelazo, P. R., Zelazo, N. A. and Kolb, S. (1972) ' "Walking" in the newborn', *Science*, **176**, 314–5.
Dataset name: `babies`.

individual points are easy to identify. One disadvantage is that, when two datapoints have the same values (as, for example, the two 10.00 values in the passive exercise group), GENSTAT plots a single point for both of them, so that the distribution of the data is not always clear. What do you conclude from the plot about possible problems with the one-way ANOVA model assumptions for these data?

(b) Perform an ANOVA on these data and describe the outcome of the test.

(c) Discuss the two datapoints flagged by GENSTAT in the light of the plot you produced in part (a).

(d) Remove these two points and repeat part (a). One way to remove them temporarily is to select the spreadsheet window, and select the rows containing the points (by clicking on the numerical label at the left-hand end of each of the rows, 5 and 12 in this case). Then choose the Spread|Restrictions|Add Selected Rows menu item. The rows disappear from the spreadsheet. (As in Exercise 4.26, you can get them back, if you need to, by choosing the Spread|Restrictions|Remove All menu item.) Update the server, rerun the ANOVA and GENSTAT will act as if the restricted rows are not in the dataset. What difference has removal made?

(e) Comment on the further flagged datapoint at this stage. (Be careful! The GENSTAT ANOVA commands number the points according to where they come in the reduced dataset, with the previous two points removed, and not according to their original numbering.)

(f) Remove this point too (number 15 in the original numbering) and produce the new ANOVA table. Now what do you see?

(g) Has the one-way ANOVA model proved robust to the presence of outliers in this case?

The results of Exercise 5.12 warn us not to be too complacent about the effects of individual points on our results. However, remember that this is an example with a very small number of individuals per treatment group, and further resolution of the effects of these treatments needs rather more data to have been collected (to determine whether the experimenters simply chose a few unusual infants or whether, for instance, infants may divide into, say, responders and non-responders under each treatment type). We cannot just throw away datapoints that upset the analysis, without further investigation.

Finally, a reminder that these standard checks on the data – producing tables of group variances, producing boxplots, looking at flagged points, looking at residual plots – are not the only ways to investigate potential problems in the data. You will learn about some more sophisticated methods in Chapter 11. But other simple methods, such as keeping a healthily sceptical attitude of mind, can help too. For example, in Exercise 5.3 you produced group means for the fats data. It might have struck you as rather strange that *all five* of these means came out to be exact integers. That is, all six group totals must have been exactly divisible by the group size (6 for each group). Possibly this is just a coincidence. But, more likely, someone has adjusted the data. Perhaps this was done by the authors of the textbook from which the data were taken, in order to make life easier for students analysing the data without a computer – but they present them as real data. Be on your guard!

5.5   Differences between treatments

The results of the analyses of Section 5.3 are unsatisfactory in the following way: a conclusion that there is evidence against the null hypothesis $H_0 : \mu_1 = \mu_2 = \cdots = \mu_k$ simply tells us that there seems to be *some* difference or differences between treatments. The question remains, how do the treatments differ from one another? The analysis of variance table can help with this question, because it provides an estimate of the residual variance σ^2 that can be used in further investigations of treatment differences. The final entries in the GENSTAT ANOVA default output are also related to this question, but will not be addressed nor used explicitly in this chapter.

5.5.1   Planned comparisons and contrasts

Quite often, certain special comparisons between treatments are of particular importance to the investigators, and are known to be so before the study is carried out. These are often referred to as *planned* comparisons.

Example 5.6
The growth medium treatments in the pea dataset in fact consist of a control (the first treatment) and four different sugar treatments (numbers 2 to 5). The question 'Does sugar (in general) affect the growth of pea sections?' can be answered by comparing μ_1 with the average of μ_2 to μ_5. A natural way to do this is by considering the quantity

$$\theta_1 = \mu_1 - \tfrac{1}{4}(\mu_2 + \mu_3 + \mu_4 + \mu_5). \tag{5.2}$$

A point estimate $\widehat{\theta}_1$ of θ_1 is immediate by replacing the treatment means in (5.2) by the sample treatment means, i.e.

$$\widehat{\theta}_1 = \overline{Y}_1 - \tfrac{1}{4}(\overline{Y}_2 + \overline{Y}_3 + \overline{Y}_4 + \overline{Y}_5).$$

The sample treatment means are given in the default GENSTAT output under the heading

Tables of means. We have

$$\widehat{\theta}_1 = 70.1 - \tfrac{1}{4}(59.3 + 58.2 + 58.0 + 64.1) = 10.2.$$

If there is any difference, then it appears to be positive, implying that the no-sugar mean is bigger than the combined sugar-treated means: sugar appears to inhibit growth of pea sections. But is there a real difference, and not one due only to random variation between experimental units? To find out, we need the standard error of $\widehat{\theta}_1$.

Since, according to the one-way ANOVA model,[4] each $Y_{ij} \sim N(\mu_j, \sigma^2)$, then $\overline{Y}_j \sim N(\mu_j, \sigma^2/n_j)$ and, in particular for this dataset, $V(\overline{Y}_j) = \sigma^2/10$. The $\overline{Y}_j$s are independent, since each is based on an independent set of Y_{ij}s, and so the variance of the linear combination of independent quantities that makes up $\widehat{\theta}_1$ is

$$V(\widehat{\theta}_1) = \frac{\sigma^2}{10} + \left(\frac{1}{4}\right)^2 \left(\frac{\sigma^2}{10} + \frac{\sigma^2}{10} + \frac{\sigma^2}{10} + \frac{\sigma^2}{10}\right) = \frac{5}{4} \times \frac{\sigma^2}{10} = \frac{\sigma^2}{8}.$$

Using the m.s.(Residual) $= 5.456$ as an estimate of σ^2, the estimated $V(\widehat{\theta}_1)$ is $5.456/8 = 0.682$, so that the estimated s.e.$(\widehat{\theta}_1)$ is $\sqrt{0.682} = 0.826$.

The value 10.2 is $10.2/0.826 = 12.35$ times its standard error and so is obviously significantly different from zero with a very tiny SP. To work the SP out properly, one needs the distribution of $\widehat{\theta}_1/(\text{estimated s.e.}(\widehat{\theta}_1))$, which turns out to be the t distribution on 45 degrees of freedom. The 45 is the number of degrees of freedom (d.f.(Residual)) associated with the estimate of σ^2 (m.s.(Residual)) and is obtained from the ANOVA table. The (two-sided) SP of the test $H_0 : \theta_1 = 0$ is therefore

$$P(|t(45)| \geq 12.35) = 2P(t(45) \geq 12.35) = 2 \times 0.2351 \times 10^{-15}.$$

To get this in GENSTAT, use the command:

        PRINT cut(12.35;45)

(In general, PRINT cut$(x; v)$ gives $P(t(v) \geq x)$.)

Indeed you can perform the entire operation of estimating θ_1 and estimating its standard error in GENSTAT – but the procedure is rather cumbersome and, for straightforward combinations of means of the sort considered in this chapter, it is quicker to do the calculations in the way shown here.

Also, when a quantity such as (5.2) is of interest, it may well be more valuable to give a confidence interval for it. For instance, a 95% confidence interval for θ_1 is $\widehat{\theta}_1 \pm t_{0.975}(45)\text{s.e.}(\widehat{\theta}_1)$. From GENSTAT (now using edt, as in Exercise 3.4, in the Input Log window in place of cut), $t_{0.975}(45) = 2.014$, yielding the confidence interval

$$(10.2 - 2.014 \times 0.826, \, 10.2 + 2.014 \times 0.826) = (8.54, 11.86).$$

Notice that this confidence interval excludes zero, and by a long way.   ∎

Combinations of treatment means like (5.2) are called *contrasts*. The defining property is that a contrast is a linear combination

$$\theta = a_1\mu_1 + a_2\mu_2 + \cdots + a_k\mu_k$$

---

[4]All this is being done under the standard assumptions of the one-way ANOVA model (so we are temporarily ignoring the evidence that these data do not fit the model very well).

of the treatment means such that

$$a_1 + a_2 + \cdots + a_k = 0.$$

(To avoid triviality we exclude the case where all the a_js are zero, but some of them may be.)

The contrast $-\theta$ makes exactly the same comparisons as θ. Indeed, so does any (positive or negative) constant times θ. This may make the definition of a contrast sound too arbitrary to be useful; but this is not so. If you need a numerical estimate of a particular contrast, then the question you wish to answer using the estimate will define the contrast. If you merely wish to test whether a particular contrast is zero, then you could legitimately multiply your contrast by any non-zero constant you like. But the standard error of a contrast depends on the actual values of the a_js in its definition. If the contrast is multiplied by a constant k, then its estimate will be multiplied by k, and so will its standard error (see below). Thus the ratio of the estimate to its standard error will be unchanged in absolute value, and the result of the test will be the same.

The contrast is, in general, estimated by

$$\widehat{\theta} = a_1 \overline{Y}_1 + a_2 \overline{Y}_2 + \cdots + a_k \overline{Y}_k,$$

which clearly has the property of unbiasedness: $E(\widehat{\theta}) = \theta$. The variance of $\widehat{\theta}$ is $a_1^2 V(\overline{Y}_1) + a_2^2 V(\overline{Y}_2) + \cdots + a_k^2 V(\overline{Y}_k)$, because of independence, which is $a_1^2 \sigma^2 / n_1 + a_2^2 \sigma^2 / n_2 + \cdots + a_k^2 \sigma^2 / n_k$, i.e.

$$V(\widehat{\theta}) = \sigma^2 \sum_{j=1}^{k} \frac{a_j^2}{n_j}.$$

(In the case of equal replications, where $n_1 = n_2 = \cdots = n_k$, this reduces to $(\sigma^2 / n_k) \sum_{j=1}^{k} a_j^2$.)

So far in this chapter, it has generally made no difference to our analyses whether the data had equal replications or not. But, as will be demonstrated in the next example, equal replications can make things more straightforward when defining and calculating contrasts.

Exercise 5.13

Consider again the `singers` dataset (Exercise 5.10). In contrast to most of the other examples we have looked at, it can clearly be argued that the explanatory variable (voice part) here is ordinal and that one might like to know, for instance, whether there is any trend in height as pitch decreases. Ignoring the ordering in the explanatory factor means ignoring some of the useful information given to us. But then how to incorporate the ordering is unclear. For instance, we could set up an explanatory variate that takes the values 1 for Tenor 1, 2 for Tenor 2, 3 for Bass 1 and 4 for Bass 2, and try linear regression. However, this labelling is fairly arbitrary – although it does respect the ordering – so linear regression is not appropriate as the data stand. Some aspects of the ordering can be investigated by looking at appropriate contrasts, such as comparing, say, a combination of the two tenor groups with a combination of the two bass groups. One possible contrast for this comparison is

$$\theta_2 = \tfrac{1}{2}(\mu_1 + \mu_2) - \tfrac{1}{2}(\mu_3 + \mu_4).$$

Note that $n_1 = n_2 = 21$, $n_3 = 39$ and $n_4 = 26$, the sample means are 68.90, 69.90, 70.72 and 71.38, respectively, and the ANOVA table for these data is given in Solution 5.10.

(a) Estimate θ_2.

(b) What is the variance of $\widehat{\theta}_2$ in terms of σ^2?

(c) Estimate s.e.$(\widehat{\theta}_2)$.

(d) Hence test the hypothesis that $\theta_2 = 0$ and interpret your result.

But is θ_2 the most appropriate contrast for this aspect of the singers data? After all, it was defined as a difference between the average of the two mean tenor heights and the average of the two mean bass heights. But why should the mean bass heights, for instance, be combined by simply taking their average? In the dataset there are many more Bass 1 than Bass 2 singers, so that $\frac{1}{2}(\mu_3 + \mu_4)$ is not even a good estimate of the mean height of basses in this choir. It could be argued that, if one wishes to compare the heights of all tenors (regardless of whether they are Tenor 1 or Tenor 2) with all basses (regardless of whether they are Bass 1 or Bass 2), one is thinking of all the tenors as being drawn from a single population of tenors and all the basses as being drawn from a single population of basses. Under this scenario, $\mu_1 = \mu_2$ and $\mu_3 = \mu_4$. Rather than using $\frac{1}{2}(\mu_3 + \mu_4)$ as a measure of the average height of all basses, one would be better off taking account that there are 39 Bass 1 singers and 26 Bass 2 singers (a total of 65 in all). Because of this, a better estimate of the mean height of all basses would be given by $\frac{39}{65}\overline{Y}_3 + \frac{26}{65}\overline{Y}_4$. (Similar difficulties do not arise with the tenors, because their group sizes happen to be equal.) These considerations suggest that the contrast $\theta_3 = \frac{1}{2}(\mu_1 + \mu_2) - \left(\frac{39}{65}\mu_3 + \frac{26}{65}\mu_4\right)$ might be more appropriate than θ_2. It can then be calculated that $\widehat{\theta}_3 = -1.584$, with an estimated standard error of 0.518.

Do not concern yourself too much with the details of this particular contrast. It serves to demonstrate two points. First, it is important when defining a contrast to be precise about the question you want the contrast to answer. Second, although for most of this chapter it has been irrelevant whether a particular dataset had equal replications or not, the difficulties here arose from the fact that there were unequal replications. This is one context in which equal replications can make things simpler. The remaining exercises in this chapter do indeed involve datasets with equal replications.

It generally happens that there is more than one planned comparison of interest. This is so for the pea data. As well as the control versus sugar comparison, differences between sucrose on the one hand and glucose and fructose on the other suggest that a second particularly interesting comparison might be between the three glucose/fructose treatments (treatment 2, glucose, treatment 3, fructose, and treatment 4, their mixture) and sucrose (treatment 5). A contrast for making this comparison is $\theta_4 = \mu_5 - \frac{1}{3}(\mu_2 + \mu_3 + \mu_4)$. Notice that μ_1 does not appear here, corresponding to $a_1 = 0$.

If one were considering several such contrasts for the same dataset, there are considerations (concerning an appropriate definition of 'separateness' of comparisons) that it can be important to take into account. For instance, in the pea data there are five treatment groups, and thus four degrees of freedom for treatments. One consequence of this is that it would not make sense (for most purposes) to consider more than four contrasts for these data, essentially because, given the estimates of four appropriate contrasts, one could work out the estimate of any other contrast without having to use the original treatment means. Speaking very loosely, the four degrees of freedom for treatments can be interpreted as meaning that there is only room for making four treatment comparisons. There is no need at this stage to go into the details of these considerations; you can assume that it is reasonable to consider all the contrasts you are asked about for any particular dataset.

Exercise 5.14
For the pea data,

(a) Estimate $\theta_4 = \mu_5 - \frac{1}{3}(\mu_2 + \mu_3 + \mu_4)$.

(b) What is the variance of $\widehat{\theta}_4$ in terms of σ^2?

(c) Estimate s.e.$(\widehat{\theta}_4)$.

(d) Hence test the hypothesis that $\theta_4 = 0$ and interpret your result.

5.5.2 Unplanned comparisons

In Section 5.5.1, care has been taken to discuss only comparisons among subsets of means that were specified as being of interest ('planned') beforehand. But when an overall ANOVA test for equality of all means shows a (strongly) significant difference, there is a strong temptation to seek out some of the biggest apparent differences in means (from the boxplots or scatterplot, or from GENSTAT's Tables of means) and claim that any overall effect is due principally to these. *This is not an illegitimate approach, but extra care has to be taken.* In the next two paragraphs, an indication of how the problem of *unplanned* comparisons can be addressed is given, but you will not be asked to make any such comparisons.

What is 'the problem of unplanned comparisons'? Consider the salinity dataset of Example 5.4. It turns out that the ANOVA test of equality of all three treatment means has $SP < 0.001$. Looking at Figure 5.4, we might be tempted to put this difference down, largely, to the apparent difference between group means 1 (Water Mass I) and 2 (Water Mass II). So can we just apply the techniques of Section 5.5.1 to $\mu_2 - \mu_1$? Not directly, because this comparison was unplanned (before the experiment took place) and, more importantly, *selected* for scrutiny because it is the largest mean difference in the data. This mean difference can be tested, but it must be tested against the null distribution of largest mean differences; i.e. the test statistic is not really $\overline{Y}_2 - \overline{Y}_1$ (ignoring its s.e. for a moment) but $\max_{i,j} |\overline{Y}_i - \overline{Y}_j|$ (which just happens to be $\overline{Y}_2 - \overline{Y}_1$ in this case). While an arbitrary (standardized) difference in sample means has the t distribution as null distribution, a maximal difference has a different null distribution; indeed, percentage points will necessarily be higher, and more difficult to surpass. Usually, however, a significant overall test will indeed correspond to a significant largest difference in means, but this is not necessarily so.

An extension of the above is to decide on which out of the $k(k-1)/2$ pairwise differences between treatments are non-zero. Several different approaches are available in this case, but again they involve (complicated) null distributions that take into account the ordering of the sample differences. A further complication relevant to both planned and unplanned comparisons also now resurfaces (having been mentioned in Section 5.1): if one performs multiple comparisons, the probability of finding a significant difference just by chance is (greatly) increased. Again, there are ways around this, but the best advice seems to be to avoid large numbers of comparisons, particularly since they suggest poorly specified research objectives.

5.6 A final example

To bring together many of the ideas of this chapter, work through the following analysis of a dataset on hemispheric dominance in the brain.

Exercise 5.15

Psychologists are interested in how different kinds of brain dominance affect recall ability of information of various types. Researchers reported an experiment in which subjects were asked to recall information presented to them in tabular form about the numbers of doctors practising in various US states. The subjects were divided into three groups of eight, depending on whether they were predominantly left-brained (active, verbal, logical; Group 1), right-brained (receptive, spatial, intuitive; Group 2) or integrative (both; Group 3). These characteristics were determined by a separate test. The data are given in Table 5.8.

Table 5.8   Scores from a recall test

Left-brained	Right-brained	Integrative
35	17	28
32	20	30
38	25	31
29	15	25
36	10	26
31	12	24
33	8	24
35	16	27

Source: Brown, T. S. and Evans, J. K. (1986) 'Hemispheric dominance and recall following graphical and tabular presentation of information', *Proceedings of the 1986 Annual Meeting of the Decision Sciences Institute*, **1**, 598. Dataset name: brain.

The data are given in the file brain.gsh. The recall variable is in the variate **recall** and the groups are defined by the factor hemi.

Does the ability to recall this kind of information depend on a subject's hemispherical dominance? If so, is there a difference between (predominantly) single-sided brain users and both-sided brain users? To get answers to these questions, follow through the analysis prompted below.

(a) Identify the response variable, treatments and experimental units in this experiment.

(b) Make a suitable plot of the data. What are your tentative conclusions? Are any of the assumptions of the one-way ANOVA model in doubt?

(c) Now assume that it is sensible to pursue an analysis of variance for these data. Obtain an ANOVA table for these data. Does the ability to recall tabular information on numbers of doctors depend on a subject's hemispherical dominance?

(d) With the aid of appropriate GENSTAT output, discuss further the appropriateness of the one-way ANOVA model for these data.

(e) Continuing to assume that the one-way ANOVA model is appropriate, set up a suitable contrast to answer the question 'Is there a difference between (predominantly) single-sided brain users and both-sided brain users?' and obtain an answer.

(f) Finally, gather together your conclusions and interpret them. Do you have any reservations about your analysis? (You might like to refer to your plot from part (b).)

Summary of methodology

1 One-way analysis of variance, or one-way ANOVA for short, concerns a normally distributed response variable and a single categorical explanatory variable which represents treatments or groups. The one-way ANOVA model for the ith individual in the jth treatment group is

$$Y_{ij} = \mu_j + \epsilon_{ij}$$

where the ϵ_{ij} are independently distributed as $N(0, \sigma^2)$.

2 The ANOVA table lays out the quantities required to test the hypothesis $H_0 : \mu_1 = \mu_2 = \cdots = \mu_k$. The method works by decomposing the total sum of squares

$$TSS = \sum_{j=1}^{k} \sum_{i=1}^{n_j} (Y_{ij} - \overline{Y})^2$$

into residual and explained sums of squares:

$$TSS = ESS + RSS$$

where

$$ESS = \sum_{j=1}^{k} n_j (\overline{Y}_j - \overline{Y})^2$$

and

$$RSS = \sum_{j=1}^{k} \sum_{i=1}^{n_j} (Y_{ij} - \overline{Y}_j)^2.$$

3 The ANOVA table has rows for Treatment, Residual and Total. Their degrees of freedom, in the d.f. column, are $k - 1$, $n - k$ and $n - 1$, respectively. ESS, RSS and TSS are given in the s.s. column. The test is based on the ratio of mean squares ($ESS/(k - 1)$ and $RSS/(n - k)$), these mean squares being given in the m.s. column, and the resulting test statistic being given under v.r.

4 The ANOVA table is completed by an SP obtained by comparing the observed v.r. with the F distribution on $k-1$ and $n-k$ degrees of freedom, denoted $F(k-1, n-k)$.

5 The quantity m.s.(Residual) estimates the variance parameter σ^2.

6 GENSTAT flags large standardized residuals as potentially outlying observations and also produces plots suitable for assessing the appropriateness of the normality and equal-variance assumptions of the model.

7 Contrasts are linear combinations $\sum_{j=1}^{k} a_j \mu_j$ of treatment means such that $\sum_{j=1}^{k} a_j = 0$.

8 Planned comparisons to test whether certain contrasts are zero can be carried out by standardizing the estimated contrast by dividing by its estimated standard error and comparing the result with the t distribution on d.f.(Residual) degrees of freedom. Unplanned comparisons require more care.

6

Multiple linear regression

In Chapter 4 you have already seen regression in its simplest form, as a technique for modelling a linear relationship between two variables, one a response variable and the other an explanatory variable. This chapter is a continuation of Chapter 4, in the sense that it introduces regression models in which there is more than one explanatory variable. There is still only one response variable, which for individual i is still denoted Y_i, but it is now assumed to be dependent on p explanatory variables, $x_{i,1}, x_{i,2}, \ldots, x_{i,p}$, where $x_{i,j}$ denotes the value of the jth variable for the ith individual. This dependence is also linear and in its basic form can be written

$$E(Y_i) = \alpha + \beta_1 x_{i,1} + \beta_2 x_{i,2} + \cdots + \beta_p x_{i,p}.$$

The variation about the mean is modelled in exactly the same way as for the linear model with one explanatory variable: add a random error ϵ_i for each individual, these errors being independent normally distributed random variables with mean 0 and constant variance σ^2. Introducing summation notation at the same time, the model becomes

$$Y_i = \alpha + \sum_{j=1}^{p} \beta_j x_{i,j} + \epsilon_i$$

with the $\epsilon_i \sim N(0, \sigma^2)$ independently of one another.

This model is called the *multiple linear regression* model, and forms the basis for many widely used statistical methods. Notice that it differs from the simple linear regression model of Chapter 4 in only one respect, the introduction of further explanatory variables into the model for the mean.

Box 6.1

The **multiple linear regression model** can be written

$$Y_i \sim N(\alpha + \sum_{j=1}^{p} \beta_j x_{i,j}, \sigma^2) \qquad (6.1)$$

where the Y_is are independent random variables and the $x_{i,j}$s, $j = 1, 2, \ldots, p$, are values of the p explanatory variables.

This chapter explores the basic properties and uses of the multiple linear regression model.

You are strongly advised to use GENSTAT to follow through, on your computer, all the analyses presented in the examples in this chapter.

6.1 Using the model

Example 6.1

This example concerns the dataset rubber, which you first met in Exercise 4.1. The experiment introduced in that exercise concerned an investigation of the resistance of rubber to abrasion (the abrasion loss) and how that resistance depended on various attributes of the rubber. In Exercise 4.1, the only explanatory variable we analysed was hardness; but in Exercise 4.19, a second explanatory variable, tensile strength, was taken into account too. There are 30 datapoints.

Figure 6.1 shows scatterplots of abrasion loss against hardness and of abrasion loss against strength. (A scatterplot of abrasion loss against hardness also appeared in Solution 4.1.) Figure 6.1(a) suggests a strong decreasing linear relationship of abrasion loss with hardness. Figure 6.1(b) is much less indicative of any strong dependence of abrasion loss on tensile strength.

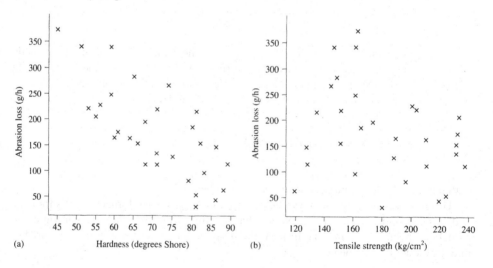

(a) Hardness (degrees Shore) (b) Tensile strength (kg/cm^2)

Fig. 6.1

In Exercise 4.5(a), you obtained the following output for the regression of abrasion loss on hardness.

```
***** Regression Analysis *****

Response variate: loss
   Fitted terms: Constant, hardness
```

```
*** Summary of analysis ***
```

	d.f.	s.s.	m.s.	v.r.	F pr.
Regression	1	122455.	122455.	33.43	<.001
Residual	28	102556.	3663.		
Total	29	225011.	7759.		

```
Percentage variance accounted for 52.8
Standard error of observations is estimated to be 60.5
* MESSAGE: The following units have high leverage:
        Unit    Response    Leverage
          1      372.0       0.182
```

```
*** Estimates of parameters ***
```

	estimate	s.e.	t(28)	t pr.
Constant	550.4	65.8	8.37	<.001
hardness	-5.337	0.923	-5.78	<.001

The output indicates strong support for a non-zero regression coefficient for the hardness variable (i.e. for a non-zero slope β), although the percentage of variance accounted for (52.8%) was not especially great.

If you now repeat Exercise 4.5(a) for the regression of abrasion loss on tensile strength, you should obtain the following output.

```
***** Regression Analysis *****
```

```
 Response variate: loss
    Fitted terms: Constant, strength
```

```
*** Summary of analysis ***
```

	d.f.	s.s.	m.s.	v.r.	F pr.
Regression	1	20035.	20035.	2.74	0.109
Residual	28	204977.	7321.		
Total	29	225011.	7759.		

```
Percentage variance accounted for 5.7
Standard error of observations is estimated to be 85.6
* MESSAGE: The following units have large standardized residuals:
        Unit    Response    Residual
          1      372.0        2.19
```

```
*** Estimates of parameters ***

                  estimate          s.e.      t(28)   t pr.
Constant             305.2          80.0       3.82   <.001
strength            -0.719         0.435      -1.65   0.109
```

Exercise 6.1

What does this output tell you about the regression of abrasion loss on tensile strength?

Individually, then, regression of abrasion loss on hardness seems satisfactory, albeit accounting for a disappointingly small percentage of the variance in the response; regression of abrasion loss on tensile strength, on the other hand, seems to have very little explanatory power. However, back in Exercise 4.19 it was shown that the residuals from the former regression showed a clear relationship with the tensile strength values, and this suggested that a regression model involving both variables is necessary. Let us try it.

Obtain the Linear Regression dialogue box once more. Change the Regression field to Multiple Linear Regression by way of the pull-down menu. Make sure that loss appears as the Response Variate. Enter both hardness and strength in the Explanatory Variates field (as hardness, strength). Click on the OK button, but do not yet click on Cancel. In the Output window, you should find the following:

```
***** Regression Analysis *****

 Response variate: loss
       Fitted terms: Constant, hardness, strength

*** Summary of analysis ***

                d.f.          s.s.          m.s.       v.r.   F pr.
Regression         2       189062.        94531.      71.00   <.001
Residual          27        35950.         1331.
Total             29       225011.         7759.
```

Percentage variance accounted for 82.8
Standard error of observations is estimated to be 36.5
* MESSAGE: The following units have large standardized residuals:
```
        Unit      Response     Residual
         19          64.0        -2.38
```
* MESSAGE: The following units have high leverage:
```
        Unit      Response     Leverage
          1         372.0         0.23
```

```
*** Estimates of parameters ***
```

	estimate	s.e.	t(27)	t pr.
Constant	885.2	61.8	14.33	<.001
hardness	-6.571	0.583	-11.27	<.001
strength	-1.374	0.194	-7.07	<.001

This is an extension of the ANOVA and regression coefficient output for the case of one explanatory variable to the case of two. Let us look at the Summary of analysis table first. The d.f.(Regression) has gone up to 2 because there are two β parameters, one for each explanatory variable; the d.f.(Residual) is consequently reduced by 1. The s.s.(Regression) is the difference between the s.s.(Total) – which, as in the case of one explanatory variable, is the sum of squares of the values remaining after the overall mean only has been fitted to the data – and the s.s.(Residual) – which, analogously to the case of one explanatory variable, is the sum of squares of the values remaining after the full two-explanatory-variable model has been fitted.[1] The mean squares are sums of squares divided by degrees of freedom; the variance ratio is, again, m.s.(Regression)/m.s.(Residual); and the SP shown is calculated by comparing v.r. with the F distribution on, here, 2 and 27 d.f. This SP pertains to the test of the null hypothesis that *both* the regression coefficient for hardness, β_1, and the regression coefficient for strength, β_2, are zero; there is strong evidence that at least one of them is non-zero.

Looking to the regression coefficient output next, the estimated model for the mean response is

$$\widehat{y} = \widehat{\alpha} + \widehat{\beta}_1 x_1 + \widehat{\beta}_2 x_2 = 885.2 - 6.571 x_1 - 1.374 x_2$$

where x_1 and x_2 stand for values of hardness and tensile strength, respectively. Associated with each estimated parameter, GENSTAT gives a standard error (details of the calculation of which need not concern us now) and hence a t statistic (estimate divided by standard error) to be compared with the t distribution on d.f.(Residual) = 27 degrees of freedom. GENSTAT makes the comparison and gives SPs, which in this case are all very small, suggesting strong evidence for the non-zeroness (and hence presence) of each parameter, α, β_1 and β_2, individually.

So, even though the single regression of abrasion loss on tensile strength did not suggest a close relationship, when taken in conjunction with the effect of hardness the effect of tensile strength is also a considerable one. A key idea here is that β_1 and β_2 (and more generally $\beta_1, \beta_2, \dots, \beta_p$ in model (6.1)) are *partial regression coefficients*. That is, β_1 measures the effect of an increase of one unit in x_1 *treating the value of the other variable x_2 as fixed*. Contrast this with the single regression model $E(Y) = \alpha + \beta_1 x_1$ in which β_1 represents an increase of one unit in x_1 *treating x_2 as zero*. The meaning of β_1 in the regression models with one and two explanatory variables is not the same.

You will notice that the percentage of variance accounted for has increased dramatically in the two-explanatory-variable model to a much more respectable 82.8%. This statistic has

---

[1] These terms have essentially the same meaning as those for the one-way analysis of variance model. See Section 5.3.

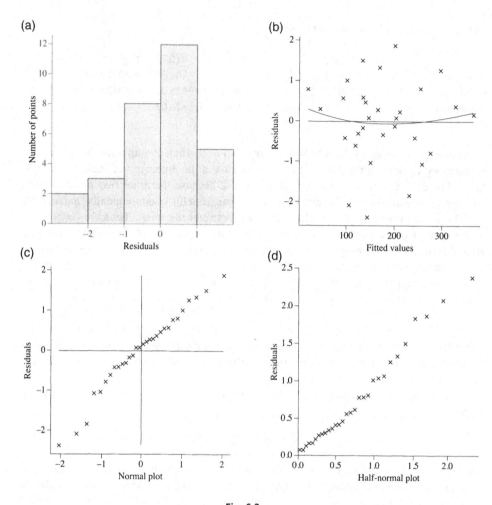

(a)

(b)

(c)

(d)

Fig. 6.2

the same interpretation – or difficulty of interpretation – as for the case of one explanatory variable (see Chapter 4).

Other items in the GENSTAT output are as for the case of one explanatory variable: the Standard error of observations is $\widehat{\sigma}$, and is the square root of m.s. (Residual); the message about standardized residuals will be clarified below; and the message about leverage will be ignored until Chapter 11.

The simple residuals are, again, defined as the differences between the observed and predicted responses:

$$r_i = y_i - (\widehat{\alpha} + \sum_{j=1}^{p} \widehat{\beta}_j x_{i,j}), \qquad i = 1, 2, \ldots, n.$$

GENSTAT can obtain these for you and produce a plot of residuals against fitted values. The fitted values are simply $\widehat{Y}_i = \widehat{\alpha} + \sum_{j=1}^{p} \widehat{\beta}_j x_{i,j}$. As in the regression models in Chapter 4, the default in GENSTAT is to use (standardized) Pearson residuals, which are the simple residuals divided by their estimated standard errors.

Return to the Linear Regression dialogue box (where the multiple regression of loss on hardness and strength should be set up). Select Further Output. Then click on Model Checking and, in the Model Checking dialogue box, leave everything in its default state. Click on the OK button. The plot shown in Figure 6.2 will be produced. These plots are of just the same sort as those in Chapter 4, and are interpreted in the same way. In this case, the normal plot and histogram show some slight suggestion of skewness in the residuals, but there is no obvious cause to doubt the model. In the regression output, GENSTAT flagged one of the points (number 19) as having a large standardized (i.e. Pearson) residual. To decide which points to flag in this way, GENSTAT uses the same rules as described in Chapters 4 and 5. In this case, it warned us about the most negative residual. However, the value of this standardized residual is not very large (at -2.38), and the plot of residuals against fitted values in Figure 6.2 makes it clear that this particular residual is not exceptionally large relative to some of the others, which are almost as large. ■

Exercise 6.2

Table 6.1 shows values for the first five datapoints, of 13, of a dataset concerned with predicting the heat evolved when cement sets via knowledge of its constituents. There are two explanatory variables, tricalcium aluminate (TA) and tricalcium silicate (TS), and the response variable is the heat generated in calories per gram (heat).

Table 6.1

heat	TA	TS
78.5	7	26
74.3	1	29
104.3	11	56
87.6	11	31
95.9	7	52
⋮	⋮	⋮

Source: Woods, H., Steiner, H. H. and Starke, H. R. (1932) 'Effects of composition of Portland cement on heat evolved during hardening', *Industrial and Engineering Chemistry*, **24**, 1207–12.
Dataset name: cemheat.

Load cemheat into GENSTAT.

(a) Make scatterplots of **heat** against each of TA and TS in turn and comment on what you see.

(b) Use GENSTAT to fit each individual regression equation (of heat on TA and of heat on TS) in turn, and then to fit the regression equation with two explanatory variables.

Does the latter regression equation give you a better model than either of the individual ones?

(c) According to the regression equation with two explanatory variables fitted in part (b), what is the predicted value of **heat** when TA = 15 and TS = 55?

(d) By looking at (GENSTAT default) residual plots, comment on the appropriateness of the fitted regression model.

Exercise 6.3

In Chapter 4, the dataset anaerob was briefly considered (Example 4.1) and swiftly dismissed as a candidate for simple linear regression modelling. The reason is clear from Figure 6.3, in which the response variable, expired ventilation (y), is plotted against the single explanatory variable, oxygen uptake (x). (The same plot appeared as Figure 4.1 in Example 4.1.) A model quadratic in x, i.e. $E(Y) = \alpha + \beta x + \gamma x^2$, was suggested instead. Since this quadratic model is none the less linear in its parameters, we can fit it using *multiple* regression of y on x plus the new variable x^2. Even though $x_1 = x$ and $x_2 = x^2$ are closely related, there is no immediate impediment to forgetting this and regressing on them in the usual way. A variable such as x_2 is sometimes called a *derived variable*.

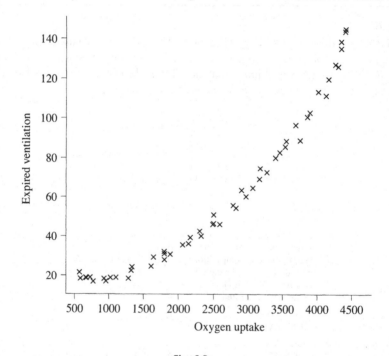

Fig. 6.3

(a) Using GENSTAT, perform the regression of expired ventilation (**ventil**) on oxygen uptake (**oxygen**). Are you at all surprised by how good this regression model seems?

(b) Now form a new variable **oxy2**, say, by squaring **oxygen**. (Choose the **Data|Transformations** menu item; in the **Transformation** dialogue box, change the **Transformation** to **Power**, and choose the appropriate values for **Data**, **Save in** and **Parameters in Equation**.) Perform the regression of **ventil** on **oxygen** and **oxy2**.

Comment on the fit of this model according to the printed output (and with recourse to Figure 4.2 in Example 4.1).

(c) Make the usual residual plots and comment on the fit of the model again.

Adding derived variables to what could already be a large number of potential explanatory variables suggests that multiple regression models could often involve a very high value of p. Is this reasonable? The next section addresses the issue of selecting appropriate explanatory variables.

6.2   Choosing explanatory variables

The more explanatory variables you add to a model, the better will be the fit to the data. To see this, think of adding a further explanatory variable to an existing single-explanatory-variable model. The particular estimate $\widehat{\beta}$ of the regression coefficient β that you already have for the smaller model, together with zero for the regression coefficient associated with the new explanatory variable, is a special case of the possibilities for the regression coefficients of the explanatory variables in the larger model. But since you will be fitting the new model by optimizing some function, you will pick as the new β estimate some value that does at least as well as the special case you already have.

But fitting the data superbly is not the be all and end all. After all, the data themselves provide a perfect model for the data, in the same way as the surface of the Earth is a perfect map of the Earth! Cutting down the number of explanatory variables is desirable from two viewpoints. The first derives from the modelling activity itself, in the sense that we are trying to understand the situation in simplified terms by identifying the *main* sources of influence on the response variable. The second viewpoint concerns prediction. Just because we can fit the data we have, with all its particular idiosyncracies, does not mean that the resulting model will be anything like such a good fit for further data that arise (under the same situation). It is better not to 'over-fit' the current data, by using large numbers of explanatory variables; rather, use just enough explanatory variables to capture its main features: this will often lead to improved prediction.

Underlying these ideas is the principle of *parsimony*: if two models explain the data about equally well, you should prefer the simpler.

Example 6.2
Table 6.2 shows the data for the first five individuals (ordered by age) from a dataset of measurements on 39 Peruvian indians. All these individuals were males, over 21, who were born at high altitude in a primitive environment in the Peruvian Andes and whose parents were born at high altitude too. However, all these people had since moved into the mainstream of Peruvian society at a much lower altitude.

Particular interest lay in the suggestion that migration from a primitive society to a ·modern one might result in increased blood pressure at first, followed by a decrease to normal levels later. For each individual, the response variable is systolic blood pressure (sbp). The explanatory variable of motivatory interest is the number of years since migration (years). But there are seven other explanatory variables also: age (in years), weight (kg), height (mm), chin (chin skin fold in mm), forearm (forearm skin fold in mm), calf (calf skin fold in mm) and pulse (pulse rate in beats per minute).

Table 6.2

age	years	weight	height	chin	forearm	calf	pulse	sbp
21	1	71.0	1629	8.0	7.0	12.7	88	170
22	6	56.5	1569	3.3	5.0	8.0	64	120
24	5	56.0	1561	3.3	1.3	4.3	68	125
24	1	61.0	1619	3.7	3.0	4.3	52	148
25	1	65.0	1566	9.0	12.7	20.7	72	140
⋮	⋮	⋮	⋮	⋮	⋮	⋮	⋮	⋮

Source: Baker, P. T. and Beall, C. M. (1982) 'The biology and health of Andean migrants: a case study in south coastal Peru', *Mountain Research and Development*, **2**, 81–95.
Dataset name: peru.

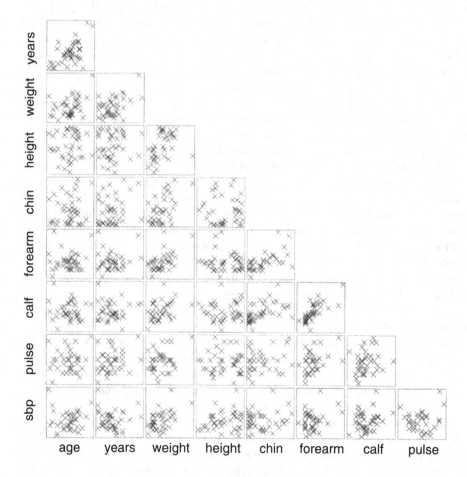

Fig. 6.4

Let us first take a graphical look at the data. Open the file peru.gsh into the spreadsheet and update the server. We would like to look, at least, at the eight scatterplots of the response variable **sbp** plotted against each explanatory variable in turn. In fact, GENSTAT allows you to produce all these scatterplots, and indeed all the scatterplots of each variable against each other variable, in one go. The result is a so-called *scatterplot matrix*, as shown in Figure 6.4. To obtain this, you need to enter the following command in the Input Log window using interactive mode:

```
DSCATTER age,years,weight,height,chin,forearm,calf,pulse,sbp
```

The ordering of the variables, separated by commas, is up to you. However, GENSTAT takes the first p of the $p + 1$ variables as the x variables, in turn, in the horizontal direction and the last p variables as the y variables, in turn, in the vertical direction. The only restrictions we should therefore impose are that our response variable be shown as a y variable and that all of its p relationships with explanatory variables are immediately visible across a single row. To ensure this, the response variable, **sbp**, must be *last* in the list. Producing the scatterplot matrix can take quite a long time on some computers. After all, it is producing $p(p + 1)/2$ plots, which is 36 in this case.

A lot of information, like this, can mean an overload of information, and you are not meant to study every single plot in detail. What should you pay attention to? First, look at the bottom row of plots, in which **sbp** is the y variable. (We shall return to the rest of the scatterplot matrix in a few paragraphs' time.) There do not seem to be many strong single dependencies between **sbp** and the explanatory variables: only the relationship with **weight** seems at all clear. In fact, the strongest visual impression in these plots is probably that given by individual points that do not conform to the main body of the data. With respect to **sbp**, one individual has a particularly high **sbp** value which seems out of step with any relationship between **sbp** and the other variables.

We shall proceed in this chapter by removing this individual's datapoint from further consideration. This point happens to be the first point in the dataset. Activate the spreadsheet window and click on the first row label. Choose the Spread|Restrictions|Add Selected Rows menu item to remove this point temporarily, and then update the server.

The scatterplot matrix with this individual removed is shown in Figure 6.5. (To save time, you need not reproduce this yourself.) Still, the individual regressions of **sbp** on individual explanatory variables fail to indicate any strong relationships, with the possible exceptions of the case of **weight** and, perhaps, some indication of a downward trend for **pulse**.

We shall not worry further about the influence of individual points here (such as perhaps the individual who has especially high values of **age, years, weight, chin, forearm** and **pulse**), nor about warnings from GENSTAT of individual points with large standardized residuals and high leverage. We shall return to the subject of detailed regression diagnostics in Chapter 11 (the peru dataset will be considered again there) and, for the time being, content ourselves with having removed the one very obvious outlier.

Exercise 6.4
Perform the multiple regression of **sbp** on all eight explanatory variables.

(a) Interpret the Summary of analysis for this overall regression.

(b) Look at the `Estimates of parameters` table given by GENSTAT and suggest, on the basis of this, which explanatory variables appear to be most important. Why is it not appropriate simply to settle on the regression model that includes only those variables with the smallest SPs in this table?

(c) Obtain residual plots and comment on how they look.

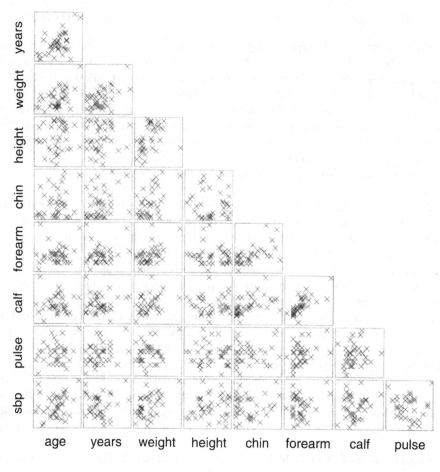

Fig. 6.5

The remainder of the scatterplot matrix may be scanned for the most obvious close relationships between explanatory variables. For instance, one might note from the three scatterplots in the top left-hand corner that **age, years** and **weight** seem fairly closely related. And several other scatterplots hint at similar relationships, e.g. **calf** and **forearm** or **calf** and **weight**.

Another way of investigating the relationships between explanatory variables is to work out their correlations. You can obtain all the correlations by choosing the Stats|Summary Statistics|Correlations menu item, and entering all the explanatory variables in the Data field in the resulting dialogue box. Make sure Print Correlations is checked, and click on OK. This produces the following correlations. (GENSTAT may split this matrix up in its output, and will print the correlations in the order in which you entered the variables.)

```
*** Correlation matrix ***
     age   1.000
   years   0.559  1.000
  weight   0.530  0.544  1.000
  height   0.116  0.112  0.434  1.000
    chin   0.215  0.259  0.552  -0.029  1.000
 forearm   0.108  0.180  0.532  -0.093  0.631  1.000
    calf   0.059  0.043  0.371  -0.032  0.505  0.729  1.000
   pulse   0.214  0.329  0.274  -0.043  0.194  0.403  0.164  1.000

           age   years  weight  height   chin forearm   calf  pulse
```

Here the correlations are given in essentially the same matrix format as the scatterplots, except that the response variable sbp has been left out of this matrix. The correlations of 1.000 are of each variable with itself and can be ignored. What we need to look for are those pairs of explanatory variables with high correlations. The highest happens to be calf and forearm at 0.729. Many others are over 0.5; in fact, in this case we can identify three (overlapping) blocks of three variables within each of which the pairwise relationships have this property: age, years and weight; weight, chin and forearm; and chin, forearm and calf.

What relevance does this have? Well, not a very great deal because these correlations are not *very* high. But had they been of the order of 0.8 or 0.9, a consideration alluded to in Chapter 1 – of the possibility of omitting one or more of a set of closely related (i.e. highly correlated) variables from the model – would have been very relevant. Remember that because the regression coefficients are partial regression coefficients, they are associated with a variable's contribution after allowing for the contributions of *other* explanatory variables. Therefore, if a variable is highly correlated with another variable, it will have little or no additional contribution over and above that of the other; and so there is a case for omitting one of the variables from the model. (This problem is called *multicollinearity*. It can even happen that if a group of such variables have mutually high correlations, *none* of them has a significant SP even though their joint contribution may be very substantial; omission of one or more such variables from the model can increase the significance of the SPs of the remaining ones. Under some circumstances GENSTAT flags this problem for you.) There is no such problem when explanatory variables are uncorrelated.

So, what do we have so far? The bottom row of the scatterplot matrix suggested sbp might depend on weight and perhaps pulse; the multiple regression suggested perhaps weight, years and maybe chin; the correlations warned us that the inclusion in the model of all three variables in each of the triples age, years and weight, weight, chin and forearm, and chin, forearm and calf might be unnecessary; and we can add that, within these triples, the most expendable on the basis of SPs would appear to be age, forearm and calf, respectively. But can we do something more formal to select explanatory variables?

Yes, we can, in very many ways, but we shall here only consider the *stepwise regression* method provided by GENSTAT. To produce a first version of this, first carry out the regression of sbp on all eight variables again. Then, in interactive mode in the Input Log window, type:

```
    STEP [MAX=16;IN=4;OUT=4] age,years,weight,height,chin,forearm,calf,pulse
```

Here, we need only specify all the explanatory variables we wish to consider – GENSTAT already knows that **sbp** is the response variable. This is what you should get.

*** Step 1: Residual mean squares ***

```
     85.84    Dropping calf
     85.85    Dropping age
     85.86    Dropping forearm
     88.12    Dropping pulse
     88.80    No change
     92.66    Dropping height
     95.13    Dropping chin
     97.97    Dropping years
    127.98    Dropping weight
```

Chosen action: Dropping calf

*** Step 2: Residual mean squares ***

```
     83.09    Dropping forearm
     83.09    Dropping age
     85.39    Dropping pulse
     85.84    No change
     88.80    Adding    calf
     89.69    Dropping height
     92.10    Dropping chin
     94.95    Dropping years
    123.94    Dropping weight
```

Chosen action: Dropping forearm

*** Step 3: Residual mean squares ***

```
     80.52    Dropping age
     83.09    No change
     83.23    Dropping pulse
     85.84    Adding    forearm
     85.86    Adding    calf
     87.62    Dropping height
     90.88    Dropping chin
     92.33    Dropping years
    131.49    Dropping weight
```

Chosen action: Dropping age

```
*** Step 4: Residual mean squares ***

     80.52    No change
     80.74    Dropping pulse
     83.09    Adding    age
     83.09    Adding    forearm
     83.12    Adding    calf
     85.26    Dropping height
     88.51    Dropping chin
     90.70    Dropping years
    135.16    Dropping weight

Chosen action: Dropping pulse

*** Step 5: Residual mean squares ***

     80.52    Adding    pulse
     80.74    No change
     82.72    Adding    forearm
     83.23    Adding    age
     83.24    Adding    calf
     84.31    Dropping height
     88.60    Dropping chin
     93.54    Dropping years
    131.55    Dropping weight

Chosen action: Dropping height

*** Step 6: Residual mean squares ***

     80.74    Adding    height
     84.31    No change
     85.26    Adding    pulse
     86.58    Adding    age
     86.81    Adding    calf
     86.83    Adding    forearm
     87.82    Dropping chin
     93.66    Dropping years
    131.48    Dropping weight

Chosen action: Dropping chin

*** Step 7: Residual mean squares ***

     84.31    Adding    chin
     87.82    No change
     88.40    Adding    pulse
```

```
 88.60    Adding    height
 89.38    Adding    forearm
 89.83    Adding    age
 89.85    Adding    calf
 95.92    Dropping years
129.52    Dropping weight
```

Chosen action: No change

***** Regression Analysis *****

Response variate: sbp
　　　Fitted terms: Constant, years, weight

*** Summary of analysis ***

	d.f.	s.s.	m.s.	v.r.
Regression	2	1596.	798.07	9.09
Residual	35	3074.	87.82	
Total	37	4670.	126.21	
Change	6	499.	83.10	0.94

Percentage variance accounted for 30.4
Standard error of observations is estimated to be 9.37
* MESSAGE: The following units have high leverage:

Unit	Response	Leverage
8	108.00	0.199
38	132.00	0.249
39	152.00	0.355

*** Estimates of parameters ***

	estimate	s.e.	t(35)
Constant	62.6	15.1	4.15
years	-0.383	0.184	-2.08
weight	1.104	0.260	4.25

At Step 1, GENSTAT lists the values of m.s. (Residual) associated with fitting the eight models each with seven explanatory variables corresponding to dropping each explanatory variable in turn, together with the m.s. (Residual) from the full model (all eight explanatory variables). The smallest m.s. (Residual) corresponds to dropping calf from the model, and GENSTAT decides to do this.

At Step 2, GENSTAT starts from the seven-variable model resulting from Step 1. It then considers nine options again: dropping each of the remaining seven variables, staying with the Step 1 model (No change), and adding calf back in (to return to the original model). The m.s. (Residual)s are computed for these models, forearm has the lowest value, and forearm is dropped. Notice that the 'order of importance' of the variables corresponding to Step 1 need not be maintained, again as a consequence of the partial nature of the regression coefficients; in particular, **age** and **forearm** were swapped.

None the less, at Step 3, **age** is dropped.

The leading option at Step 4 is No change. So does GENSTAT come to a halt and select the model with the five remaining variables **pulse**, **height**, **chin**, **years** and **weight**? No: it drops **pulse**! This indicates that GENSTAT is not simply taking the minimum m.s.(Residual) option at each step. What it in fact does is: first, it looks at the minimum m.s.(Residual) 'dropping' option and drops this variable if it satisfies a criterion discussed below; then, second, if it is decided not to drop this variable, GENSTAT looks at the minimum m.s.(Residual) 'adding' option and adds this variable if it satisfies a further criterion described below; finally, if neither criterion is satisfied, or if the number of steps has reached the number given by MAX, the selection mechanism is stopped and GENSTAT finishes by producing the full regression output for the model it has finished up with. The default value of MAX (short for MAXCYCLE) is 1 (i.e. one step), so it must be changed; we suggest using $2p$, here 16 (as in the STEP command above).

Let us consider the criterion for dropping a variable. Minimizing the m.s.(Residual) turns out to be exactly equivalent to maximizing the absolute values of the t statistics associated with each $\hat{\beta}_j$ or, equivalently again, maximizing appropriate variance ratios (these are just the squares of the t statistics) in an ANOVA set-up (which you need not worry about constructing). The basic idea is to drop the variable which has the 'least significant' value of $\hat{\beta}_j$. However, one has to decide whether to drop that variable or not; to this end, it turns out that it is not appropriate to use standard SP considerations but it is better to employ a fixed value, the OUT (short for OUTRATIO) option: if the variance ratio in question is smaller than OUT, drop the variable. A similar argument in the variable addition case leads to our adding the variable with the greatest variance ratio (the 'most significant' variable) provided that the variance ratio is greater than another threshold, IN (short for INRATIO). The GENSTAT default options for OUT and IN are not the most usual choices, so they have been changed to more popular values, namely OUT = IN = 4, in the STEP command used above. It is suggested that you stick to these more popular values as default choices for this book.

For the peru data, the model resulting from this stepwise regression contains just two explanatory variables, **weight** and **years**. Given our earlier exploration of which variables might have been important, this choice seems pretty sensible. Certainly, **weight** appeared important both from looking at individual regressions and from looking at the eight-variable multiple regression; **years** was suggested in the full multiple regression also. Note that the correlation between weight and years is 0.544, but this level of correlation need not seriously preclude the two variables from appearing together.

The resulting fitted model is

$$\text{sbp} = 62.6 - 0.383 \text{ years} + 1.104 \text{ weight}.$$

Notice how blood pressure goes up with **weight** and down with **years**.[2] Recall too that the main interest of the researchers who collected the data was in the relationship of **sbp** and **years** (in the presence of other variables). The researchers' expectation was that blood

---

[2] You can obtain the residual plots associated with this two-variable model if you wish: you will find that they do not differ greatly from those you produced for the full model in Exercise 6.4.

pressure would go (back) down as the number of years at low altitude increased, and this is borne out, when proper adjustment is made for weight, by this model. It is interesting that the usual increase of blood pressure with age does not appear in this model, quite likely because of age's correlation with both years and weight.

What if years hadn't appeared in the final model? First, there would simply be less evidence than expected for a relationship between sbp and years when other variables are taken into account. But there may not be much less evidence: we have selected a single model, but perhaps other models are almost as good (or, in fact, perhaps better).

One way of exploring this would be to select variables while forcing years to be in the model. (Or, because it is known that in other populations sbp increases with age, you might wish to force age into the model. It turns out that this does not make a useful difference in this case: as was already suggested, in the presence of weight and years, age has little further effect.) To ensure that years, say, would be in the final model suggested by stepwise regression, you would simply not include years in the list of variables attached to the STEP command. (You need not do anything like this now.)

Let us look back at some more of the GENSTAT output. The percentage of variance accounted for has increased very slightly from the 29.6% of the eight-variable model to the 30.4% of the selected two-variable model. Some people might find this surprising! Many authors simply define the percentage of variance accounted for to equal the R^2 statistic given by $R^2 = $ s.s. (Regression)/s.s. (Total). (R^2 is also called the *multiple correlation coefficient*.) If this were the case, the percentage of variance accounted for must always be greater for a model with more variables than for one with less; this reflects what was alluded to in the very first sentence of this section. However, GENSTAT uses as its definition of the percentage of variance accounted for, the *adjusted* R^2 *statistic*, given by

$$R_a^2 = 1 - \frac{n-1}{n-p-1}(1 - R^2).$$

This formula involves p, the number of explanatory variables in the model, and is designed such that comparisons between adjusted R^2s for models with different p are meaningful and such that R_a^2 is not automatically greater for larger p. What the stepwise regression procedure actually does, thanks to close connections between R_a^2 and the sums of squared errors in the ANOVA table, is to seek a model that balances a relatively small number of variables with a good fit to the data by seeking a model with high R_a^2. Stepwise regression (which need not necessarily find the maximum R_a^2 model) is used rather than a direct maximization of R_a^2 over all possible subsets of variables because, even with modern computers, the latter can sometimes take too much time.

The method employed above is not the only sensible way of performing stepwise regression in GENSTAT. Instead of starting with the full model, as we just did, you could start with the null model, i.e. with no explanatory variables fitted. Choose the Stats|Regression Analysis|Linear menu item. Make sure the Regression field contains Multiple Linear Regression. Enter sbp as the Response Variate. In the Explanatory Variates field, type a space: this tells GENSTAT to fit the null model. You have to type this space because otherwise GENSTAT demands to be given some explanatory variates! Click on OK. If you look in the Output window, you will see that the null model has been fitted (the only regression coefficient is the constant).

By the way, if you wanted to force, say, years, into the model, start with the model containing years only and do not include years in the STEP command.

Now, repeat the STEP command exactly as used before. This time, the procedure goes through three steps, adding **weight**, then **years**, then making no change. It thus ends up with the same model found before. Achieving the same result with both methods often happens and is reassuring when it does. It is recommended that, in performing stepwise regression, you use both methods, because there can also be useful information in the results when they differ. You could also try setting OUT and IN to different values. For example, starting from the null model, they could be set such that variables can only be added to the model (this is called 'forward selection'); likewise, starting from the full model, they could be set so that variables can only be removed (this is called 'backward elimination').   ■

Exercise 6.5

The dataset **crime** concerns the crime rates (offences per 100 000 residents) in 47 states of the USA for the year 1960 and contains 13 possible explanatory variables. We shall try to find a good multiple regression model for these data, using a suitably selected subset of explanatory variables. The full set of explanatory variables is as follows.

crime	Crime rate
malyth	Number of adult males aged 14–24 per 1000 of state population
state	Codes for southern states (1) and the rest (0)
school	Mean number of years of schooling ($\times 10$) for the state population aged 25 years and over
pol60	Police expenditure (in $) per person by state and local government in 1960
pol59	Police expenditure (in $) per person by state and local government in 1959
empyth	Labour force participation rate per 1000 civilian urban males in the age group 14–24
mf	Number of males per 1000 females
popn	State population size in hundred thousands
race	Number of non-whites per 1000 of state population
uneyth	Unemployment rate of urban males per 1000 in the age group 14–24
unemid	Unemployment rate of urban males per 1000 in the age group 35–59
income	Median value of family income or transferable goods and assets (in units of 10$)
poor	Number of families per 1000 earning below one-half of the median income

Source: Vandaele, W. (1978) 'Participation in illegitimate activities: Ehrlich revisited' in Blumstein, A., Cohen, J. and Nagin, D. (eds) *Deterrence and Incapacitation*, Washington DC, National Academy of Sciences, 270–335.

Dataset name: crime.

Perusal of the scatterplot matrix[3] for these data (not shown) suggests that **crime** might be particularly dependent on **pol60**, **pol59** and **income** and less so on **malyth** and **school**. Linear relationships do not seem unreasonable. Plots involving the variable **state** catch the eye because **state** is binary; none the less, we can treat a binary explanatory variable on the same footing as the quantitative ones (such a cavalier attitude is not appropriate for

---

[3]Do not bother to produce this matrix unless your computer is fast.

categorical variables with more than two categories, as was mentioned in Chapter 5). There are also some strong dependencies among the explanatory variables.

(a) Carry out a regression on all 13 explanatory variables and decide whether or not a linear regression model is plausible (you should include a check of the residuals). Which variables does this full multiple regression suggest are most important?

(b) Obtain the correlation coefficients for pairs of explanatory variables. Which variables are most highly correlated? Do the two most highly correlated pairs make intuitive sense? Should some variables be dropped and, if so, which might they be?

(c) Perform a stepwise regression starting from the full model (using the book default: IN = OUT = 4). Which explanatory variables does this procedure suggest should be in the model? Does this seem reasonable on the basis of your exploration of the data? Also, perform a stepwise regression starting from the null model. Does this lead to the same model?

(d) Write down the finally fitted model, discuss how well the model fits, and draw conclusions about the factors that influenced crime in the USA in 1960.

Variable selection, as developed in Example 6.2 and Exercise 6.5, was a particularly useful exercise because, in both cases, interest centred on an understanding of which variables mainly affected the response. (In either case, there are many different methods for selecting the best model.)

Had the purpose been prediction, however, this kind of variable selection approach has its opponents. Two alternatives are worthy of brief mention. The first is an argument that says that, for prediction purposes, it may be better to retain all the explanatory variables but to change the estimates of the β_js. Instead of setting some of the $\widehat{\beta}_j$s to zero, as we have just been doing by selecting a subset of variables, 'shrinkage' estimators that modify all the $\widehat{\beta}_j$s without reducing any of them to zero have been proposed. The second alternative for prediction argues against selecting a single best model at all. Instead, there may well be several plausible models which each fit the data reasonably well (for example, they might all have fairly similar values of R_a^2). In the absence of compelling subject-matter knowledge that allows one to choose between them, it may be better to average one's predictions over all the reasonable models, perhaps giving different models different weights according to how good they are. Neither of these important approaches will be pursued further in this book.

6.3 Parallels with the case of one explanatory variable

In Section 4.2, a few details were given about the parameter estimates and their properties in the case of simple linear regression with one explanatory variable. In this section, you will learn that you have been using just the same kinds of properties in the $p > 1$ case, and you will get to produce confidence intervals for the mean and prediction intervals corresponding to those of Section 4.3.

There are explicit formulae for the parameter estimates and explicit distributional results in the multiple linear regression case as in simple linear regression, and when $p = 1$ the former reduce to the latter. You may ignore the next two paragraphs at no disadvantage whatsoever for what follows if they mean nothing to you (and restart after the forthcoming

(*) sign), but they are included so that you will see something familiar if you read other textbooks on multiple regression and so that you can be reassured that Chapters 4 and 6 provide a coherent approach to regression.

Assume the standard multiple linear regression model as in Box 6.1. Put α and $\beta_1, \beta_2, \ldots, \beta_p$ in a vector $\boldsymbol{\beta}$ of length $p + 1$; write $\mathbf{y} = (y_1, y_2, \ldots, y_n)$; and let $\mathbf{X}$ be the $n \times (p + 1)$ matrix whose ith row is $(1, x_{i,1}, x_{i,2}, \ldots, x_{i,p})$. Using superscripts T and -1 for matrix transpose and inverse, respectively, the estimated parameters can be written in matrix–vector form as

$$\widehat{\boldsymbol{\beta}} = (\mathbf{X}^T \mathbf{X})^{-1} \mathbf{X}^T \mathbf{y}.$$

Because of the normality assumption for the random errors, it turns out that $\widehat{\boldsymbol{\beta}}$ has a multivariate normal distribution with mean $\boldsymbol{\beta}$ and variance $\sigma^2 (\mathbf{X}^T \mathbf{X})^{-1}$. The multivariate normal distribution is an extension of the univariate normal distribution and the 'univariate normal distribution' is just another name for the usual normal distribution for one random variable, as in Section 2.1. Its important properties are that each $\widehat{\beta}_j$ is normally distributed with mean β_j and variance $\sigma^2 v_j$, where v_j is the jth diagonal element of the matrix $(\mathbf{X}^T \mathbf{X})^{-1}$; also, for each j and k, $\widehat{\beta}_j$ and $\widehat{\beta}_k$ are correlated, and their correlation can be derived from the (j, k)th element of $(\mathbf{X}^T \mathbf{X})^{-1}$. For inferential purposes, t and F distributions follow from the normality.

When $p = 1$, the ith row of $\mathbf{X}$ is $(1, x_i)$ so that

$$\mathbf{X}^T \mathbf{X} = \begin{bmatrix} n & \sum_i x_i \\ \sum_i x_i & \sum_i x_i^2 \end{bmatrix}.$$

It follows that

$$\widehat{\boldsymbol{\beta}} = \frac{1}{n \sum_i x_i^2 - (\sum_i x_i)^2} \begin{bmatrix} \sum_i x_i^2 & -\sum_i x_i \\ -\sum_i x_i & n \end{bmatrix} \begin{bmatrix} \sum_i y_i \\ \sum_i x_i y_i \end{bmatrix}$$

$$= \frac{1}{n \sum_i x_i^2 - (\sum_i x_i)^2} \begin{bmatrix} \sum_i x_i^2 \sum_i y_i - \sum_i x_i \sum_i x_i y_i \\ -\sum_i x_i \sum_i y_i + n \sum_i x_i y_i \end{bmatrix}.$$

Here, $\widehat{\boldsymbol{\beta}}$ is the two-element vector containing $\widehat{\alpha}$ and $\widehat{\beta}$, in the notation of Section 4.2; and, with a little more algebraic manipulation, the formulae given there for $\widehat{\alpha}$ and $\widehat{\beta}$ appear. Sampling distributions for $\widehat{\alpha}$ and $\widehat{\beta}$ also follow as special cases of the general results for any number p of explanatory variables. (*)

You should remember that: the s.s.(Total) is just the sum of squares of differences between the y_is and the overall mean $\overline{y}$, that is, the total sum of squares; the s.s.(Residual) is the sum of squares of differences between the y_is and the fitted regression model, i.e. the residual sum of squares; and the s.s.(Regression) is the difference between the two, the sum of squares explained by the regression.

In Section 4.3, a confidence interval for the mean and a prediction interval were produced for the simple linear regression model. As a further consequence of the above, corresponding confidence intervals for the mean and prediction intervals can be obtained for multiple linear regression.

Example 6.3

For the data in rubber on abrasion loss and its dependence on hardness and tensile strength, let us find both a confidence interval for the mean and a prediction interval for the abrasion loss for a rubber sample with hardness = 60, strength = 205. First run the multiple regression of loss on hardness and strength. Then make the Input Log active and enter the command:

```
PREDICT [pred=yhat;se=se] hardness,strength;60,205.
```

Note carefully where the commas and semicolons go. This is related to what you did in Exercise 4.14; it was mentioned in Solution 6.2. GENSTAT does not give either of the intervals we want directly, but it does yield much of the necessary information. In fact, it says that the predicted value of loss (according to the fitted regression equation, loss = 885.2 − 6.571 hardness − 1.374 strength) is 209.18, with an estimated standard error of 9.27. As in Exercise 4.14, this standard error is (directly) appropriate to forming a confidence interval for the mean, but cannot be used directly for prediction intervals for individual values.

To calculate, say, a 95% confidence interval for the mean loss at the given values of hardness and strength, use the point prediction plus or minus the 97.5% point (the 0.975 quantile) of the t distribution on 27 degrees of freedom times the standard error. 27 is d.f.(Residual), as also used for assessing individual regression coefficients in Example 6.1. To remind you, the t quantile can be found using the following command.

```
PRINT edt(0.975;27).
```

Alternatively, you can calculate upper and lower confidence limits directly in GENSTAT by using the following commands.

```
CALC lower=yhat-se*edt(0.975;27)
CALC upper=yhat+se*edt(0.975;27)
PRINT lower,upper.
```

Either way, the 95% confidence interval for the mean is

$$(209.18 \pm 2.052 \times 9.27) = (190.2, 228.2).$$

To make a prediction interval for an individual value instead of a confidence interval for the mean, you have to add the estimated variance to the square of the standard error stored in se and then take the square root in order to get the appropriate analogue of the standard error. This is just the same as in Exercise 4.14 for simple linear regression, and just the same commands in GENSTAT are needed:[4]

```
RKEEP deviance=rss;df=df
CALC s2=rss/df
CALC pse=sqrt(se**2+s2)
CALC plower=yhat-pse*edt(0.975;df)
CALC pupper=yhat+pse*edt(0.975;df)
PRINT plower,pupper
```

This analogue accounts for the ϵ component of Y in addition to its estimated mean. The 95% prediction interval is thus

$$(131.9, 286.4).$$

This is, of course, much larger than the 95% confidence interval for the mean.　■

---

[4] Alternatively you can calculate pse by using ;scope=new inside the brackets in the PREDICT command.

Exercise 6.6

Consider the `crime` dataset once more. Suppose that a certain US state had values of malyth = 140, school = 110, pol60 = 98, unemid = 20 and poor = 180. Using the regression model based on these five variables, obtain a 95% prediction interval for the crime rate.

A word of warning: the prediction interval obtained in Exercise 6.6 takes no account of the variable selection procedure that we went through in Exercise 6.5 to obtain this five-variable model. The prediction interval, wide as it is, is therefore too narrow and too optimistic. This is because the sampling distribution used accounts only for estimating the $p + 1 = 6$ regression parameters in the chosen model and does not account for uncertainty in the appropriateness of the model in the first place.

6.4   Using indicator variables I: comparing regression lines

In this and the next section, the notion of *indicator variables* will be introduced as a means of using multiple regression to perform two types of analysis that you have met before. The first, in this section, concerns the comparison of two regression lines, as in Section 4.6; the second, in Section 6.5, concerns one-way analysis of variance (Chapter 5). Multiple regression affords both a unified analysis of these problems and the opportunity to analyse (later in the book) more complicated versions thereof.

Example 6.4

This is a continuation/reworking of Example 4.3 in Section 4.6, which concerned the dataset `water`, whose spreadsheet contains three columns (plus one with the town names): mortalty is the response variable of interest, actually the annual mortality rate per 100 000 for males; calcium is a quantitative explanatory variable, the calcium concentration (parts per million) in the drinking water supply; and north is what we shall now call an indicator variable, a binary variable that indicates whether or not the town in question is in the south (0) or north (1) of England and Wales. While other labels for the south/north grouping could equally well have been provided (e.g. south/north or s/n or 1/2), the values 0 and 1 were chosen for good reasons, as you will shortly see. (There remains an arbitrariness in which is 0 and which is 1.) In GENSTAT terms, north has been set up as a factor (with levels 0 and 1), for reasons that will become apparent later.

Figure 6.6 is the same as Figure 4.6. To obtain a GENSTAT version of this plot, having loaded water, choose the Graphics|Point Plot menu item, select mortalty as the Y Coordinates, calcium as the X Coordinates, and north as the Groups. This will result in the points for the northern and southern towns being plotted on the same axes, with the two sets of points having different colours. In Exercise 4.26, separate regression lines were fitted to the data for the northern and southern towns, with estimated intercepts 1692.3 and 1522.8 and estimated slopes -1.931 and -2.093, respectively. (The fitted lines are shown in Figure 4.7.) It was then argued that there seemed to be no evidence to suggest that the regression slopes were not the same, i.e. that there seemed to be no evidence to suggest that the lines were not parallel. Let us look again at the comparison of regression parameters, this time using multiple regression.

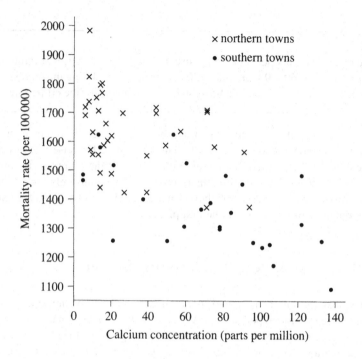

Fig. 6.6

You will now see the power of distinguishing the groups by means of a 0, 1 indicator variable. First, we could form a new variable called **cnprod** which is the product of **calcium** and **north**. This would give us a variable which is 0 for southern town data and has the same value as **calcium** for northern town data. (This extends Section 4.7.) You could now use multiple regression to fit the model

$$\text{mortalty} = \alpha + \beta_1 \text{ north} + \beta_2 \text{ calcium} + \beta_3 \text{ cnprod} + \epsilon$$

(where ϵ is the random error as usual). This is equivalent to fitting the two equations

$$\text{mortalty} = \alpha + \beta_2 \text{ calcium} + \epsilon$$

for the southern towns (**north** = **cnprod** = 0) and

$$\text{mortalty} = (\alpha + \beta_1) + (\beta_2 + \beta_3) \text{ calcium} + \epsilon$$

for the northern towns (**north** = 1, **cnprod** = **calcium**). The second equation reduces to the first when $\beta_1 = 0 = \beta_3$, i.e. β_1 and β_3 represent corrections to the coefficients for the southern data that account for towns being in the north.

In GENSTAT you don't have to go to the trouble of calculating **cnprod**. GENSTAT will fit the combined model and produce output giving us information about β_1 and β_3 (as well as about α and β_2). In particular, we can obtain, with comparative ease, tests of the null hypotheses of equal slopes and equal intercepts (and confidence intervals for such parameters).

Choose the Stats|Regression Analysis|Linear menu item. In the Linear Regression dialogue box, change the **Regression** field to **Simple Linear Regression with Groups**. This choice enables GENSTAT to make appropriate use of the fact that north is a factor. Choose mortalty as the **Response Variate**, calcium as the **Explanatory Variate** and north as the **Groups**. Click on **OK**. Let us concentrate on the final Estimates of parameters part of the GENSTAT output for now.

	estimate	s.e.	t(57)	t pr.
Constant	1522.8	48.9	31.11	<.001
calcium	-2.093	0.610	-3.43	0.001
north 1	169.5	58.6	2.89	0.005
calcium.north 1	0.16	1.01	0.16	0.874

Both the north and the calcium.north terms are labelled with a 1 because they correspond to the level 1 of the factor north.

GENSTAT has denoted what was called cnprod as calcium.north. The coefficient for this term has an SP of 0.874, giving no evidence against $\beta_3 = 0$. It is therefore reasonable to conclude that the lines in the fitted model are parallel. The coefficient for the north variable has an SP of 0.005, strongly suggesting that $\beta_1 \neq 0$ and hence that the intercepts are not equal.

The Simple Linear Regression with Groups command actually does the following. First, it fits the **Explanatory Variate** (calcium) alone, then it adds the **Groups** factor to the model, and finally it fits the full model including the product variable which we called cnprod. In the **Output** window, GENSTAT records the output for each of these fitted models in turn.

Exercise 6.7
By looking at the GENSTAT output, comment on the goodness of fit of the model without the product variable. What is the equation of the fitted model? What are the corresponding equations for the northern and southern data separately?

It should be stressed that the above analysis assumes a single variance parameter σ^2 for all the errors. If you were just contemplating two separate regressions, as in Section 4.6, you would tend to include different error variances for the two regressions. The single-variance assumption is preferable if it is tenable – and the residual plot given as Figure 6.7 suggests that it is an acceptable assumption in this case. To obtain this plot: fit the calcium + north regression model using **General Linear Regression** in the **Regression** field; save the residuals and fitted values using the **Save** button in the **Linear Regression** dialogue box; produce the plot using **Graphics|Point Plot** with the residuals as the **Y Coordinates**, the fitted values as the **X Coordinates** and north as the **Groups**. (Notice that the residuals for north = 1 correspond to higher fitted values than those for north = 0, but their distributions are much the same.) ∎

Exercise 6.8
The data in Table 6.3 are the first five rows of a dataset gathered to see if the presence of urea formaldehyde foam insulation (represented by the indicator variable uffi, with levels 0 for

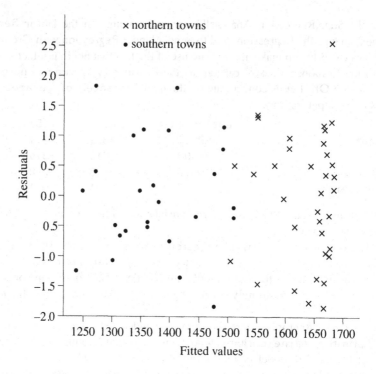

Fig. 6.7

no uffi present, 1 for uffi present) has an effect on the formaldehyde concentration in homes. The response variable is the formaldehyde concentration, **conc**, in parts per billion measured over a week. As well as uffi, a further explanatory variable is a measure of the airtightness of the home (**airtight**, in arbitrary units). You can take it that the linear regression model is adequate, at least providing a good enough approximation to give sensible answers to the questions below.

Table 6.3

conc	airtight	uffi
50.52	1.56	1
60.80	6.09	1
65.06	5.55	1
51.79	4.18	1
59.67	6.05	1
⋮	⋮	⋮

Source: data of R. J. MacKay in Jorgensen, B. (1993) *The Theory of Linear Models*, London, Chapman and Hall. Dataset name: uffi.

Load uffi into GENSTAT.

(a) Plot **conc** against **airtight**, distinguishing between the two groups on your scatterplot, and comment on what you see.

(b) Fit the full two-straight-line model to these data. Test whether the lines are parallel.

(c) Comment on how well the model with parallel lines fits the data. Is there a need for separate intercepts?

6.5   Using indicator variables II: analysis of variance

As tools in data analysis, indicator variables are far more versatile and widely applicable than we have seen so far and have particular application to problems involving the analysis of variance. In this section, some of the work of Chapter 5 will be reformulated using indicator variables so that it can be reanalysed using multiple regression. As well as giving an alternative – but equivalent – way to solve the one-way ANOVA problems you have seen so far, the multiple regression approach will prove to be the more versatile later in that it can generally cope more easily with deviations from the basic set-up such as having missing data or wishing to adjust one's analysis to account for additional explanatory variables.

Example 6.5
In Chapter 5 you analysed the pea dataset, concerned with the effect of different sugars on the growth of peas. Pea section lengths were measured for growth under five different treatments: a control, glucose, fructose, glucose and fructose, and sucrose.

Load pea into GENSTAT and use the Linear Regression dialogue box to carry out a simple linear regression of length on sugar. Because sugar is a factor not a variate, it does not appear in Available Data; but sugar can be typed into the Explanatory Variate field. The output is reproduced below.

```
***** Regression Analysis *****

  Response variate: length
    Fitted terms: Constant, sugar

*** Summary of analysis ***

                 d.f.        s.s.         m.s.      v.r.   F pr.
Regression        4        1077.3      269.330     49.37   <.001
Residual         45         245.5        5.456
Total            49        1322.8       26.996

Percentage variance accounted for 79.8
Standard error of observations is estimated to be 2.34
* MESSAGE: The following units have large standardized residuals:
         Unit     Response     Residual
           5        65.00       -2.30
           9        76.00        2.66
```

```
* MESSAGE: The error variance does not appear to be constant:
            large responses are more variable than small responses
```

*** Estimates of parameters ***

	estimate	s.e.	t(45)	t pr.
Constant	70.100	0.739	94.91	<.001
sugar glucose	-10.80	1.04	-10.34	<.001
sugar fructose				
	-11.90	1.04	-11.39	<.001
sugar g&f	-12.10	1.04	-11.58	<.001
sugar sucrose	-6.00	1.04	-5.74	<.001

The output looks like the standard multiple regression output for a model involving a constant and *four* explanatory variables (d.f.(Regression) = 4), namely the four sugars: glucose, fructose, glucose and fructose (g&f) and sucrose. But there is only one explanatory variable, **sugar**. And what has happened to the control?

Here's how indicator variables come into it. Essentially, a categorical variable (or factor) z having say q values (or levels), such as **sugar** with $q = 5$, is 'expanded' into a series of $q - 1$ binary indicator variables. First the values of z are numbered from 1 to q. Then the $q - 1$ variables are defined as follows:

$$x_2 = \begin{cases} 1 & \text{when } z = 2 \\ 0 & \text{otherwise} \end{cases}$$

$$x_3 = \begin{cases} 1 & \text{when } z = 3 \\ 0 & \text{otherwise} \end{cases}$$

$$\vdots$$

$$x_q = \begin{cases} 1 & \text{when } z = q \\ 0 & \text{otherwise.} \end{cases}$$

Always with q factor levels, only $q - 1$ indicator variables are needed. These are the $q - 1$ explanatory variables that GENSTAT recognized. The pea data, when changed to this format, appears as in Table 6.4.

The regression model we can fit to data in this form is

$$E(Y) = \alpha + \beta_2 x_2 + \beta_3 x_3 + \beta_4 x_4 + \beta_5 x_5.$$

It follows that an individual in the control group has $x_2 = x_3 = x_4 = x_5 = 0$ and so the mean value for the control group is modelled as α. For the glucose treatment group, $x_2 = 1$, $x_3 = x_4 = x_5 = 0$, and so β_2 is the usual regression coefficient corresponding to the glucose indicator variable x_2, but with the special interpretation that it is the extra mean response due to glucose over and above the control: the mean response in the glucose treatment group is $\alpha + \beta_2$. And so on.

Table 6.4

length	x_2, glucose	x_3, fructose	x_4, g&f	x_5, sucrose
75	0	0	0	0
67	0	0	0	0
$\vdots$	$\vdots$	$\vdots$	$\vdots$	$\vdots$
68	0	0	0	0
57	1	0	0	0
58	1	0	0	0
$\vdots$	$\vdots$	$\vdots$	$\vdots$	$\vdots$
61	1	0	0	0
58	0	1	0	0
61	0	1	0	0
$\vdots$	$\vdots$	$\vdots$	$\vdots$	$\vdots$
58	0	1	0	0
58	0	0	1	0
59	0	0	1	0
$\vdots$	$\vdots$	$\vdots$	$\vdots$	$\vdots$
59	0	0	1	0
62	0	0	0	1
66	0	0	0	1
$\vdots$	$\vdots$	$\vdots$	$\vdots$	$\vdots$
67	0	0	0	1

Exercise 6.9

You will find the data in the arrangement given here in pea1. Load this version into GENSTAT and carry out a multiple regression of length on the indicator variables. How do your results compare with those given above?

This, then, is what GENSTAT automatically does for you when it sees that an explanatory variable is a factor. It always takes the lowest level of the factor as the level corresponding to the constant; so if, as here, one value of the factor is a control, it helps interpretation to ensure that the control takes the lowest level, in which case effects are measured relative to the control.

From the ANOVA table in Solution 5.11(b), we concluded from the $SP < 0.001$ that there is strong evidence that different treatments result in different mean lengths. The ANOVA table obtained above is (almost) identical to the table in Solution 5.11(b), and so leads to the same conclusion. However, the Estimates of parameters above suggest, additionally, that there is strong evidence that *each* of the non-control treatments gives results different from the control. This is what the SPs associated with the β coefficients are telling us. In the figures in Solutions 5.11(a) and (d) there was some evidence of a non-constant variance. (The residual plot in Solution 5.11(d) can be reproduced now from the Linear

Regression dialogue box (but with a different vertical scale because of standardization).) GENSTAT warns about this non-constancy of variance in the output above.

As in Chapter 4, it is worth experimenting with a few transformations of the length variable to rectify this. Notice that you should be concerned here only with the issue of constancy of variance. You need not worry about the plot looking like a straight line: as the explanatory variable is a factor, a single straight line is not being fitted. The easiest way to explore transformations is to choose the Graphics|Point Plot menu item and, in the resulting dialogue box, enter the transformation of length in the Y Coordinates field (e.g. type in log(length) or sqrt(length) or 1/length); then enter sugar in the X Coordinates field and click on OK. A log transformation does not seem to do enough, and we could settle on the reciprocal 1/length. Actually, more manageable numbers occur if we use 1000/length. The GENSTAT output for Linear Regression with 1000/length typed in the Response Variate field is as below. Notice that all warnings about possible outliers and unequal variances have disappeared.

***** Regression Analysis *****

Response variate: 1000/ length
   Fitted terms: Constant, sugar

*** Summary of analysis ***

	d.f.	s.s.	m.s.	v.r.	F pr.
Regression	4	64.76	16.1908	53.44	<.001
Residual	45	13.63	0.3030		
Total	49	78.40	1.6000		

Percentage variance accounted for 81.1
Standard error of observations is estimated to be 0.550

*** Estimates of parameters ***

	estimate	s.e.	t(45)	t pr.
Constant	14.306	0.174	82.19	<.001
sugar glucose	2.569	0.246	10.44	<.001
sugar fructose				
	2.892	0.246	11.75	<.001
sugar g&f	2.944	0.246	11.96	<.001
sugar sucrose	1.305	0.246	5.30	<.001

As you can see, none of our conclusions change as a result of this transformed output, but we may be happier with the fit, and the percentage of variance accounted for has risen a little from 79.8% to 81.1%. It can be shown that the residuals look rather more reasonable, in terms of constancy of variance, and that their normal probability plot is more acceptable. It is not surprising that in this case conclusions are unaffected by the better modelling, because there are very clear effects of sugars inhibiting growth. ∎

Exercise 6.10

In Chapter 5, Exercise 5.10, you used one-way analysis of variance to examine data on the heights of singers in the New York Choral Society, where the explanatory variable was voice part with values Tenor 1, Tenor 2, Bass 1 and Bass 2. The data are in singers.

(a) Carry out the regression of height on voice. Discuss the ANOVA part of the output. Produce residual plots and comment on them.

(b) From the GENSTAT output you obtained in part (a), what is the SP for the test of the null hypothesis that the contrast between Tenor 2 and Tenor 1 is zero? What do you conclude?

Contrasts in general can be formulated in terms of regression coefficients, but this will not be done here.

Summary of methodology

1 The methods of Chapter 4 for regression of a response variable on a single explanatory variable have been extended to regression on more than one explanatory variable. The multiple linear regression model is

$$Y_i = \alpha + \sum_{j=1}^{p} \beta_j x_{i,j} + \epsilon_i$$

where the ϵ_is are independent $N(0, \sigma^2)$ random errors.

2 It was shown that there is a strong parallel between multiple linear regression and simple linear regression. There is an analogous ANOVA table. Transformations are sometimes appropriate and (standardized) residuals can reveal features. The several explanatory variables in a multiple regression model may include some extra variables derived (by transformation) from others.

3 The coefficients in multiple regression are to be interpreted as *partial* regression coefficients, reflecting the effect of the corresponding explanatory variable in the presence of the other explanatory variables with their values held fixed.

4 GENSTAT's Percentage variance accounted for is actually an adjusted R^2 statistic which attempts to make allowance for different numbers of variables in the regression model.

5 The question of selecting a subset of explanatory variables in arriving at a final regression model was addressed. A pragmatic approach was suggested, involving the exploratory use of scatterplot matrices (to identify, *inter alia*, strong single dependencies between the response variable and the explanatory variables), the full multiple regression fit (to identify which appear to be the most important variables) and the matrix of correlations between explanatory variables (to identify strongly correlated variables which may not add information one over the other), along with (two versions of) a stepwise regression method provided by GENSTAT.

6 The two main aims of regression analysis, for prediction and for understanding relationships, were stressed.

7 Indicator variables were used (a) to compare regression lines for two different groups and (b) to provide a regression approach to one-way analysis of variance.

7

The analysis of factorial experiments

This chapter is largely concerned with the important concept of *factorial analysis of variance*. In the analyses of variance you met in Chapter 5, the treatments were defined by a categorical variable, which in GENSTAT is represented as a factor. So, to that extent, you have already been doing factorial analysis of variance! But the term is usually used for analyses in which the treatments differ in terms of two or more factors. For instance, in an experiment on animal feeding, the diets involved may differ in terms of the substances they contain (one factor) as well as the quantity of food (another factor). In a pharmaceutical experiment, the people involved may differ in terms of the drug they receive (one factor), the way the drug is administered (a second factor) and their sex (a third factor).

This last example indicates that, in a sense, there is nothing in this chapter that could not be analysed by methods you have already met in Chapters 5 and 6. Suppose there are three different drugs (A, B and C), two different methods of administration (I and II), and two sexes (M and F). Then there are in all $12 = 3 \times 2 \times 2$ different types of person in the experiment, from males receiving drug A by method I (whom we might label A–I–M) to females receiving drug C by method II (C–II–F). The resulting data could be analysed by a one-way analysis of variance with twelve different treatments (as in Chapter 5), and we could investigate, for instance, differences between the methods of administration by calculating appropriate contrasts. Really, all that this chapter involves is an equivalent analysis, done by a slightly different version of analysis of variance (which is very like some of the analyses of variance you met in Chapter 6). The new analysis is worthwhile because it often provides more insight and easier interpretation. In some cases, because of the way GENSTAT works, these analyses have to be done directly using regression methods very similar to those you used in Chapter 6.

7.1 Two-way factorial analysis of variance

7.1.1 The basics: main effects and interactions

We begin by looking at a very simple factorial experiment, and analysing it using methods you have already met.

Example 7.1

An experiment was carried out to investigate the effect of six different diets on the gain of weight in rats. For now, to keep things simple, we shall look at just four of these diets. They differ in terms of two different factors, the source of the protein (beef or cereal) and the amount of protein (low or high). The data are given in Table 7.1.

Table 7.1   Weight gain (in grams) of rats under four diets.

Low protein beef protein	Low protein cereal protein	High protein beef protein	High protein cereal protein
90	107	73	98
76	95	102	74
90	97	118	56
64	80	104	111
86	98	81	95
51	74	107	88
72	74	100	82
90	67	87	77
95	89	117	86
78	58	111	92

Source: Snedecor, G. W. and Cochran, W. G. (1967) *Statistical Methods*, 6th edition, Ames (IA). Iowa State University Press. Dataset name: `weight`.

This is an experiment with four treatments (which happen to be defined by two different factors). It can be analysed using the methods of Chapter 5.

Exercise 7.1

Open the datafile `weight.gsh` as a GENSTAT spreadsheet. In the spreadsheet, you will see that there are four columns. The first is the weight gain (a variate called **weight**). The second is a factor with four levels, called **treatmnt**, distinguishing the four treatments. The third and fourth are also factors, called **amount** and **source**, but each has only two levels. Between them, the two factors also distinguish the four different treatments. (The reason that the treatments are distinguished in two different ways is that the one-way analysis of variance methods you used in Chapter 5 require the treatments to be defined by a single factor in GENSTAT, whereas the new methods you will meet in this chapter require a separate factor in GENSTAT for each of the separate factors in the data.)

Update the GENSTAT server, and carry out a one-way analysis of variance on the response variable (**weight**) using the factor **treatmnt** to define the treatments. You should save the output and plots from this analysis; you will need them again soon. Is there evidence that different diets result in different average gains in weight? Plot the residuals to investigate whether the model you have fitted is appropriate.

Exercise 7.1 indicates that the diets differ in their effect on the response variable, weight gain. But the fact that the treatments are defined in terms of two factors indicates that the experimenters were particularly interested in certain comparisons between treatments. For instance, was there any overall difference in weight gain for rats on a cereal diet compared with rats on a beef diet? Let the mean response for rats on the low beef protein diet be denoted by μ_{LB}, that for rats on the low cereal diet by μ_{LC}, and those for the high beef and high cereal diets by μ_{HB} and μ_{HC} respectively. Consider first the rats on the low protein diets. For them, the contrast $\mu_{LC} - \mu_{LB}$ is a measure of the difference between weight gains on the two sources of protein (cereal and beef). For the rats on high protein diets, the contrast $\mu_{HC} - \mu_{HB}$ measures the same thing. So it seems reasonable to use the average of these two contrasts,

$$\theta_1 = \tfrac{1}{2}((\mu_{LC} - \mu_{LB}) + (\mu_{HC} - \mu_{HB})) = \tfrac{1}{2}(\mu_{LC} + \mu_{HC} - \mu_{LB} - \mu_{HB}),$$

as an overall measure of the difference between weight gains from cereal and beef sources.

Exercise 7.2

(a) Estimate the contrast θ_1 from the data. Estimate the standard error of this estimated contrast. What can you conclude about the average effect of changing the source of protein in the diet of rats?

(b) Write down a similar contrast θ_2 that measures the average effect of changing the amount of protein in the diet of rats from low to high. Estimate this contrast, and estimate its standard error. What do you conclude?

There are four treatments in this experiment, and so three degrees of freedom for treatments. Thus it should be possible to write down a useful third contrast. Such a contrast can be found as follows. In defining the contrast θ_1, we just averaged two contrasts comparing weight gain on different protein sources, one for the rats on low protein amounts and the other for the rats on high protein amounts. But is this the whole story? It might be the case that the mean change in weight gain when the diet changes from beef to cereal is different in rats fed on low protein from that in rats fed on high protein. How can we investigate if this is the case? Well, if the two contrasts $\mu_{LC} - \mu_{LB}$ and $\mu_{HC} - \mu_{HB}$ are the same, then their difference

$$\theta_3 = (\mu_{HC} - \mu_{HB}) - (\mu_{LC} - \mu_{LB}) = \mu_{HC} + \mu_{LB} - \mu_{LC} - \mu_{HB}$$

will be zero. Is this the case for these data?

Exercise 7.3

(a) Estimate the contrast θ_3 from the data. Estimate the standard error of this estimated contrast. What do you conclude?

(b) The contrast θ_3 was derived by investigating whether the effect of changes in protein source differed for different protein amounts. Write down a contrast that you could use to investigate whether the effect of changes in protein amount differs for different protein sources. Check that your new contrast is actually equal to θ_3 (or to $-\theta_3$ if you did the subtractions the other way round).

There is some very weak evidence, then, that the effect of changing protein source is different, depending on the protein amount. Also, Exercise 7.3(b) showed that this implies there is the same very weak evidence that the effect of changing protein amount is different, depending on the protein source. In general, if this sort of thing happens, i.e. if the effect of one factor depends on the level of another factor, the two factors involved are said to *interact*, and a contrast such as θ_3 is called their *interaction*. The concept of interaction will continue to be discussed throughout this chapter. However, in this case the evidence that the two factors interact is very weak (and indeed does not quite reach the conventional 5% significance level).

So far, you have simply used the methods you already know from Chapter 5. Now let us analyse the same data using a different GENSTAT method, one designed explicitly for factorial data.

Exercise 7.4

Load the weight data into GENSTAT again, if they are not already there. Choose the Stats|Analysis of Variance menu item. In the resulting dialogue box, in the Design field, choose Two-way ANOVA (no Blocking). The method is called *two-way* ANOVA because there are two factors; you will learn what blocking is in Chapter 8. Enter weight as the Y-Variate (i.e. as the response variable). Enter source and amount in the Treatment 1 and Treatment 2 fields, respectively. Check that the Interactions field reads All interactions, and that the Covariates check box is unchecked. Click on OK.

Look at the results in the Output window and compare them with the results of the one-way ANOVA you performed in Exercise 7.1. If you can, leave GENSTAT running until the next exercise.

The main difference between the results of the two analyses is that there is more detail in the output from the two-way analysis.

Here are the ANOVA tables from the analyses.

One-way analysis

Source of variation	d.f.	s.s.	m.s.	v.r.	F pr.
treatmnt	3	2404.1	801.4	3.58	0.023
Residual	36	8049.4	223.6		
Total	39	10453.5			

Two-way analysis

Source of variation	d.f.	s.s.	m.s.	v.r.	F pr.
source	1	220.9	220.9	0.99	0.327
amount	1	1299.6	1299.6	5.81	0.021
source.amount	1	883.6	883.6	3.95	0.054
Residual	36	8049.4	223.6		
Total	39	10453.5			

Unsurprisingly, the total sum of squares and the total degrees of freedom are the same in both analyses, because the same response variable is being analysed. More interestingly, the residual sum of squares and the residual degrees of freedom are also the same. This is because, as you will see shortly, the model being fitted is essentially the same. In the two-way analysis there are rows for each of the two factors (source and amount) individually, and one labelled source.amount which corresponds to the interaction between the two factors.

Each of these rows has one degree of freedom. Together they correspond to the treatmnt row in the one-way analysis, which has three degrees of freedom. In fact, the sums of squares for the two factors and their interaction in the two-way analysis add up to 2404.1, the sum of squares for treatmnt in the one-way analysis. This is because, essentially, all the two-way analysis is doing is splitting up (or 'decomposing') the information on differences between treatments so that it can be related more easily to the two treatment factors.

Since the ANOVA table for the two-way analysis has three rows relating to treatments, it also has three different F tests in relation to treatments. The SP in the row for the source factor is given as 0.327. This is exactly the same SP as for the t test you carried out in Exercise 7.2 of whether the contrast θ_1 was zero. Again, this is no coincidence. The F test is exactly equivalent to the t test; indeed the variance ratio (v.r.) is exactly the square of the t statistic you calculated in Exercise 7.2. Similarly, the SPs for the amount and source.amount rows are the same as those for θ_2 and θ_3 (ignoring rounding error). The analysis of variance table is, in this case, providing the same information as the contrasts did.

Looking through the remaining output from the two analyses, you will see some further differences in the Tables of means and in the Standard errors of differences of means. Again there is more information in the results from the two-way analysis. The new Tables of means are particularly useful. As well as means for the four different treatments, they give means relating to each factor, so that (for instance) the mean responses for all the rats on beef protein and all those on high protein levels are given.

As with the analyses of variance you have already met, further output can be generated by GENSTAT.

Exercise 7.5

(a) Return to the Analysis of Variance window and click on Further Output. In the resulting window, choose Residual Plots, and click on OK. Compare the residual plots with those you obtained in Exercise 7.1.

(b) Again click on Further Output in the Analysis of Variance window. This time choose Means Plots, and in the resulting window select amount as the Factor for X-axis, select source as Groups, and select the Lines button under Method. Click on OK. Examine the resulting plot.

The plot resulting from Exercise 7.5(b) is shown in Figure 7.1. In the printed diagrams here, the different lines are shown drawn in different styles. The default for GENSTAT is to produce them in different colours, which unfortunately cannot be reproduced here. Four of the small crosses indicate the means of the four different treatments. They are joined with lines to make it clearer how they relate to the factors. The two levels of the amount factor are arranged along the x-axis (as requested in the dialogue box that produced the plot). The steeper line corresponds to beef protein source, and the other line to cereal protein source. The use of lines to link the means corresponding to the same *source* of protein is because we chose source as the Groups factor. The vertical bar at the left of the diagram has total length equal to the size of the standard error of the difference between any *pair* of treatment means (in this case $\sqrt{2\sigma^2/10}$, estimated to be 6.69). It is centred on a fifth cross, indicating the overall mean weight gain.

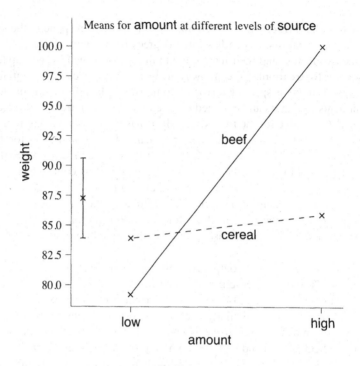

Means for amount at different levels of source

Fig. 7.1

Since the vertical distance between the two crosses for low protein amounts is less than the length of the standard error bar, the difference between these two treatment means is less than one standard error, so there is no significant difference between them. However, the difference between the two means for high protein amounts is much larger than the standard error. Without doing a formal test, we can't really conclude from the diagram alone that this difference is significant, but it looks like it might be! In fact the SP turns out to be 0.042.

The steeper line indicates that, for beef protein, changing from low to high protein amount corresponds to a relatively large increase in mean weight gain. On the other hand, for cereal protein (the other line), changing from low to high protein amount corresponds to a much smaller (and not significant) increase in mean weight gain. That is, the plot indicates that a change from low to high protein amount has a different effect for different protein sources; in other words, there appears to be an interaction between amount and source of protein. Our earlier analysis, in Exercises 7.3 and 7.4, showed that this difference between effects is scarcely significant, and that it would not be totally unreasonable to take it as zero; the plot merely indicates that the difference *might* be quite substantial. In general, the greater the difference between the slopes, the more likely it is that the two factors interact; conversely, the closer the two lines are to being parallel, the less likely it is that the two factors interact.

It would be quite possible to produce a similar means plot, but with source as the Factor for X-axis and amount as the Groups factor. You might like to try it. Again, the lines do not look parallel, and again the standard error bar looks long in relation to one of the differences in means. (In light of Exercise 7.3(b), this is only to be expected.) ∎

Let us take stock at this point, by defining some jargon and generalizing what has been demonstrated in Example 7.1.

A *factorial experiment* is one where the treatments are defined by more than one factor. A *factor* is simply a categorical variable, usually taking a small number of possible values. Its possible values, or labels, can be assigned numerical values, often called *levels*.[1] (Thus, in the rat experiment, the **source** factor had two levels, labelled **beef** and **cereal**.) A set of factor levels, uniquely defining a single treatment, is sometimes called a *cell*. A *two-way* factorial experiment involves two factors. It can be analysed using *two-way analysis of variance (two-way ANOVA)*. This essentially involves considering contrasts comparing treatment differences of different kinds: first, the average difference in response corresponding to a difference in levels of just one factor and, second, differences between such differences when a second factor is changed. The first type of contrast is called the *main effect* of the factor in question, and the second type is called the *interaction* of the two factors involved. (In this case, it is a *two-factor* interaction because two factors are involved. Later you will meet interactions involving more than two factors.)

Although two-way ANOVA is frequently used in the context of experiments, it can be used in other types of situation where there is a single, quantitative, response variable that depends on two categorical variables. You will see examples later in this chapter, and elsewhere in the book.

In Example 7.1, it was claimed that the reason for many similarities between one-way and two-way analyses of the same data was that the models involved were the same. Why is this?

You already know from Chapter 5 that, for one-way analysis of variance, the response Y_{ij} of the ith experimental unit in the jth treatment group is modelled as

$$Y_{ij} = \mu_j + \epsilon_{ij}$$

where each $\epsilon_{ij} \sim N(0, \sigma^2)$ and all the ϵ_{ij}s are independent of each other. In this model, if there are k treatments in all, there are k parameters involved (namely the k μ_js – ignoring, for the moment, the parameter σ^2 describing the error variance).

Later in Chapter 5, a slightly different way of expressing the same model was introduced:

$$Y_{ij} = \mu + \tau_j + \epsilon_{ij}$$

subject to the constraint that

$$\sum_j \tau_j = 0.$$

(This constraint only applies to the case of equal replications; as was mentioned in Chapter 5, a different constraint is needed in the case of unequal replications.) Here, again assuming k treatments, there are $k + 1$ parameters (μ and the k τ_js); but the constraint on the sum of the τ_js means that effectively there are only k of them, as in the original model.

For two-way analysis of variance, write Y_{ijl} for the response of the ith experimental unit from the treatment group where the first factor is at level j and the second factor is at level l. The model is

$$Y_{ijl} = \mu + \alpha_j + \beta_l + \gamma_{jl} + \epsilon_{ijl} \qquad (7.1)$$

---

[1] We shall sometimes, for convenience, refer to levels by their labels rather than by their numerical values.

where again each $\epsilon_{ijl} \sim N(0, \sigma^2)$ and all the ϵ_{ijl}s are independent of each other. Suppose that the first factor has J levels and the second factor has L levels. This is called a $J \times L$ factorial experiment.[2] The total number of treatments involved is the total number of possible combinations of levels of the two factors; that is, JL. Now, equation (7.1) introduces a lot of parameters – one μ, J α_js, L β_ls, and JL γ_{jl}s, or in all $1 + J + L + JL$ of them. There are only JL treatments, so we need some constraints – indeed we need $1 + J + L$ of them. The usual constraints (used by GENSTAT) are as follows:

$$\sum_j \alpha_j = 0, \quad \sum_l \beta_l = 0;$$

$$\sum_j \gamma_{jl} = 0 \quad \text{for } l = 1, 2, \dots, L;$$

$$\sum_l \gamma_{jl} = 0 \quad \text{for } j = 1, 2, \dots, J.$$

(These are the forms of the constraints when there are equal replications, i.e. equal numbers of experimental units in all combinations of factor levels.) This appears to be a set of $2+L+J$ constraints, which is one more than is needed; but in fact the constraints $\sum_j \gamma_{jl} = 0$ for $l = 1, 2, \dots, L$ imply that the overall sum of all the γ_{jl}s is zero, which makes one of the constraints $\sum_l \gamma_{jl} = 0$ for $j = 1, 2, \dots, J$ unnecessary.

Thus the model amounts to setting a different mean ($\mu + \alpha_j + \beta_l + \gamma_{jl}$) for each treatment group, and using a normal distribution with a constant variance σ^2 to describe the variability (represented by ϵ_{ijl}) of data within each treatment group. That is exactly the model for one-way analysis of variance; the only difference here is that the parameters are set up in a rather different, more complicated, but also more meaningful way. The parameters α_j correspond to the main effects of the first factor; the parameters β_l correspond to the main effects of the second factor; and the parameters γ_{jl} correspond to the interaction of these two factors.

An important point, when there are equal replications, is that the estimates of these parameters are independent (which is not the case for the parameters of regression models in general). This means that the parameters do not have a 'partial' interpretation (as do the partial regression coefficients of multiple regression models), i.e. their values remain the same regardless of the presence or absence of other parameters.

How does this two-way ANOVA model apply to the rats example? In that case there were four treatments, arising from a 2×2 factorial structure, which we labelled LB and so on. Alternatively, they could be labelled as follows. Regarding protein sources as the first factor, number its levels 1 for beef and 2 for cereal. For the second factor, protein amount, the levels are numbers 1 for low and 2 for high. Thus, according to the model in equation (7.1), the means for the four treatments (i.e. for the four cells) are:

beef, low	(1, 1)	$\mu + \alpha_1 + \beta_1 + \gamma_{11}$
cereal, low	(2, 1)	$\mu + \alpha_2 + \beta_1 + \gamma_{21}$
beef, high	(1, 2)	$\mu + \alpha_1 + \beta_2 + \gamma_{12}$
cereal, high	(2, 2)	$\mu + \alpha_2 + \beta_2 + \gamma_{22}.$

---

[2] $J \times L$ is said as 'J by L'.

The constraints are:

$$\alpha_1 + \alpha_2 = 0, \qquad \beta_1 + \beta_2 = 0;$$
$$\gamma_{11} + \gamma_{21} = 0, \qquad \gamma_{12} + \gamma_{22} = 0;$$
$$\gamma_{11} + \gamma_{12} = 0, \qquad \gamma_{21} + \gamma_{22} = 0.$$

Thus $\alpha_2 = -\alpha_1$, $\beta_2 = -\beta_1$, and $\gamma_{11} = -\gamma_{12} = -\gamma_{21} = \gamma_{22}$, so that there is really only one independent value each for the α_js, β_ls and γ_{jl}s. This corresponds to the fact that, in this case, where each factor has only two levels, the main effects for each factor and their interaction have only one degree of freedom each.

In analyses where the factors have more than two levels, things get a little more complicated, as you will see in the next subsection.

7.1.2   Developing the methods

This subsection considers the extension of the methods of the last subsection to two situations: first, where the factors do not both have two levels and, second, where the response variable needs to be transformed.

The model in equation (7.1) was introduced in the context of a two-way analysis of variance where each of the two factors had two levels – a so-called 2×2 analysis of variance. However, the model is more general than that, and applies to any two-way analysis of variance, however many levels there are. Again, let us see what happens via an example.

Example 7.1 continued
As mentioned previously, in the experiment on rat diets there were really six treatments. There were in fact three levels of the protein source factor. As well as beef and cereal protein, some rats were fed on protein derived from pork – again, ten of them on a low amount of protein and ten on a high amount. Thus the experiment should be analysed using a 3×2 analysis of variance, because the first factor (protein source) has three levels. The extra data for the rats fed on pork protein are given in Table 7.2, and the data for all the rats in the experiment are in the datafile `weight3.gsh`.

Table 7.2   Weight gain (in grams) of rats under two more diets.

Low protein pork protein	High protein pork protein
49	94
82	79
73	96
86	98
81	102
97	102
106	108
70	91
61	120
82	105

Source as for Table 7.1.
Dataset name: `weight3`.

Exercise 7.6

(a) Load the `weight3` file into GENSTAT. Carry out a two-way analysis of variance with **source** and **amount** as the two treatment factors just as you did in Exercise 7.4. Look at the resulting analysis of variance table. Can you explain why the numbers of degrees of freedom are as given? (Think about what the degrees of freedom would be in a one-way analysis of the same data.) How do you interpret the results of the F tests?

(b) Plot the residuals and comment on what you see.

(c) Plot the means (as in Exercise 7.5(b)) and comment on what you see.

∎

The discussion of degrees of freedom in Solution 7.6(a) can be generalized. In a two-way factorial ANOVA where the numbers of levels of the two factors are J and L respectively, there will be $J - 1$ and $L - 1$ degrees of freedom respectively for the two factors, and $(J - 1)(L - 1)$ degrees of freedom for their interaction. Since there are JL treatments in all, there should be $JL - 1$ degrees of freedom for treatments in total. You might like to check that $(J - 1) + (L - 1) + (J - 1)(L - 1) = JL - 1$.

By the way, in deciding, as in Solution 7.6(a), that only one main effect was important, there is no need to rerun the analysis using only the corresponding factor, because the estimate of the effect would not change. This is a consequence of the independence of parameter estimates in this context, as mentioned in the previous subsection.

So far in the datasets that you have analysed, the factors have not had a significant interaction. If there is a significant interaction, the interpretation of the results will need a little more care, as the next example shows.

Example 7.2

In a laboratory working on influenza viruses, there were three different operators of photoelectric titration equipment and two different methods of performing the titration. The methods involve several dilutions of the virus preparation; either a single pipette could be used for every dilution, or a fresh pipette could be used for each dilution. What was being studied was not the virus preparations themselves, but the operators and the measurement methods. In fact, apart from measurement variability caused by having different operators and different methods, there was no reason for the responses to differ, since all the measurements were made on samples drawn from the same virus preparation. The experimental units were single measurements, and they differed in two ways: first, in terms of the operator involved (A, B or C) and, second, in terms of the method used (single pipette or multiple pipettes).

Thus there are two factors, which we shall call **operator** and **pipette**. The first has three levels and the second has two. There are $3 \times 2 = 6$ different treatments (i.e. six cells), and four measurements were made for each treatment group. Table 7.3 gives the data.

Do the different operators and measurement methods have an effect on the measurements obtained? If so, what sort of effect? These questions are addressed in Exercise 7.7.

Table 7.3

operator A		operator B		operator C	
single pipette	multiple pipettes	single pipette	multiple pipettes	single pipette	multiple pipettes
290	282	285	310	269	261
277	255	209	286	331	255
282	233	252	282	266	249
243	251	265	320	289	255

Source: Osborn, J. F. (1979) *Statistical Exercises in Medical Research*, Oxford, Blackwell Scientific Publications.
Dataset name: influ.

Exercise 7.7

Load the datafile influ.gsh into GENSTAT.

(a) Use GENSTAT's Stats|Summary Statistics|Summaries of Groups (Tabulation) menu item to calculate the variance of the response variable (titrate) for each of the treatment groups. (You should enter *both* the factors, operator and pipette, into the Groups field, so that GENSTAT will deal with the six treatment groups separately.) What do you observe from the resulting table?

The table you obtained in part (a) should have given you some cause for concern about the appropriateness of the ANOVA model for these data; but, as the concern is not too great (see the solution), continue with the analysis.

(b) Perform a two-way analysis of variance. As part of your analysis, produce appropriate plots of means and of residuals. Which main effects and interactions are significant? Do the residual plots show anything untoward?

You will have found in part (b) that the interaction between the two factors is highly significant. This means, for instance, that the effect of changing the pipette factor from single to multiple depends on the level of the operator factor. Now, the contrast for the main effect of the pipette factor can be thought of as an average of two contrasts between single and multiple pipettes, measured at different levels of the operator factor; and, because the interaction is significant, these two contrasts are different, with the result that their average – the pipette main effect – does not tell us anything useful. Similar conclusions apply in the case of the operator main effect. In general, if an interaction is significant, the corresponding main effects are of little interest.

(c) Examine the plot of means and that part of the Tables of means which gives separate means for all six treatments. Report your conclusions.

■

We now turn to situations where the raw data do not satisfy the usual analysis of variance assumptions. Actually, there is nothing new to say here. You have already seen that the model for two-way analysis of variance is basically the same as that for one-way analysis of variance – the only difference is that the parameters that describe the treatment means are set up in a different way. In essence the model still says that each observation is the sum of a treatment mean and a random, normally distributed error, and that all the errors are independent and have the same variance. You saw in Chapter 5 how certain problems with the assumptions in one-way ANOVA could be identified by looking at residual plots, and in some cases they could be solved by transforming the response variable. Exactly the same is true for two-way ANOVA.

Example 7.3

This example concerns an investigation into the effectiveness of different kinds of psychological treatment on the sensitivity of headache sufferers to noise. There were two groups of 22 subjects each, those suffering from a migraine headache and those suffering from a tension headache. Half of each of the two groups was chosen at random to receive a psychological treatment. The other half remained as an untreated control group. Afterwards, all the subjects listened to a tone which gradually increased in volume. The response variable represents the volume level at which the subject found the tone unpleasant.

The subjects are thus classified according to two factors, each at two levels – headache type (migraine or tension) and treatment type (active treatment or control). There are, then, $2 \times 2 = 4$ groups of subjects (i.e. four cells), with 11 subjects in each group. The responses for the first five cases in each treatment group are given in Table 7.4.

Table 7.4

migraine headache		tension headache	
active treatment	control	active treatment	control
5.70	2.80	2.70	2.10
5.63	2.20	4.65	1.42
4.83	1.20	5.25	4.98
3.40	1.20	8.78	3.36
15.20	0.43	3.13	2.44
⋮	⋮	⋮	⋮

Source: Hand D. J. and Taylor. C. C. (1987) *Multivariate Analysis of Variance and Repeated Measures*, London, Chapman & Hall. Dataset name: headache.

Exercise 7.8

Load the data from headache.gsh into GENSTAT.

(You may notice that GENSTAT reports in the Output window that the response variable, score, is skewed. This is a strong indication that the assumption of normal residuals is going to be inappropriate. However, with some sorts of regression data, it is possible for the distribution of the response variable to be skewed simply because the distributions of some of the explanatory variables are skewed, even though the residuals do have a normal distribution. But here, the 'explanatory variables' are simply the factors that define the four groups of subjects; there are equal numbers in each group, so no sign of skewness there, and so the only way the response variable can be skewed is if the residuals are not normal. But, for now, ignore this and carry on.)

(a) Analyse the data using two-way analysis of variance and plot the residuals. What do you see? How would you suggest transforming the response variable?

(b) Take a log transformation of the response variable. Analyse the resulting data, producing plots of means and residuals. What do you conclude?

In this example, transforming the data produced an analysis in which the assumptions were apparently still not entirely satisfied, in that there seemed to be two outliers. The departure from the assumptions was not as great as for the untransformed data, so the conclusions are likely to be more reliable. It is possible to investigate what happens if these outliers are omitted; but things become a little more complicated because then there are no longer equal numbers of subjects in each of the groups. How to deal with unequal replications is covered in Section 7.3, so we shall leave this example until then. ∎

Let us look briefly at what the model can look like when a log transformation of the response variable is used. Suppose, for example, both main effects are important, but the interaction is not. Denoting the (untransformed) response by Y, using equation (7.1) the model is of the form

$$\log Y = \mu + \alpha + \beta + \text{error}$$

where α relates to one factor and β to the other. Thus

$$Y = e^{(\mu+\alpha+\beta+\text{error})} = e^\mu e^\alpha e^\beta e^{\text{error}}.$$

This has a multiplicative form rather than the additive form we have been using so far. This means that, for instance, a change in the factor corresponding to α from level 1 to level 2, say, does not add a quantity $\alpha_2 - \alpha_1$ to the mean response (as it would in an additive model), but instead increases or decreases the mean response by multiplying it by a quantity $e^{\alpha_2 - \alpha_1}$: e.g., if $\alpha_2 - \alpha_1 = 0.5$, the mean has 0.5 added to it in the additive case, but is multiplied by $e^{0.5} = 1.649$ in the multiplicative case. In practice, many datasets show this kind of multiplicative structure and taking logs allows them to be analysed using a linear model.

7.2 More than two factors

In this section you will see how factorial analysis of variance can be extended to cover experiments with more than two factors. Again let us start with an example.

Example 7.4

The data in Table 7.5 come from a study carried out on a pilot chemical plant. The response variable was the yield of chemical produced, and there were eight different treatments, defined in terms of three different factors. These factors were the temperature of the reaction (160° C or 180° C), the concentration (20% or 40%) and the catalyst used (A or B). The number of possible different treatments is thus $2 \times 2 \times 2 = 2^3 = 8$. An experiment like this is said to have a 2^3 factorial structure. Two different runs were carried out for each of the eight treatments. The treatments are each given a number in the left-hand column of Table 7.5 to make it easier to refer to them later.

Table 7.5

Treatment	Temperature (°C)	Concentration (%)	Catalyst	Yield run 1	run 2
1	160	20	A	59	61
2	180	20	A	74	70
3	160	40	A	50	58
4	180	40	A	69	67
5	160	20	B	50	54
6	180	20	B	81	85
7	160	40	B	46	44
8	180	40	B	79	81

Source: Box, G. E. P., Hunter, W. G. and Hunter, J. S. (1978) *Statistics for Experimenters*, New York, John Wiley. The authors report that the data have been simplified somewhat from the original, for illustrative purposes. Dataset name: `plant`.

To analyse data like these in GENSTAT is no harder than to analyse a 2×2 experiment; but there is one further concept that you need to understand. There are, in all, eight treatments, so there are seven degrees of freedom for treatments. All the factorial analysis of variance does is to split up these seven degrees of freedom. Each of the three main effects, one for each factor, will have one degree of freedom, because there are two levels of each factor. Each of the interactions between pairs of factors (such as the temperature.catalyst interaction) will have one degree of freedom. There are three such pairs, so we have so far six degrees of freedom. Where does the seventh degree of freedom go to? Well, it belongs to a *three-factor interaction*, the interaction temperature.concentration.catalyst. (In larger factorial experiments, there are more degrees of freedom and things are a little more complicated.) As with the main effects and two-factor interactions, this corresponds to a single contrast (because it has one degree of freedom). What contrast is it? It is easier to start by defining the other contrasts. They are defined qualitatively in the same way as in two-factor experiments, but, because there are more treatments, the definitions look more complex.

A contrast for the main effect of temperature can be defined as the average of *four* simpler contrasts, each of which measures the difference in response when the temperature is at one level (say 180° C) and the response when the temperature is at the other level (160° C) within a single set of values of the other two factors. So, in terms of the treatment means $\mu_1, \mu_2, \ldots, \mu_8$ (numbered to correspond to the treatments in Table 7.5), the contrast for the main effect of temperature is

$$\tfrac{1}{4}[(\mu_2 - \mu_1) + (\mu_4 - \mu_3) + (\mu_6 - \mu_5) + (\mu_8 - \mu_7)]$$
$$= \tfrac{1}{4}(\mu_2 + \mu_4 + \mu_6 + \mu_8 - \mu_1 - \mu_3 - \mu_5 - \mu_7).$$

The contrast for the interaction of temperature and concentration can be found by writing down a contrast to measure the effect of changing temperature from 160° C to 180° C when concentration is at one level (say 40%) and subtracting a similar contrast for measuring the effect of changing temperature from 160° C to 180° C when concentration is at the other level (20%). What makes this more complicated than in the two-factor case is that we have to average things out over the different levels of the third factor (catalyst). For instance, a contrast to measure the effect of changing temperature from 160° C to 180° C when concentration is at 40% is $\frac{1}{2}[(\mu_4 - \mu_3) + (\mu_8 - \mu_7)] = \frac{1}{2}(\mu_4 + \mu_8 - \mu_3 - \mu_7)$. A similar contrast when concentration is at 20% is $\frac{1}{2}(\mu_2 + \mu_6 - \mu_1 - \mu_5)$. Thus the appropriate contrast for the temperature.concentration interaction is the difference of these two, i.e.

$$\frac{1}{2}[(\mu_4 + \mu_8 - \mu_3 - \mu_7) - (\mu_2 + \mu_6 - \mu_1 - \mu_5)].$$

Other two-factor interactions can be defined in a corresponding but equally tedious way.

Now the three-factor interaction is, in a sense, just more of the same.[3] It is there to answer questions such as the following. Does the amount of interaction between two factors, say temperature and concentration, depend on the level of the third factor, catalyst? To investigate this, a contrast is calculated to measure the temperature.concentration interaction when catalyst is at one level, say B, and a similar contrast is calculated for the temperature.concentration interaction when catalyst is at its other level, A. If the temperature.concentration interaction does not depend on the catalyst, these two contrasts should be the same. So we look at their difference and see if it could plausibly be zero. A contrast for the temperature.concentration interaction when the catalyst is B is $(\mu_8 - \mu_7) - (\mu_6 - \mu_5)$, and a contrast for the same interaction when the catalyst is A is $(\mu_4 - \mu_3) - (\mu_2 - \mu_1)$. Thus the contrast for the three-factor interaction is

$$(\mu_8 - \mu_7) - (\mu_6 - \mu_5) - (\mu_4 - \mu_3) + (\mu_2 - \mu_1).$$

The definition of this contrast did not seem to treat the three factors on an equal footing – we calculated it as the difference between the interaction of two particular factors at different levels of the third – but in fact (as with two-factor interactions) if you swap the roles of the three factors in the definition you will end up with exactly the same contrast.

In experiments with more than three factors, one can go on to define four-factor interactions and so on in increasingly horrendous ways. But for interpreting an analysis, the exact definitions hardly ever matter. One way to think of things is to interpret interactions in a negative way. If the interaction between two factors is zero, then you can think of the effects of the two factors as simply being added together. Thus, if the interaction is not zero, the effects do not simply add together – it's more complicated than that. Similarly, if a three-factor interaction is non-zero, you cannot think of the effects of the three factors involved simply in terms of adding together their main effects and their two-factor interactions. Again, things must be more complicated.

In practice, things tend to be a lot more straightforward than the last few paragraphs might have indicated. Honest!

Exercise 7.9

Load the data from the file plant.gsh into GENSTAT.

Choose the Stats|Analysis of Variance menu item. In the resulting dialogue box, choose General Treatment Structure (no Blocking) in the Design field. (There is not a three-way ANOVA option, as there was a two-way ANOVA option.) Choose yield as the Y-Variate.

In the Treatment Structure field, enter temperat*concentr*catalyst. You can pick up symbols like * from the Operators list in the same way that you can pick up names from

---

[3] All these contrasts look beastly, but you will never have to work them out from scratch in this book (or, if you are lucky, anywhere else).

the Available Data list – or you can just type them in. In GENSTAT, the symbol * appearing between two factors indicates that GENSTAT will fit a model including the main effects of the two factors and their interaction; so temperat*concentr would fit the main effects of temperature and of concentration, together with their interaction. (Another way of telling GENSTAT the same thing would be to type temperat + concentr + temperat.concentr; you have seen that GENSTAT denotes an interaction between two factors with a dot, so that temperat.concentr is the interaction of the temperature and concentration factors.) Extending this, temperat*concentr*catalyst means 'fit all three main effects, all three two-factor interactions, and the three-factor interaction'. (This could be written instead as temperat + concentr + catalyst + temperat.concentr + temperat.catalyst + concentr.catalyst + temperat.concentr.catalyst, but the shorter form is a lot easier to type!)

Check that the Interactions field is set to All Interactions, and click on OK.

The resulting output looks much like the output from a two-way factorial analysis, only longer. What effects and interactions are reported as significant? Produce a means plot for any significant two-factor interaction. Plot the residuals and comment.

Let us interpret these results more fully. The concentration factor has a significant main effect, and none of the interactions involving it is significantly different from zero. This makes it easy to interpret. The (additive) effect of changing the concentration from 20% to 40% can be taken to be the same, regardless of the level of the other two factors. The mean yield at 40% concentration is 61.75, and at 20% concentration it is 66.75; so we can estimate that changing the concentration from 20% to 40% on average reduces the yield by 5 units. To obtain a standard error for this estimate, note that the estimated contrast we have calculated is

$$\tfrac{1}{4}(\widehat{\mu}_3 + \widehat{\mu}_4 + \widehat{\mu}_7 + \widehat{\mu}_8 - \widehat{\mu}_1 - \widehat{\mu}_2 - \widehat{\mu}_5 - \widehat{\mu}_6),$$

and each treatment mean is the mean of two values, so that the variance of the contrast is $\left(\tfrac{1}{4}\right)^2 8 \left(\sigma^2/2\right) = \sigma^2/4$. The residual mean square is 8, so the estimated variance is $8/4 = 2$, and the estimated s.e. of our contrast is $\sqrt{2} = 1.414$.

The other two factors have a significant interaction, so their interpretation is a little more complicated. It cannot be summarized just by giving the main effects for these two factors (and so the fact that the temperature main effect is significant is not particularly useful). The plot of means (see Solution 7.9) indicates that increasing the temperature increases the mean yield on both catalysts, but by far more on catalyst B than on catalyst A. Indeed, at the lower temperature, catalyst A produces the higher yield, but at the higher temperature catalyst B does. We can quantify this by reporting the relevant table of means:

temperat catalyst	A	B
160	57.00	48.50
180	70.00	81.50

■

Example 7.5

A researcher interested in water pollution and related matters carried out a study to investigate the effect of individual bathers on the faecal and total coliform bacterial populations in water. Coliform bacteria are generally present in water that has been used by humans, and need to be reduced to a low level in drinking water supplies. Faecal coliform bacteria are a subgroup that are commonly found in the human gut (among other places). The researcher was interested in the effects of the following factors on bacterial populations: the time since the subjects' last bath, the vigour of the subjects' activity in the bath water, and the subjects' sex. Each of these factors was studied at two levels – 1 hour and 24 hours since bathing, lethargic or vigorous bathing activity, and male or female sex. Thus the study had a 2^3 factorial structure. Two measurements were made for each combination of factors. The experiments were performed in a 100-gallon polyethylene bathtub, using dechlorinated tap water at 38° C. (Chlorine in the water would tend to kill the bacteria.)

Four response variables were measured: the bather's contribution to the faecal and the total coliform bacterial concentration after 15 minutes and after 30 minutes. The unit of measurement is the number of organisms present per 100 ml of water. Each contribution was measured by subtracting the initial bacterial concentration from that measured after 15 or 30 minutes. The data for the response variable faecal coliform concentration after 30 minutes are given in Table 7.6; the data for all four response variables are in the datafile bath.gsh.

Table 7.6

Time since last bath (hours)	Vigour of bathing activity	Bather's sex	Faecal coliform contribution after 30 minutes (organisms per 100 ml water)	
1	lethargic	female	1	4
24	lethargic	female	15	39
1	vigorous	female	10	21
24	vigorous	female	6	5
1	lethargic	male	170	67
24	lethargic	male	148	360
1	vigorous	male	170	377
24	vigorous	male	217	250

Source: Drew, G. D. (1971) *The Effects of Bathers on the Fecal Coliform, Total Coliform and Total Bacterial Density of Water*, Master's thesis, Department of Civil Engineering, Tufts University, Medford, Mass. Dataset name: bath.

One matter to take account of when analysing these data is that, for technical reasons to do with the method of measurement of bacterial populations in water, errors of measurement do not have a constant variance. Instead, the variance increases with the value of the mean. In situations like this, it is often appropriate to transform the response variable using a transformation down the ladder of powers, and a logarithmic transformation is often appropriate.

Exercise 7.10
Load the data from the file bath.gsh into GENSTAT.

(a) In this part of the exercise, you will be concerned only with the data on faecal coliform contribution after 30 minutes, which is in the variate fc30. Transform this response variable using a log transformation, and analyse the transformed response variable using a three-way factorial ANOVA. Report your conclusions. Was the log transformation appropriate? Can you find a better one?

(b) Repeat part (a) using the data on total coliform contribution after 30 minutes (variate tc30).

∎

In situations like this, where a transformation does not entirely enable the ANOVA assumptions to be satisfied, an alternative may be to carry out a weighted analysis of variance, the ANOVA equivalent of a weighted least squares regression. This methodology is beyond the scope of this book.

7.3   Using regression

So far, all the data you have analysed have had a particular property: the number of observations on each treatment (i.e. in each cell) has been constant within an experiment – in other words, there have been equal replications. In the last example (Example 7.5), for instance, there were eight different treatments (in a 2^3 factorial experiment), and there were two observations on each treatment. If an experiment has equal replications, some aspects of its analysis are simpler than if it has unequal replications. But not all data come in this simple form.

Example 7.6
The data in the file foster.gsh come from an experiment on foster feeding in rats. The rats were classified into one of four different genotypes, labelled A, B, I and J. The experimental unit was a litter of rats, and the response variable is the litter weight (littwt, in grams) after a trial feeding period. The foster mother did not necessarily have the same genotype as the litter she was fostering, so the litters are classified according to two factors, litter genotype and mother's genotype, each at four levels. This is therefore a 4 × 4 factorial experiment.

What makes this different from the examples you have met so far is that the number of observations in each cell is not constant, because the numbers of rats and litters available to the experimenters made this impossible. For instance, there were five litters of genotype A fostered by mothers also of genotype A, but there were only three litters of genotype A fostered by mothers of genotype B. The number of litters in each cell is shown in Table 7.7.

Table 7.7

		Genotype of mother			
		A	B	I	J
Genotype of litter	A	5	3	4	5
	B	4	5	4	2
	I	3	3	5	3
	J	4	3	3	5

Source: Scheffé, H. (1959) *The Analysis of Variance*, New York, John Wiley.
Dataset name: foster.

Exercise 7.11

Load the data from the file `foster.gsh` into GENSTAT and attempt to analyse them using two-way analysis of variance. What goes wrong?

■

The GENSTAT ANOVA command, which is invoked by the Stats|Analysis of Variance menu item, can only deal with data that possess a property known as 'first-order balance'. In general, this is a little complicated to define; lack of first-order balance can arise only when the number of observations in each cell is not constant, but it does not always happen even then, as you will see later. But it did happen in this case.

However, we are not stuck! In Chapter 5 you saw that data of the sort that could be analysed by one-way ANOVA could also be analysed using regression. This is also true for two-way ANOVA, and indeed for any factorial ANOVA, and using regression has the advantage that it works with unbalanced data (which in the current context results from unequal replications) as well. To see how to do this, let us start with an example we have looked at before.

Example 7.1 continued

Exercise 7.12

Load the (full) data on the rat-feeding experiment from the file `weight3.gsh` into GENSTAT. Unless you happened to keep a full printed copy of the output from your previous analysis, reanalyse the data using the two-way ANOVA command. This analysis should include producing and printing the usual residual plots.

Now click on the Stats|Regression Analysis|Linear menu item. In the resulting dialogue box, choose General Linear Regression in the Regression field. (This choice makes it easier to use factors in the model, because they are in the Available Data list.) Choose weight as the Response Variate. Ignore the Maximal Model field, and in the Model to be Fitted field enter source*amount. The * notation here is as for ANOVA models: entering source*amount in the Model to be Fitted field means 'fit a model involving main effects for source and for amount, and also their interaction'. Click on the Options button, and in the resulting dialogue box make sure the Accumulated item is checked. (The Model, Summary, F-probability, Estimates, t-probability and Estimate Constant Term boxes should also be checked by default.) Click on OK in this dialogue box, and then on OK in the Linear Regression dialogue box.

If you omitted to check the Accumulated item, you can do it later as a choice under the Further Output button in the Linear Regression dialogue box. In this case, you should also check the F-probability option in the Linear Regression Further Output dialogue box, to ask GENSTAT to calculate the appropriate SPs. This option is not available until after you have checked the Accumulated option.

Compare the outputs from the analysis of variance and the regression analysis in the output window. Can you work out what is happening in the regression analysis?

Click on the **Further Output** button in the **Linear Regression** dialogue box and then on the **Model Checking** button. In the resulting dialogue box, change the **Type of Residual** to Simple, retain **Composite** under **Type of Graph**, and compare the resulting plots with the residual plots from the analysis of variance. (The reason for choosing simple residuals is that the ANOVA residual plots use simple rather than standardized residuals.)

Your work on Exercise 7.12 should have given you some confidence that the regression model fitted by the above procedure is the same, effectively, as the two-way ANOVA model. This is indeed true; let us attempt to see why.

As for one-way ANOVA, when asked to fit a model where some or all of the explanatory variables are factors, GENSTAT effectively begins by setting up binary indicator variables to represent each level of each factor. In this case, you can think of it as setting up three indicator variables, x_1, x_2 and x_3, to represent the three levels (**beef**, **cereal** and **pork**) of the first factor, protein source, such that each variable takes the value 1 for an experimental unit receiving protein from that source and 0 otherwise. It sets up two more indicator variables, z_1 and z_2, to represent the two levels (**low** and **high**) of the second factor, protein amount, in a similar way. Finally (and this is what is new about the two-way case) it sets up further variables to represent the interaction of the two factors, and it sets these up as products of the indicator variables for the factors. So, for instance, there is a binary variable $w_{21} = x_2 z_1$ that takes the value 1 when, and only when, $x_2 = 1$ and $z_1 = 1$, i.e. for rats receiving cereal protein at low level.

If GENSTAT proceeded to fit a regression model with all the x_js, z_ls and w_{jl}s, there would be problems because of dependencies between the explanatory variables; this is because, for instance, if you know the values of x_2 and x_3 for a particular rat, you can work out exactly what its value of x_1 is. Thus, as usual, constraints are imposed by missing out some of the explanatory variables. For the variables corresponding directly to factors, GENSTAT omits the first for each factor, so that x_1 and z_1 are omitted. Then, for the product variables (the w_{jl}s), GENSTAT omits any that are defined in terms of variables it has already left out. Thus w_{11}, w_{12}, w_{21} and w_{31} are omitted, and the only ones left are w_{22} and w_{32}. Finally, then, GENSTAT fits by regression the model in which the mean response for a rat is

$$E(Y) = \mu + \beta_2 x_2 + \beta_3 x_3 + \gamma_2 z_2 + \delta_{22} w_{22} + \delta_{32} w_{32}.$$

The variation about this mean is, as usual, supposed to be normal with constant variance. If you compare this equation with equation (7.1), bearing in mind that at most one of the x_js, one of the z_ls and one of the w_{jl}s is non-zero for any one rat, you should be persuaded that the two models match, allowing for the fact that the ANOVA model involves sum-to-zero constraints and the regression model involves different constraints (e.g. omitting x_1).

Because the constraints differ in the two cases, the parameter estimates will not agree. In fact, the ANOVA commands do not output their parameter estimates unless you specifically ask them to. However, the regression output does give parameter estimates, and if you check back on the output you produced in Exercise 7.12 you will see that they are there, identified by rather more helpful but also rather more longwinded labels than the x_js, z_ls and w_{jl}s used here.

However, the analyses of variance will be the same, the fitted values will be the same, and the residuals will thus be the same, so that the conclusions from the models should be the same. ∎

Now that you have learned why the regression model is appropriate for factorial data, let us go back and use it for the rat-fostering data, for which ANOVA did not work.

Example 7.6 continued

Exercise 7.13

Load the data from foster.gsh into GENSTAT once more, if necessary.

(a) Analyse them by regression, as you did for the rat weight-gain data. Use littwt as the **Response Variate** and litter*mother as the **Model to be Fitted**. Do not forget to produce the accumulated analysis of variance table.

You will find that one of GENSTAT's warning messages refers to a large residual. This is an outlier; there may be very good reasons why its response has the value it does, but, for now, let us simply leave it out. Using the **Restrictions** command in the Spread menu, temporarily remove the row for that case. Reanalyse the data. What do you conclude? Are there any problems with the residual plots?[4]

(b) Reanalyse the data (still with the outlier removed), this time using mother*litter as the **Model to be Fitted**. How does the accumulated analysis of variance table differ from that in part (a)? Are the residual plots the same?

The reason for the difference between the two accumulated analysis of variance tables for parts (a) and (b) of Exercise 7.13 (with the outlier removed) lies in the lack of balance in this experiment. To understand this, we need to investigate what is going on in the accumulated analysis of variance table. This effectively reports the results from fitting a succession of regression models. The table for the regression of Exercise 7.13(a) is as follows.

```
*** Accumulated analysis of variance ***
```

Change	d.f.	s.s.	m.s.	v.r.	F pr.
+ litter	3	105.37	35.12	0.87	0.466
+ mother	3	745.57	248.52	6.12	0.001
+ litter.mother	9	1263.48	140.39	3.46	0.003
Residual	44	1785.60	40.58		
Total	59	3900.02	66.10		

The first row sums up how the regression results change when the model changes from one where only a constant term is included to one where the litter main effect is included as well. What happens is that the regression d.f. increases by 3 and the regression sum of squares increases by 105.37. (Since the total sum of squares remains fixed, this means that the residual sum of squares is reduced by 105.37.) The second row shows the changes

---

[4]Because you are not comparing these residual plots with residual plots from ANOVA commands, there is no need to change the residual type from Pearson to simple.

on adding the mother main effect to the model as well as the litter main effect. The regression sum of squares increases by another 745.57. Finally the third row shows the changes resulting from adding the interaction to the model that already contains the two main effects.

It turns out that SPs for these changes in sums of squares can be calculated using an F test in much the same way as those in the ANOVA tables discussed before. GENSTAT does these calculations using the residual mean square from the final model (with interaction), and reports the SPs in the accumulated ANOVA table. Thus, in the table for Exercise 7.13(a), the first row has an SP of 0.466 – there is no evidence that adding the litter main effect to a model with only a constant term improves the fit. The second row shows an SP of 0.001 – strong evidence that adding the mother main effect to a model that already includes the litter main effect *does* improve the fit. Finally the third row shows that adding the interaction to a model that already includes both main effects also improves the fit.

In a two-way ANOVA with equal replications (whether the calculations are done using ANOVA or regression methods), the estimates of the main effects are independent of each other. Therefore it does not matter what order they are entered in the analysis of variance; the results will always be the same. (If you do n‹ believe this, try it.) Thus, in the case of equal replications, the fact that the accumulated ANOVA table tests the effects of adding terms to the model in a particular order makes no difference. Any order would do. However, this is not necessarily true if there are unequal replications. Basically what happens is that the binary indicator variables representing the main effects become correlated in a particular way, so that the estimates of the corresponding regression coefficients are correlated too. Therefore the order in which the effects are fitted can make a difference. In this example using the foster data, the numbers in the cells did not differ very much, so the correlations involved were small and the difference between the two tables was not important. But in other examples, the difference can be greater. In such cases, the regression coefficients have the same sort of 'partial' interpretation as in the multiple regression examples of Chapter 6. Typically one would often want to test whether adding some effect to the model improves the fit in comparison with a model that already includes all the other effects. This means you have to fit the effect being tested *last*.   ∎

Any factorial experiment can be analysed (in GENSTAT) using regression methods. You might even be wondering why there are separate analysis of variance commands. One answer is that some more complicated experiments, which you will meet in Chapter 8, cannot easily be analysed by regression methods. Another is that, in GENSTAT, the analysis of variance commands are sometimes easier to use, and generally produce output that is easier to interpret when the data come from a factorial experiment.

Under certain circumstances, there are other ways that can be used to analyse factorial data with unequal replications using GENSTAT. One such circumstance is when the reason for the unequal replications is that data from some experimental units in a two-way experiment have somehow gone missing. The topic of how to deal with such missing values is a tricky one, and the details are beyond the scope of this book. Despite this, the following example illustrates two basic methods employed by GENSTAT for dealing with missing values.

Example 7.3 continued
In your study of the data on headaches, you found that, after transforming the response variable by taking logs, the usual ANOVA model fitted reasonably well. However, there

were two large outliers, corresponding to the cases numbered 5 and 18 in the GENSTAT spreadsheet. These have responses of 15.2 and 11.5 respectively, far above all the other responses. (The next largest is 8.78.) This does not, of course, mean that these data are in error, but we cannot go back and check with the experimenters. One possibility is to leave them out and see whether the conclusions from the analysis are affected. Let us do this, by two different methods.

Exercise 7.14
Load the data from the file headache.gsh into the GENSTAT spreadsheet.

(a) Update the server and take a log transformation, lscore, of the response variate, score. Unless you kept a copy of the output from your previous analysis of these data, perform a two-way analysis of variance using the logged responses.

We are now going to replace the values of lscore for cases 5 and 18 by the symbol *. The * is GENSTAT's symbol for missing data. In doing this, we shall be telling GENSTAT that the experiment was designed to include data from these individuals, but that the data have got corrupted or gone missing, so that we do not know what they are. To do this, enter * in place of the values for score in rows 5 and 18 of the spreadsheet, pressing <Return> after you have entered each one, and then recalculate lscore. Repeat the analysis of variance. How do the results differ? What do the residual plots look like?

(b) Now make the spreadsheet active and use the **Restrictions** command in the **Spread** menu to tell GENSTAT to ignore rows 5 and 18 temporarily. This amounts to telling some parts of GENSTAT to forget about these cases entirely – the values of the two factors as well as the responses. Update the server. Choose the **Data|Display** menu item. In the resulting dialogue box, select the row for lscore by clicking on it. Click on the **Delete** button, and then on **Cancel**. Now recalculate the log transformation of score. (This is one of several ways in which to ensure that **Restrictions** applies to lscore as well as to the original variables.)

You might expect that GENSTAT cannot analyse the resulting data using analysis of variance, since we no longer have equal replications. However, essentially because we have deleted two observations with the same headache type but different treatment groups, the data still possess what GENSTAT refers to as first-order balance, so that it *can* analyse them using ANOVA.

Perform an analysis of the restricted data. Do the conclusions differ from those in part (a)? When you have finished this exercise, do not save the modified spreadsheet.

If you like, you could analyse the headache data with the two outliers completely removed using regression. You would find that the results match those from the second ANOVA in Exercise 7.14(b). In this case, because of the first-order balance, the estimates of the two main effects remain independent, so the order in which you enter the effects makes no difference. ∎

You may be wondering how you can tell whether data from a particular factorial experiment with unequal replications can be analysed using the GENSTAT ANOVA commands.

There are two ways to do this. One is to learn and understand the definition of first-order balance used in GENSTAT, and to know how to apply it in particular cases. The other is simply to use trial and error, and this is recommended! GENSTAT knows whether it can analyse a particular dataset, and will tell you if it cannot.

7.4   Factorial ANOVA without replication

It is very often the case in the analysis of data with several factors that three-factor interactions are smaller and less important than two-factor interactions, four-factor interactions are smaller still, and so on. Therefore, in some contexts, data analysts *assume* that certain interactions are zero at the outset. If such an assumption is justified, it can permit the analysis of data that could not otherwise be dealt with.

Example 7.7
An experiment was carried out in which the response variable was the number of cycles of loading applied to a sample of worsted yarn before it broke. The treatments were defined by three factors each at three levels: length of test specimen (250 mm, 300 mm, 350 mm); amplitude of loading cycle (8 mm, 9 mm, 10 mm); load (40 g, 45 g, 50 g). There are thus $3 \times 3 \times 3 = 27$ different treatments. One observation was made for each treatment. The first six cases are given in Table 7.8.

Table 7.8

Length (mm)	Amplitude (mm)	Load (g)	Cycles
250	8	40	674
250	8	45	370
250	8	50	292
250	9	40	338
250	9	45	266
250	9	50	210
⋮	⋮	⋮	⋮

Source: Box, G. E. P. and Cox, D. R. (1964) 'An analysis of transformations (with discussion)', *Journal of the Royal Statistical Society, Series B*, **26**, 383–430.
Dataset name: wool.

Because only one observation was made for each treatment, it seems obvious that there is no way of measuring the variability within treatments. Therefore there cannot be a row for the residual in the ANOVA table (or, if there is, it will have no degrees of freedom and zero sum of squares), so the usual F tests cannot be performed. Another way of thinking of it is that there are 27 observations and so 26 degrees of freedom in all, and there will be three main effects with two degrees of freedom each, three two-factor interactions with four degrees of freedom each, and one three-factor interaction with eight degrees of freedom, making a total of $3 \times 2 + 3 \times 4 + 8 = 26$ for the main effects and interactions, leaving none for the residual. One way to get round this is to make the assumption that all the contrasts making up three-factor interactions are zero. (Experience with similar data in other contexts indicates that this may well be reasonable.) In that case, the eight degrees of freedom for the three-factor interaction can instead be assigned to the residual, and the data can be analysed successfully.

Experience with data like these (and the plotting of these data) indicates that a log transformation of the response variable is appropriate.

Exercise 7.15
Load the data from `wool.gsh` into GENSTAT.

(a) Take a log transformation of the response variable (**cycles**). Attempt a three-way analysis of variance, by choosing the Stats|Analysis of Variance menu item, choosing General Treatment Structure (no Blocking) in the Design field of the resulting dialogue box, choosing the log-transformed **cycles** as the Y-variate, and entering length*ampli*load in the Treatment Structure field. Click on OK. What happens?

(b) Go back to the Analysis of Variance dialogue box. For the Interactions field, choose Specify level of interaction from the pull-down menu. Another dialogue box should appear, in which you are asked to specify the highest order of interactions to be fitted. Fill in this field with a 2. This means that two-factor interactions will be included, but the three-factor interaction will not be (which effectively means you are assuming that all the relevant contrasts are zero). Click on OK to close this dialogue box, and then OK in the Analysis of Variance dialogue box. Compare the results with what you found in part (a). Plot the residuals and comment.

The validity of this analysis depends on the validity of the assumption that there are no three-factor interactions. In this case, it would be possible to check the assumption by making different assumptions instead. The factors defining the treatments are all, in this case, quantitative really. The ANOVA treated them simply as nominal variables; it did not even use the fact that they are ordered. We could instead have fitted a regression model, treating them as variates rather than factors. In this case, each would have one degree of freedom for the 'main effect' instead of two. We could define other variates as products of these, and use them as regressors as well; again, each would have only one degree of freedom. Working in this way, we could fit a complicated model with fewer degrees of freedom, leaving more over for the residual, and in this model it would be possible to test whether the equivalent of the three-factor interaction really is zero. However, for this regression analysis to be valid, assumptions need to be made about the nature of the relationship between the (quantitative) explanatory variables and the response variable — in particular we would need to assume that the relationship is linear. Such an analysis in fact gives no reason to doubt that the three-factor interaction really is zero. This kind of approach does not deal with every problem, though. ∎

Example 7.8
The data in the file `soil.gsh` form part of the data from an experiment to determine the effectiveness of blast furnace slags as agricultural liming materials on three types of soil: sandy loam, sandy clay loam and loamy sand. (Source: Johnson, D. E. and Graybill, F. A. (1972) 'An analysis of a two-way model with interaction and no replication', *Journal of the American Statistical Association*, **67**, 862–868. Dataset name: `soil`.) There were seven

soil treatments: none, coarse slag, medium slag, agricultural slag, agricultural limestone, agricultural slag plus minor elements, and agricultural limestone plus minor elements. Each was applied at 4000 lb per acre. The response variable was the corn yield in bushels per acre. The data involve two factors in a 3 × 7 structure, but there is only one response per cell (i.e. each soil treatment was applied only once to each soil type). Thus there are no degrees of freedom for the residual.

Here, because there are only two factors, the only interaction is that between the two factors, soil type and soil treatment. It seems risky to assume that this is non-existent. (Indeed, as the title of the source paper indicates, there is some evidence that the two factors *do* interact.) You might like to try loading the data into GENSTAT and analysing it. Even without any degrees of freedom for the residual, GENSTAT will perform an analysis of a sort (as you saw in Exercise 7.15(a)), so you can use it to produce a means plot (though of course there will be no standard error bar). The means plot certainly *looks* as if the factors interact, but without any idea of the standard error of differences between treatment means, one cannot be definite! These data cannot therefore be analysed without making yet more assumptions of some sort. There are methods for doing this sort of thing, but they are complicated and not without controversy, and you are spared the details. (One such method is described in the paper from which these data came. As you might expect from the title of that paper, it finds that the two factors interact. But that paper also reports that another, better-known, way of analysing the data in cases like this finds no interaction. Who knows what the true state of affairs is?) ■

In the next chapter, you will see how the analysis of variance model can be extended further to cope with data from experiments where one needs to take account of differences between experimental units other than those caused by treatment factors.

Summary of methodology

1 In some data, the explanatory variables are all categorical (i.e. factors) and the response variable is quantitative. In cases like this when there are two or more explanatory factors, the data are said to be factorial. The different possible values of the factors are often assigned numerical values known as levels.

2 In experimental and similar situations where the treatments are defined in terms of factors, one can define contrasts called main effects, measuring the change in the mean response when there is a change between the levels of one factor, two-factor interactions, which are differences between such changes in the mean response when there is a change between the levels of a second factor, three-factor interactions, which are differences between two-factor interactions according to the level of a third factor, and so on.

3 It is possible to write down a model involving terms directly related to main effects and interactions, together with an additive error term that has a normal distribution. It is possible to fit such models using GENSTAT analysis of variance and regression commands.

4 Since such a model is simply a reparametrization of linear models from Chapters 5 and 6, the residual plotting methods described there are still appropriate.

5 Plots of treatment means are helpful also.

6 GENSTAT'S ANOVA commands deal readily with data in which there are equal replications per cell. The same commands can also deal with certain patterns of unequal replication, provided first-order balance (a concept not explained in detail) is maintained; otherwise, regression can be used. Experiments with no replication can be analysed if higher-order interactions can be assumed to be zero.

8

Experiments with blocking

Chapters 5 and 7 concerned experimental situations in which the researcher has control over the allocation of experimental units to the levels of certain explanatory factors (treatments), and this chapter continues in similar vein. Chapter 5 considered experiments in which experimental units were allocated to treatments in a completely randomized way, while Chapter 7 extended the ideas to cases where the treatments have a factorial structure. This is the last chapter to concentrate solely on designed experiments.

Those earlier chapters had in common the fact that the experimental units were considered to be reasonably *homogeneous*. This means that we thought of the experimental units as being similar to one another for the purposes of the experiment: of course, units cannot be identical (else there'd be no need for statistics!), but homogeneous units will not differ widely from each other in ways that could be important determinants of values of the response variable (with the result that differences between responses can be put down solely to the different treatments applied to the experimental units).

Often, finding enough homogeneous experimental units is not practical. For example, in an experiment concerned with applying different fertilizers to improve the growing of a cereal crop, perhaps the field in which the experiment is to be done slopes. Parts of the field at the top of the hill will tend to be drier than parts of the field down the slope, and the crop might naturally grow better or worse in different parts of the field depending on its liking, or otherwise, for wet conditions. This will happen regardless of which fertilizer is applied. The way round this heterogeneity of field areas with respect to moisture is to compare fertilizers with each other within each of the wet and dry areas of the field. Such an experimental design provides an example of a technique known as *blocking*, which is the central new idea of this chapter.

8.1 Blocking

Blocking is, then, a general way to cope when the available experimental units are not homogeneous. A *block* is a set of experimental units – a subset of all the available experimental units – that can be considered to be homogeneous. *Blocking* is the grouping of experimental units into blocks of homogeneous units. By performing blocking, the experimenter is explicitly incorporating the relevant aspect of the heterogeneity of experimental units into the model, as will be seen later.

As an example, imagine an experiment to compare the growth of piglets fed on two different diets. Each piglet is an experimental unit. The piglets will vary in many ways. For the purposes of the experiment, the most important differences are:

- when the piglets were born;

- the health and age of each piglet's mother;

- and, perhaps, the initial weight and any inherited tendency to grow rapidly or slowly.

These points should lead us to choose blocks based on different litters of piglets. This would give blocks of piglets which were born at about the same time and had the same parents, in particular the same mother, and which, for the purpose of the experiment, could be assumed to form a homogeneous group. (A sow will usually give birth to about six piglets per litter.) It does not seem reasonable to assume that piglets from different litters form a homogeneous group.

Blocking is carried out before the experiment is done; it is part of the design of an experiment. The designing of experiments will be discussed briefly in Section 8.4. Until then, let us concentrate on how to analyse the results from a variety of experiments involving blocking. You have met all the basic ideas that will be needed already, in Chapters 5 and 7.

Section 8.1.1 covers the simplest form of blocking, namely paired data. In paired datasets, each block contains two experimental units. You have met this idea before (Section 2.3.1) but without thinking of it in experimental design terms. Section 8.1.2 moves on to blocks containing more than two experimental units.

8.1.1   Paired data

Pairing is used in many statistical studies. Experiments using paired experimental units are usually easy to analyse and many practical problems are solved by using *paired data*. Section 2.3.1 used the term matched pairs data for what we shall here refer to as paired data. We shall only consider pairing in experiments, but it is also used in observational studies.

Example 8.1
The data in Table 8.1 are the responses of the first five subjects (out of 25) in a study to assess the effect of sunlight on the solid content of grapefruit. One half of each (uncut) grapefruit (the upper half) was exposed to sunlight, the other half was shaded. The response variable was the percentage of solids in each grapefruit half. Interest was in whether or not exposure to sunlight affects the percentage of solids in (the exposed half of the) grapefruit.

It is clear that these are paired data, being pairs of observations arising from individual grapefruit, and comparisons between the responses under shaded and exposed conditions are possible within each grapefruit. Now think of this experiment in blocking terms.

Exercise 8.1
What are the blocks in this experiment? What are the experimental units within blocks? Why was the experiment designed in this way?

Table 8.1

Fruit	Shaded	Exposed
1	8.59	8.49
2	8.59	8.59
3	8.09	7.84
4	8.54	7.89
5	8.09	8.19
⋮	⋮	⋮

Source: data of P. L. Harding in
Croxton, F. E., Cowden, D. J.
and Klein, S. (1968) *Applied
General Statistics*, 3rd edition,
London, Pitman.
Dataset name: grapeft.

In the terminology of designed experiments, then, it might be said that this paired data experiment concerned 25 blocks each of size two.

Provided normality of the differences between the responses for each block is a reasonable assumption, the conventional way to analyse paired datasets is to use the *paired t test*. The method for performing this test is described in Section 2.3.1. It is precisely the one-sample *t* test applied to the differences between responses. Performing tests of this type in GENSTAT is described in Section 3.3. You can therefore go ahead, in Exercise 8.2, and analyse the data from this experiment, after first checking the relevant assumption.

Exercise 8.2

Load the dataset stored in grapeft.gsh into GENSTAT.

(a) Produce a preliminary plot of the data in the following way. Click on the **Graphics** menu; click on **Point Plot**. In the resulting dialogue box: type in the **Y Coordinates** field as exposed,shaded; fill in the **X Coordinates** field as fruit; click on the **OK** button; click on the **Cancel** button.

How does this plot differ from an ordinary scatterplot of **exposed** against **shaded** and why is this plot more useful in the current circumstances? Look at the plot: do you think the amount of solids differs between exposed and shaded halves?

(b) Calculate the differences between **exposed** and **shaded** responses. Produce a histogram of these differences. Does a normality assumption seem appropriate?

(c) Assume a normality assumption is tenable. Choose the **Stats|Statistical Tests|Two-sample tests** menu item. In the resulting dialogue box: change the **Test** to t-test (paired samples); fill in the **Data Set 1** field as exposed; fill in the **Data Set 2** field as shaded; click on the **OK** button; click on the **Cancel** button.

Look at the information in the **Output** window.

(i) Why is the output for the *t* test headed One-sample T-test?

(ii) What are the estimates of μ_Q and σ_Q^2, the mean and variance of the differences?

(iii) What is the SP and what do you conclude from the result of the test? Is this a reasonable conclusion?

8.1.2 More than two units per block

In Chapter 5, we saw that two groups could be compared using the two-sample t test, and that an equivalent test could be done by performing the one-way ANOVA. There is a similar relationship between the paired t test and the type of ANOVA developed here, which is called the *randomized block ANOVA*. The randomized block ANOVA gives us the equivalent of the paired t test, but can be used for blocks of any size, not just of size two.

The datasets that you will meet in this section are examples arising from a *randomized block design*. A randomized block design is a type of experiment in which the blocks are all of the same size and each of the treatments appears the same number of times within each block. Strictly speaking these are randomized *complete* block designs, in contrast to incomplete block designs that you will meet later. The 'randomized' aspect refers to random allocation of treatments to experimental units *within* blocks. Often, each treatment appears just once in each block; for simplicity, we shall confine our attention to this type of randomized block design.

Exercise 8.3

The data in Table 8.2 are from an experiment that had a randomized block design. The experiment concerned the comparison of five different diets and their effect on the weight gain of young rats in the four weeks after weaning. The response variable is the weight gain in grams. For precisely the reasons described for piglets at the beginning of this section, different litters of rats were to be considered as different blocks. To accommodate one rat within each litter getting one of the five experimental diets, rat litters of size five were used as the blocks. Within each litter (block), the allocation of each of the five diets to one of the five rats was performed at random, this being the 'randomized' aspect of the randomized block design. (Notice that, in Table 8.2, the data have been tidied up so that diet a results are in the first column, diet b results in the second column, and so on. This is not to be misread as suggesting any kind of systematic allocation of diets to rats within litters.)

Table 8.2

Litter	Diet				
	a	b	c	d	e
1	57.0	64.8	70.7	68.3	76.0
2	55.0	66.6	59.4	67.1	74.5
3	62.1	69.5	64.5	69.1	76.5
4	74.5	61.1	74.0	72.7	86.6
5	86.7	91.8	78.5	90.6	94.7
6	42.0	51.8	55.8	44.3	43.2
7	71.9	69.2	63.0	53.8	61.1
8	51.5	48.6	48.1	40.9	54.4

Source: data of M. Chalmers reported in John,
J. A. and Quenouille, M. A. (1977) *Experiments:
Design and Analysis*, London, Griffin.
Dataset name: wtgain.

(a) Load the wtgain data into the GENSTAT spreadsheet. Notice that the format of the data differs from Table 8.2; ensure that you understand how the two correspond. Update the server.

Click on the Graphics menu; click on Point Plot. In the resulting dialogue box, enter wtgain in the Y Coordinates field, diet in the X Coordinates field and litter in the Groups field. Click on the OK button; click on Cancel. What information is there in the resulting plot?

(b) To analyse these data, click on Stats|Analysis of Variance to produce the familiar dialogue box. From the pull-down menu, choose One-way ANOVA (in Randomized Blocks) in the Design field. Put wtgain in the Y-Variate field, diet in the Treatments field, and litter in the Blocks field. Click on OK. An ANOVA table will appear as part of the output in the Output window. How much of this ANOVA table appears to correspond with ANOVA tables you have seen before? Using the row labelled diet, what do you think is the result of a test that there is no difference between the different diets?

(c) To check the appropriateness of the model GENSTAT is using, in the Analysis of Variance dialogue box, choose Further Output|Residual Plots|OK. Does the model seem a suitable one? (You will be told about this model after this exercise.)

The model used to analyse data from a randomized block design with each treatment appearing once per block is

$$Y_{jl} = \mu + \gamma_j + \tau_l + \epsilon_{jl} \tag{8.1}$$

where each $\epsilon_{jl} \sim N(0, \sigma^2)$ and all the ϵ_{jl}s are independent of one another. It was this normality assumption that was particularly considered in Exercise 8.3(c).

The terms in (8.1) have the following meanings:

Y_{jl} is the response for the experimental unit in block j which received treatment l;

μ is the overall mean response, taken over all the treatments and all the blocks;

γ_j is the difference between the overall mean and the mean response for block j, the *block effect*;

τ_l is the difference between the overall mean and the mean response for treatment l, the *treatment effect*.

As in earlier chapters, there are constraints placed on the γ_js and τ_ls. The meanings for Y_{jl}, μ, γ_j and τ_l given above are based on the constraints $\sum_j \gamma_j = 0$ and $\sum_l \tau_l = 0$. If we were to use different constraints, then the meanings of μ, γ_j and τ_l would be different. The ANOVA, the fitted values and the residuals are not affected by the choice of constraints, so the choice is not a big issue.

Does model (8.1) strike you as familiar? Well, you haven't seen precisely this formula written in this book before, but you have seen something just a little grander that includes the same ingredients. The model given in Chapter 7 at (7.1) for the two-way analysis of variance is closely related to this one. The special case of (7.1) in which we take only one experimental unit per treatment combination (enabling us to drop the subscript i, which always takes the value 1) is

$$Y_{jl} = \mu + \alpha_j + \beta_l + \gamma_{jl} + \epsilon_{jl} \tag{8.2}$$

where the α_js and β_ls are the effects of two different treatments and the γ_{jl}s are their interactions. Now, set all the interaction parameters γ_{jl} to 0, and replace α in (8.2) by γ and β by τ, and we have precisely (8.1). The constraints associated with the two models coincide too.

What model (8.1) does, then, is to deal with block and treatment effects as we dealt with two treatment effects in Chapter 7. A crucial difference, however, is that, while in Chapter 7 we were interested in both of the treatment effects α and β, here only the treatment effects τ are of real interest to us, and the block effects γ are in the model to account for the extraneous response variability associated with the heterogeneity of experimental units.

And there's a second major difference. Recall that, in the two-way analysis of variance, not having replication (Section 7.4), i.e. having just one response per treatment combination, gave us a distinct problem; this is because, when two treatments are being considered on an equal footing, we are rarely happy to assume that their interactions can be ignored. However, when one of the two 'treatments' is a block, this is precisely what we are happy to do! It would be a badly designed experiment in which we chose blocks that interacted with the treatment of interest, because we would have introduced an element that simply muddies the waters. (If an experiment were designed to have more than one response for each block/treatment combination, one could use the methods of Chapter 7 to *check* whether this assumption of no interaction is indeed valid. This is sometimes done.)

In fact there's also a third difference, namely that the block effects γ_j in (8.1) are often taken to be random variables (with a normal distribution) rather than fixed values. (The treatment effects τ_l are still taken to be fixed.) However, this makes no difference to any of the analyses in this book, so you can ignore it.

The reanalysis, in Exercise 8.4, of the weight gain data for rats should help clarify these points.

Exercise 8.4

(a) If a randomized block experiment is a special case of two-way analysis of variance, we should be able to analyse it by the latter route too. To try this, in the Analysis of Variance dialogue box, change the Design field to Two-way ANOVA (no Blocking). Let the Y-Variate be wtgain, and enter diet and litter in the Treatment fields. Change the Interactions field to No interactions, (ie. main effects.). Click on OK. How does the ANOVA table thus produced differ from that seen in Solution 8.3? Do your conclusions change?

(b) Repeat part (a) having switched the Interactions field back to All interactions. What difference does this make?

The next exercise considers another randomized block experiment, which you should analyse using the One-way ANOVA (in Randomized Blocks) option in the Design field of the Analysis of Variance dialogue box.

Exercise 8.5

Consider an experiment in which a drug is added to the feed of chicks in an attempt to promote growth. The comparison is between three treatments: standard feed (control), denoted by C; standard feed plus low dose of drug, L; and standard feed plus high dose of drug, H. The experimental unit is a group of chicks, reared and fed together in the bird house. The response variable is the average weight per bird (in pounds), at maturity, for the group of birds in an experimental unit. Because of variation in lighting and ventilation, the position of each experimental unit in the bird house is thought to influence the response, so the 24 experimental units were grouped three to a block, with adjacent units going into the same block. That is, in this case the physical proximity of units is thought to make for a homogeneous block, with heterogeneity between units some distance from each other. Within each block, the three experimental units were randomly allocated to the three treatments; this avoids any possibility of systematic differences between locations within blocks (guarding against any remaining differences within blocks, even though they are thought to be reasonably homogeneous). The data are given in Table 8.3.

Table 8.3

Block	Control	Low dose	High dose
1	3.93	3.99	4.08
2	3.78	3.96	3.94
3	3.88	3.96	4.02
4	3.93	4.03	4.06
5	3.84	4.10	3.94
6	3.75	4.02	4.09
7	3.98	4.06	4.17
8	3.84	3.92	4.12

Source: data of S. M. Free in Snee, R. D. (1985) 'Graphical display of results of three treatment randomized block experiments', *Applied Statistics*, **34**, 71–7.
Dataset name: chicks

(a) Perform a randomized block analysis of the chicks data. Is there evidence of a difference between treatments?

(b) Produce residual plots. Do these cast any doubt on the analysis?

You may have noticed that a question possibly of more interest to the researcher than the one you have just answered, namely 'Are there any differences between the treatments at all?', is 'Does addition of the drug to the chicks' feed make any difference?'. As in Chapter 5, this calls for a planned comparison between μ_1, the mean of the chicks' weights under the standard feed, and the average of μ_2 and μ_3, the mean weights of the low-dose and high-dose groups respectively. Note that, here, there is no indeterminacy about what the contrast should be (as there was for the singers data in Section 5.5.1) because now we have equal replications. Interest therefore centres on the contrast

$$\theta_1 = \mu_1 - \tfrac{1}{2}(\mu_2 + \mu_3).$$

As in the completely randomized design of Chapter 5, the way ahead is to estimate the contrast in the natural way using the treatment means:

$$\widehat{\theta}_1 = \overline{Y}_1 - \tfrac{1}{2}(\overline{Y}_2 + \overline{Y}_3).$$

Here, the means are averages across blocks. Had this been a completely randomized experiment, we would then say that

$$V(\widehat{\theta}_1) = \frac{\sigma^2}{8} + \left(\frac{1}{2}\right)^2 \left(\frac{\sigma^2}{8} + \frac{\sigma^2}{8}\right) = \frac{3}{2}\frac{\sigma^2}{8} = \frac{3\sigma^2}{16}.$$

Here we have given the formula specific to the case of the chicks experiment (where each treatment is replicated eight times). To see that this formula still holds true for a randomized block experiment, notice from (8.1) that

$$Y_{jl} \sim N(\mu + \gamma_j + \tau_l, \sigma^2).$$

Because the Y_{jl}s are independent,

$$\overline{Y}_l \sim N\left(\mu + \overline{\gamma} + \tau_l, \frac{\sigma^2}{b}\right)$$

where $\overline{Y}_l$ is the mean for the lth treatment, $\overline{\gamma} = b^{-1} \sum_j \gamma_j$ (= 0 under the sum-to-zero constraint) and b is the number of blocks. It follows that $\widehat{\theta}_1$ has mean θ_1 (provided we write $\mu_l = \mu + \overline{\gamma} + \tau_l$) and variance as given because the $\overline{Y}_l$s are independent. And notice that $\overline{\gamma}$ cancels out of any contrast.

So if all this is essentially the same as in the completely randomized design, where does the randomized block nature of the design come in? The answer is in the nature of σ^2. Take the estimate (and its d.f.) from the Residual line of the randomized block ANOVA table. This estimate is an estimate of a smaller σ^2 than in the completely randomized case, because the block effects have been explicitly taken account of by being introduced into the mean structure (and hence taken out of the random error). To see this, think of the block effects γ_j as being $N(0, \sigma_\gamma^2)$ random variables, say, independent of the ϵ_{jl}s, which remain $N(0, \sigma^2)$ random variables. If the experiment is in randomized blocks, then the m.s.(Residual) simply estimates σ^2. However, if the experiment is randomized differently, as a completely randomized experiment ignoring the blocks, then the m.s.(Residual) estimates the variance of the response for a randomly chosen plot from the whole experiment, and this variance is the sum of error and block variances, $\sigma^2 + \sigma_\gamma^2$, which is larger than σ^2.

Exercise 8.6

Taking information from the GENSTAT output arising in the previous exercise, but otherwise using only pen, paper and calculator, what is the numerical value of $\widehat{\theta}_1$ for the chicks data? Give a 95% confidence interval for this difference (you may need to use GENSTAT to obtain the appropriate percentage point). Does this give evidence for a difference between control and treatment results?

At the start of this subsection, it was claimed that a paired t test was effectively the same as doing a randomized block analysis for blocks of size two. If that is so, running the grapeft data through a randomized block ANOVA should yield the same results as in Section 8.1.1.

Exercise 8.7
Load the dataset grapeft1 (not grapeft) into GENSTAT.

(a) Check that you understand how grapeft1.gsh gives the same data but in different format to that in grapeft.gsh (which looks much like Table 8.1). (The different data format is required to allow GENSTAT to cope with paired data in its ANOVA menu.)

(b) Use the One-way ANOVA (in Randomized Blocks) option in the Design field, having first chosen the Stats|Analysis of Variance menu item. Run this randomized block ANOVA for the grapeft1 data by making the appropriate choices in the other fields. Does the ANOVA F pr. value coincide with that in Solution 8.2? How else is the paired t test information apparent from the ANOVA output?

Have you spotted any possible objection to considering the grapefruit experiment as a randomized block experiment? It is certainly OK in having blocks with two treatments in each, but the element of randomization is missing: which half of each grapefruit is shaded and which half of each grapefruit is exposed was not allocated at random by the experimenter. This, and other deviations from the strict randomized block design happen in many other contexts too, and yet it remains appropriate to perform the paired t test or its ANOVA counterpart. Why so? Well, in the absence of randomization, extra care has to be taken to ensure that any systematic differences are eradicated in some other way. For instance, consider an experiment in which differences between patient response before and after some treatment are of interest. Provided one can be sure that nothing else changed for the patient between the times at which the two responses were taken (e.g. the patient may have had time to get better anyway!), the model underlying the paired t test remains applicable. For the grapefruit, one would have to be convinced that it is only the effect of amount of sunlight that affects the percentage of solids and not some other factor. One objection might be, 'What about gravity?': perhaps it is this and not the effects of the sun that makes solids have a tendency to drop to the bottom (shaded half) of the grapefruit. It would be for the plant physiologist to convince us that this is not so, else we cannot disentangle the effects of gravity and the sun.

The point is that, if at all possible, randomization should be performed to avoid just this kind of uncertainty about any effects observed. With it, we can be confident that other systematic effects are 'averaged out'; without it, a convincing argument for putting observed effects down to treatment is much more difficult to make.

Before we leave this experiment, it is worth commenting on the fact that the estimated mean difference in the percentage of solids between the two halves of a grapefruit is only about 0.2%. The difference is statistically significant; but whether such a small difference is of any practical significance is another question.

Finally, in this section, as our analysis can be thought of as separating out block and treatment effects, there should be no great difficulty in introducing more structure into the treatment effects part while retaining a blocking element. This is the case in the following example in which an experiment on factors affecting turnip growth is analysed. There are blocks and also a factorial structure to the treatments (as in Chapter 7).

Example 8.2

The data in Table 8.4 are from an experiment which had a randomized block design. There were 64 plots, arranged in four blocks each of size sixteen. Each block was a rectangular piece of land, measuring 3 metres by 32 metres. Each block was divided into sixteen plots by splitting the long side of the block into sixteen two-metre pieces. So, each plot was a 3 m × 2 m rectangle of land.

Table 8.4

variety	sowing date	sowing density (kg/ha)	label	I	II	III	IV
Barkant	21/8/90	1	A	2.7	1.4	1.2	3.8
		2	B	7.3	3.8	3.0	1.2
		4	C	6.5	4.6	4.7	0.8
		8	D	8.2	4.0	6.0	2.5
	28/8/90	1	E	4.4	0.4	6.5	3.1
		2	F	2.6	7.1	7.0	3.2
		4	G	24.0	14.9	14.6	2.6
		8	H	12.2	18.9	15.6	9.9
Marco	21/8/90	1	J	1.2	1.3	1.5	1.0
		2	K	2.2	2.0	2.1	2.5
		4	L	2.2	6.2	5.7	0.6
		8	M	4.0	2.8	10.8	3.1
	28/8/90	1	N	2.5	1.6	1.3	0.3
		2	P	5.5	1.2	2.0	0.9
		4	Q	4.7	13.2	9.0	2.9
		8	R	14.9	13.3	9.3	3.6

Source: unpublished data supplied by P. C. Taylor.
Dataset name: `turnip`.

The River Thames runs along one edge of the field used in this experiment, and usually floods part of the field each year. The blocks were designed so that the long side of each block was parallel to the river-bank. The blocks were at different distances from the river-bank.

Exercise 8.8

(a) With pencil and paper, sketch a diagram to show the positions of the blocks relative to the river and how each block is split into plots.

(b) Write down a brief explanation of why the experiment was designed in this way.

The experiment was about growing turnips for fodder. The turnips would not normally be harvested because they are grown to provide food for farm animals in winter: the farmer simply releases animals into the field and the animals graze on the turnips. The turnips are not even the main crop in the field during the growing season; the turnips are sown after the main crop has been harvested.

There were sixteen treatment combinations in the experiment. These combinations are formed from: two different varieties of turnip – Barkant or Marco; two different sowing dates – one as soon as possible after the main crop has been harvested, the other a week later; and four different sowing densities – 1, 2, 4 or 8 kilograms per hectare (kg/ha). Each of the sixteen treatment combinations formed from this $2 \times 2 \times 4$ factorial arrangement appears once in each block. Treatment combinations were allocated to plots within blocks at random, and labelled with letters of the alphabet (see Table 8.4). Treatment L, for example, corresponds to Marco seeds, sown on 21 August 1990, at a density of 4 kilograms of seed per hectare of land. The letters I and O are not used, because they tend to get confused with the numbers 1 and 0. The response variable is the yield of turnips in kilograms.

The questions of interest are the following.

- Does one variety produce more fodder than the other?
- Can the farmer afford to wait an extra week before sowing?
- Bearing in mind that seed is sold by weight, what sowing density should the farmer use?

The turnips were harvested and weighed by staff and students of the Departments of Agriculture and Applied Statistics of the University of Reading, in October 1990. This group included one of the current authors.

Exercise 8.9

Load the `turnip` dataset into the GENSTAT spreadsheet. Ensure that you understand how the columns correspond to blocks, treatments and responses.

Update the GENSTAT server. Click on the **Stats** menu; click on **Analysis of Variance**. In the **Analysis of Variance** dialogue box, select **General Analysis of Variance** in the **Design** field.[1] (This is necessary in place of, for example, **One-way ANOVA (in Randomized Blocks)**, because the latter does not allow a factorial treatment structure; the former does.) Fill in the Y-Variate as weight, the **Treatment Structure** as density*sowing*variety, and the **Block Structure** as block. Click on **OK**.

(a) Look at the ANOVA table in the **Output** window. As in the earlier ANOVA tables, the block 'stratum' is taken out as a separate row at the top of the table. Why are there 3 d.f. for blocks? With the block effect taken out, the remainder of the table, under the heading `block.*Units* stratum`, gives all the information about treatment effects. Why are there 45 d.f. for `Residual`?

(b) This latter part of the ANOVA table can be treated as for any other factorial experiment (without blocking). Which effects seem to be important and which unnecessary to the model?

(c) Using appropriate tables of means, relate your answer to part (b) back to the bulleted questions asked just before this exercise. (Treat the **sowing.variety** interaction term as if unimportant here.)

(d) Obtain residual plots of the data (in the usual fashion). What effects do you notice in the residual plots that might make you wish to delve more deeply into the analysis of these data? (Attempt no further analysis.)

∎

---

[1] There is a **General Treatment Structure (in Randomized Blocks)** option as well, which would also work in this context.

It may be worth stressing that blocking is not some extra complication that one has, unfortunately, to incorporate into the model when there is clear heterogeneity between experimental units. Instead, it is positively advantageous to introduce a blocking structure if possible; by arranging experimental units in the most homogeneous way, we are eliminating some of what would otherwise be extraneous error variability. (And even if there turns out to be little variability between blocks, rarely will anything be lost by blocking.)

8.2   More complicated blocking

The blocking in the randomized block design is the simplest form of blocking possible. Now we shall look at more complicated types of blocking.

We saw, in Chapter 7 and in Exercise 8.9, that there are experiments whose treatment combinations have a factorial structure. In such an experiment there are two or more factors that define the treatments. There are blocking schemes which are, in a sense, similar to factorial experiments in that there are two or more ways of dividing the set of available experimental units into blocks. (In this case, one says that there are two or more *blocking factors*.) A widely used form of two-factor blocking is called the *latin square*. The latin square is presented in Section 8.2.1.

The second type of blocking that we shall look at is called an *incomplete block* design. The main difference between incomplete block designs and the randomized block design is that each block contains only some, not all, of the treatment combinations. Incomplete block designs are the subject of Section 8.2.2, and are considered further in Section 8.3.

8.2.1   Latin squares

A latin square design is simpler to explain once you have seen an example; Example 8.3 is based on a dataset from an experiment that used such a design.

Example 8.3
An experiment was carried out to compare the wear resistance of four different types of rubber-covered fabric, labelled Material A to Material D. The data are given in Table 8.5.

Table 8.5   Losses in weight (in units of 0.1 mg) from an experiment to study the resistance to abrasion of four types of rubber-covered fabric

Run	\multicolumn{4}{c}{Position in machine}			
	4	2	1	3
2	A 251	B 241	D 227	C 229
3	D 234	C 273	A 274	B 226
1	C 235	D 236	B 218	A 268
4	B 195	A 270	C 230	D 225

Source: Davies, O. L. (ed.) (1956), *The Design and Analysis of Industrial Experiments*, 2nd edition, London, Longman.
Dataset name: `martinda`.

The apparatus used for testing the wear resistance was a device called a Martindale wear tester. This machine has four abrading surfaces made of emery paper. Four test samples of the fabrics rest on the emery paper surfaces, and the samples are mechanically moved over the surfaces, thus rubbing them away. The response variable is the weight loss of a fabric sample after a fixed number of cycles of the machine. (Fabrics that are more resistant to wear lose less weight.)

Clearly more than one run of the machine should be used to compare the four fabrics, to allow for random variability. Measurements made on the same run tend to be more alike than measurements made on different runs, because it is not possible to set up and run the machine in exactly the same way each time. So one possible design for the experiment would be to use randomized blocks, with a machine run as the block and the four materials allocated at random to the four available positions on the machine on each run.

However, it is also known that, because of imperfections in the construction of the wear testing machine, measurements made in the same position on the machine (on different runs) tend to be more alike than measurements made in different positions (on different runs). Thus another design for the experiment would be to use machine positions as blocks; that is, to carry out four runs of the machine in which the four materials all occupy Position 1 once each, all occupy Position 2 once each, and so on, with the actual allocation of materials to positions on any run being done at random. This is a feasible design but it does not take account of differences between runs (just as the design in the previous paragraph did not take account of differences between positions).

The latin square design that was actually used took into account *both* the possible methods of blocking (runs and positions), in a way described below. ■

In a latin square design there are two blocking factors. In Example 8.3 the blocking factors are, first, the different machine runs and, second, the different positions on the machine.

The blocking factors in a latin square are assumed to have no interactions with each other. For Example 8.3 this means that each run produces different responses, that the different machine positions produce different responses, but that the pattern of responses across the four machine positions is the same for each run. (Experience with the machine indicated that this was in fact the case.)

As with the randomized block design, the blocking factors are assumed not to have interactions with the treatment factors. For Example 8.3 this means that, for instance, the mean difference in response between Material A and Material B is the same for any machine run and any position. The same assumption is made for differences between other materials.

Both blocking factors in a latin square design have the same number of levels, which is the same as the number of treatment combinations. This translates into there being four different runs, four different machine positions and four different fabrics in Example 8.3. (In this example, the number of positions is fixed by the design of the Martindale wear tester. If the number of fabrics to be compared was not four, a different design would have to be used.)

The main characteristic of a latin square design is that each treatment combination appears once at each level of each of the two blocking factors. This tells us that the number of experimental units must be equal to the square of the number of treatment combinations. For Example 8.3, this means that each material is used once in each run and once in each position, and that there are sixteen experimental units; each experimental unit is the combination of a run and a position.

Finally, the design is called a latin *square*, because you can write down the treatment combinations in a square grid, where the rows represent one blocking factor, and the columns the other, to form a square. It is called a *latin* square because the treatment combinations are usually labelled with letters taken from the Roman alphabet, which comes from Latin! You can see this latin square pattern in Table 8.5; each letter (material) appears once in each row (run) and once in each column (position).

There are similar designs involving two different sets of treatments, called *graeco-latin squares*, in which one set of treatments is usually labelled using the Roman alphabet and the other using the Greek alphabet. Graeco-latin squares will not be covered in this book. Nor will other more complicated latin square designs.

As with all experiments, randomization is important with latin squares. The process of randomization is rather more complicated with latin squares than with simpler designs. You might well have wondered why the rows and columns in Table 8.5 appear in the order they do; this is to demonstrate the randomization. To randomize a latin square design, begin with any latin square of letters (i.e. any square of letters in which each letter appears once in each row and once in each column), then allocate the levels of one blocking factor to rows at random, do the same with the levels of the second blocking factor and the columns, and finally allocate the treatments to the letters at random. In Table 8.5, for example, the first row was randomly allocated to the second run and the first column to position 4 in the machine. (Also the four types of fabric were allocated randomly to the letters A, B, C, D.)

The model used to analyse data from a latin square design is

$$Y_{jkl} = \mu + \gamma_j + \beta_k + \tau_l + \epsilon_{jkl} \tag{8.3}$$

where each $\epsilon_{jkl} \sim N(0, \sigma^2)$, and all the ϵ_{jkl}s are independent of one another. The terms in (8.3) have the following meanings:

Y_{jkl} is the response for the experimental unit in row j and column k, which received treatment l;

μ is the overall mean response, taken over all the treatments and all the blocks;

γ_j is the difference between the overall mean and the mean response for row j;

β_k is the difference between the overall mean and the mean response for column k;

τ_l is the difference between the overall mean and the mean response for treatment l.

As for the randomized block design, we could use the methods of Chapter 7 to fit this model. We shall not do this. Instead, we shall go straight to the analysis which makes a distinction between blocking factors and treatment factors.

Exercise 8.10

(a) Load the dataset stored in `martinda.gsh` into the spreadsheet. Compare the spreadsheet with Table 8.5, so as to ensure that you know how the data in the spreadsheet correspond to the table.

(b) Click on the **Graphics** menu; click on **Point Plot**. In the resulting dialogue box: fill in the Y Coordinates field as **wear**; fill in the X Coordinates field as **material**; click on the **OK** button; click on the **Cancel** button. Do you think that there are systematic differences in the amount of wear produced in each fabric? Do you have any other comments on the pattern in the plot?

(c) Click on the **Stats** menu; click on **Analysis of Variance**. In the resulting dialogue box: choose **Latin Square** as the **Design**; fill in the **Y-Variate** field as wear; fill in the **Treatments** field as material; fill in the **Rows** field as run; fill in the **Columns** field as position; click on the **OK** button. Find the ANOVA table in the Output window. What does it tell you about differences in wear resistance between the four fabrics? Produce the standard ANOVA residual plots. Does the model seem to be appropriate?

(d) Scroll up in the **Input Log** window until you find the BLOCK command that tells GENSTAT how the blocks are set up in this experiment. Note what it says.

8.2.2 Incomplete block designs

The blocks in all the experiments with a randomized block design that we have studied so far contain enough experimental units for each treatment combination to appear once in each block. However, this is not always practical. If the number of experimental units in each block is smaller than the number of different treatment combinations, then it is impossible to have each treatment combination appearing once in each block. An *incomplete block design* is a design in which there are blocks that are incomplete, in that they have some treatment combinations missing.

Example 8.4
The data in Table 8.6 are from an incomplete block design. Identical plates were made dirty using some dirt prepared specially for the experiment. This was so that all the plates would be in the same state of dirtiness at the start of the experiment. The plates were then

Table 8.6

Session	Detergent								
	1	2	3	4	5	6	7	8	9
1	19	17	11	—	—	—	—	—	—
2	—	—	—	6	26	23	—	—	—
3	—	—	—	—	—	—	21	19	28
4	20	—	—	7	—	—	20	—	—
5	—	17	—	—	26	—	—	19	—
6	—	—	15	—	—	23	—	—	31
7	20	—	—	—	26	—	—	—	31
8	—	16	—	—	—	23	21	—	—
9	—	—	13	7	—	—	—	20	—
10	20	—	—	—	—	24	—	19	—
11	—	17	—	6	—	—	—	—	29
12	—	—	14	—	24	—	21	—	—

Source: John, P. W. M. (1961) 'An application of a balanced incomplete block design', *Technometrics*, **3**, 51–54.
Dataset name: dishwash.

washed using nine different detergents to see which detergent would wash most plates. Three people and three basins were available for washing the plates. This meant that it was only possible to use three different detergents simultaneously.

During a washing session, the washers were instructed to take care to make their pace of scrubbing plates the same as each other; however, there was no means of checking that plates were washed at the same speed in different sessions. Therefore it was believed that there might be differences between the test conditions for different sessions, but that conditions would be much more similar for sets of washing done at the same time (even though three different basins and three different people were involved). The experimenters therefore decided to block on session. Each block consisted of the three stacks of plates to be washed in a single session. The experimental unit, therefore, was a stack of plates. The experimenters could perhaps have used an even more complicated design, in which there was another type of block corresponding to the individual people–basin combinations, but this was not done.

Each plate was washed individually until it was clean. The number of plates washed before the foam disappeared is the response variable. Thus the response variable is discrete. This might lead you to think that it is not appropriate to assume a normal distribution. There is something in this, but we shall see later that such an assumption is not too wide of the mark in this case.   ■

The experiment described in Example 8.4 has blocks of size three. There are nine different treatments. The blocks are incomplete because each contains only three out of the nine detergents.

Because the blocks are not all alike in experiments (like this) with incomplete block designs, it is important to allocate the blocks at random, as well as allocating the experimental units to treatments at random within the blocks. Thus, although Table 8.6 is laid out very systematically, in practice it would be important to perform the session patterns numbered 1 to 12 in the table in a random order, rather than just working through them in the order listed. Also, within a session, one would allocate the three detergents involved at random to the three washers.

The design of the dishwashing experiment is also *balanced*. Consider Table 8.6. Notice that if you choose any pair of detergents, such as 5 and 9, there is a block in which both detergents of the pair appear together. For Detergents 5 and 9, the block in which they appear together is Session 7. In fact, not only does every pair of detergents appear within at least one block, every pair appears the same number of times (once in this case). This is what is meant by saying that an incomplete block design is balanced. There are $(9 \times 8)/2 = 36$ unique pairs of detergents overall and $(3 \times 2)/2 = 3$ pairs of detergents within each block. That is why there have to be $36/3 = 12$ blocks.

Note that this is only one of several meanings of the word 'balance' in the context of the design of experiments. You saw in Chapter 7 that the GENSTAT ANOVA command cannot be used unless the data possess a property called 'first-order balance'. In the next section, you will meet other types of incomplete block designs which do possess the property of first-order balance but which are *not* balanced in the sense discussed here. However, balanced incomplete block designs do possess first-order balance; so, as we shall see, GENSTAT can analyse them with its ANOVA command. Also the word 'balanced' is sometimes used incorrectly to mean that each treatment combination appears the same number of times. This is actually 'equal replication'.

Balance (in the sense of 'balanced incomplete block designs') is important for the following reason. If a design is balanced, then we have the same amount of information about the difference between each pair of treatments. Three practical consequences of using a balanced incomplete block design are:

- the standard error of the difference between the effects of two treatments is the same regardless of which two treatments we are looking at;

- GENSTAT can use a fast way to fit the model and produce the ANOVA;

- as in many other (but not all) designs that are balanced in the first-order sense, the estimates of the block parameters in the model are independent of the estimates of the treatment parameters.

The first of these points may not be relevant if there is a good reason why you want to estimate some treatment differences more accurately than others. The second point is becoming less and less important as computers get faster; nowadays it is hardly ever crucial. The third point sounds rather technical and uninteresting, but arguably it is the most important. Usually nobody is very interested in the values of the block parameters; but, if there are strong dependences between the block and treatment parameter estimates, a consequence is that the treatment parameters are estimated less accurately than they would be otherwise. Slight lack of balance in this respect is not important, but severe imbalance is. You saw in Exercise 7.13 a dataset where there was slight lack of balance. (There were no blocks but the principle is the same.)

If the design is not balanced (in the first-order sense), then, as you saw in Chapter 7, it cannot be analysed using GENSTAT's **Analysis of Variance** menu item. We shall see an example of what to do when the design is not balanced in Exercise 8.12. Now let us return to the dishwashing experiment, and use GENSTAT to perform the ANOVA.

Exercise 8.11

(a) Load the dataset stored in `dishwash.gsh` into the spreadsheet. Compare the spreadsheet with Table 8.6, so as to ensure that you know how the data in the spreadsheet correspond to the table.

(b) Click on the **Graphics** menu; click on **Point Plot**. Fill in the **Y Coordinates** field as plates; fill in the **X Coordinates** field as detergen; click on the **OK** button; click on the **Cancel** button.

Assume for the moment that there are not huge differences between the sessions. Do you think that there are systematic differences between the numbers of plates that can be washed by each detergent? If so, which detergent do you think washes most plates, and which least? Are there several good, or bad, detergents or are most detergents similar, with one or two being very different from the rest? If not, what makes you think that there are no systematic differences between detergents?

One worry with these data is that there may indeed be considerable differences between sessions. In this case, your conclusions from the plot of **plates** against **detergen** may be misleading, because not every detergent was used in every session. You might like to explore the data further by producing a plot where the different sessions are distinguished by a colour code, or perhaps a plot of the number of plates washed against the washing session, with the detergents distinguished by a colour code.

(c) Click on the Stats menu; click on Analysis of Variance. Check that the General Analysis of Variance option is selected in the Design field. Fill in the Y-Variate field as plates; fill in the Treatment Structure field as detergen; fill in the Block Structure field as session; click on the OK button. (You could alternatively use the One-way ANOVA (in Randomized Blocks) option in the Design field. For balanced designs, this works even if the blocks are incomplete.)

Find the ANOVA table in the Output window. What is odd about this ANOVA table compared with the other ANOVA tables that you have seen before? Looking at the session.*Units* stratum, what does the ANOVA table tell you about differences between detergents?

(d) Check the residuals, by plotting them in the usual way. This is particularly important for this dataset, because the data are counts. This means that they cannot really have come from a normal distribution, but we are hoping that their distribution is very similar to a normal one. Do the residuals look as if they satisfy the usual ANOVA assumptions?

In the following example, you will look at a dataset which has incomplete blocks but is not balanced.

Example 8.5

The dataset presented in Table 8.7 was designed as a latin square, with the rows and columns corresponding to the two blocking factors. The aim was to compare the yield of turnips of six different varieties, labelled A–F. Each plot was 15 ft × 15 ft. The experiment was vandalized. This is why three plots in the bottom right of Table 8.7 have no weights recorded. The data are stored in vandal.gsh, with the vandalized plots omitted.

Table 8.7 Fresh weights (in pounds) of six varieties of turnips.

E 29.0	F 14.5	D 20.5	A 22.5	B 16.0	C 6.5
B 17.5	A 29.5	E 12.0	C 9.0	D 33.0	F 12.5
F 17.0	B 30.0	C 13.0	D 29.0	A 27.0	E 12.0
A 31.5	D 31.5	F 24.0	E 19.5	C 10.5	B 21.0
D 25.0	C 13.0	B 31.0	F 26.0	E 19.5	A –
C 12.2	E 13.0	A 34.0	B 20.0	F –	D –

Source: Rayner, A. A. (1969). *A First Course in Biometry for Agriculture Students*, Pietermaritzburg, University of Natal Press.

Dataset name: vandal.

Exercise 8.12

(a) Load the dataset stored in vandal.gsh into the spreadsheet. Compare the spreadsheet with Table 8.7.

(b) Produce a point plot of the data, with weight on the vertical axis and variety on the horizontal axis. Do you think that the plot suggests systematic differences between the fresh weights for different varieties? If so, summarize the differences. If not, what makes you feel that there are no systematic differences between the weights for different varieties?

(c) Use GENSTAT to fit the model for a latin square design. What does the Output window contain now? (Do not be alarmed by the appearance of the warning window about faults; just click on the OK button and look in the Output window.)

(d) As the method that we used for a latin square design does not work for this dataset, we shall have to do something else. Choose the Stats|Regression Analysis|Linear menu option. Choose General Linear Regression in the Regression field; fill in the Response Variate field as weight; fill in the Model to be Fitted field as row,column,variety (it is important to enter these three terms in this order). (GENSTAT will then fit the blocks part of the model (row and column) first and then see if there are any remaining differences between varieties, i.e. apart from those that are due to the rows and columns.) Leave the Maximal Model field blank. Now, click on the Options button and click on Accumulated; click on the OK button. Finally, click on the remaining OK button. In the Output window, you should be able to see an ANOVA table. What does it tell you about differences between varieties?

(e) Above the ANOVA table there is a set of regression coefficients. This set of coefficients contains values for Variety B up to Variety F, but no value for Variety A. What do these coefficients tell us about the differences between the varieties?

You could also check the residuals; if you did, you might feel that the variance is not constant and/or that the distribution is not normal. However, the discrepancies are not large and for most purposes it is reasonable to take the results of this analysis as they stand. An alternative would be to analyse a transformation (perhaps log or reciprocal) of the fresh weight. However, with transformed data it would be more complicated to produce meaningful quantitative statements about differences between varieties.

The residual plots highlight the fact that the weight of 29.0 pounds, in the top left of Table 8.7, is exceptionally high for Variety E.

An alternative way of analysing these data would be to add rows to the spreadsheet corresponding to the vandalized plots, but to enter the missing value code * for the response on those plots. You could then use the GENSTAT ANOVA command to analyse the data. (This would give slightly different results.)   ∎

8.3   Factorial experiments with incomplete blocks

In Example 8.4, we looked at an experiment with a balanced incomplete block design. In that experiment, we concentrated entirely on what was happening in the stratum of experimental units; in other words, we compared treatments on the basis of comparisons between experimental units within single blocks. But the ANOVA showed that there was information about the treatments in the blocks stratum, i.e. in the differences between block totals. In this section, we look at two different types of incomplete block design for experiments with factorial treatment structure. In each case, one of the sets of comparisons between treatments (a main effect or an interaction) is estimated in the blocks stratum.

The first such design that we shall study is called a *split plot* design. In an experiment with a split plot design, there is a factorial treatment structure, but it is not possible to use different levels of one of the treatment factors within a block. For an example of this, consider the piglet experiment discussed at the beginning of Section 8.1. Suppose that, as well as investigating the effect of varying the piglets' diets, you wished to study simultaneously the effect of adding vitamin supplements to the sows' diets during pregnancy. Now there are two treatment factors, sow diet and piglet diet, but sow diet must be the same for all the piglets in a litter. We discussed using litters as the blocks in this sort of experiment; so you would be confronted with a treatment factor, namely sow diet, which it is impossible to vary within a block. Thus this factor can only be investigated in the blocks stratum. A split plot design is how you could solve this problem. The split plot design is covered in Section 8.3.1.

In the second type of design covered in this section, the experimenters deliberately choose to arrange things so that one of the interactions between treatment factors is tested only in the blocks stratum. The interaction in question is then said to be *confounded* with the blocks, because its contrasts are inextricably mixed up with the differences between blocks. By doing this, smaller blocks can be used. Such designs are dealt with in Section 8.3.2.

8.3.1   Split plot designs

A split plot design is a design for a factorial experiment in which, for some reason, one cannot vary one of the factors within a block, so that factor is applied to entire blocks rather than to individual experimental units within a block. The term 'split plot design' comes from the use of this kind of design in agricultural field experiments; Example 8.6 discusses such an experiment.

Example 8.6

The data in Table 8.8 come from an experiment carried out many years ago to investigate the yield of oats. The treatment structure was factorial. There were two factors, the amount of nitrogen fertilizer applied (at four levels, labelled N0 (no fertilizer), N1, N2 and N3) and the variety of oats (three different varieties, labelled X, Y and Z). Different varieties of oats were sown on plots (of size $\frac{1}{80}$ acre). These plots were grouped into blocks of three plots each (labelled I, II, ..., VI). Each plot was then divided into four (equal) smaller areas, and each of the four areas was given a different level of nitrogen fertilizer. Thus the basic experimental unit was one of these small areas, or *subplots*. Each of these subplots ($\frac{1}{320}$ acre or about 13 m^2) was large enough to have its own level of fertilizer, but really too small for oats to be sown on it by the usual mechanical method; thus the subplots are grouped into *whole plots* (the original $\frac{1}{80}$ acre plots), each consisting of four subplots, on which the same variety of oats was sown. These whole plots form a sort of block (not the same as the larger blocks of three whole plots mentioned above).

To summarize, the experimental units (subplots) in this experiment are grouped into a type of block called whole plots, and the whole plots are grouped into larger blocks (which we shall actually call blocks). This nomenclature is really a historical accident. It would arguably make more sense to call the experimental units 'plots', which are grouped into 'blocks' and then into 'hyperblocks' (or some such term), but these are not the usual names.

Keep in mind that, in a split plot experiment, a whole plot is a kind of block, not a kind of experimental unit. In fact, it makes sense to think of the whole plots in an experiment with a split plot design as incomplete blocks. Each unit (subplot) in a whole plot has the

Table 8.8 Yields of oats (in quarter pounds) from an experiment with a split plot design. Different varieties of oats (X, Y and Z) were planted on whole plots and different levels of nitrogen fertilizer (N0, N1, N2 and N3) were applied to subplots.

Block	Variety	Nitrogen level and yield				Nitrogen level and yield				Variety	Block
	Z	N3	156	N2	118	N2	109	N3	99	Z	
		N1	140	N0	105	N0	63	N1	70		
I	X	N0	111	N1	130	N0	80	N2	94	Y	II
		N3	174	N2	157	N3	126	N1	82		
	Y	N0	117	N1	114	N1	90	N2	100	X	
		N2	161	N3	141	N3	116	N0	62		
	Z	N2	104	N0	70	N3	96	N0	60	Y	
		N1	89	N3	117	N2	89	N1	102		
III	X	N3	122	N0	74	N2	112	N3	86	X	IV
		N1	89	N2	81	N0	68	N1	64		
	Y	N1	103	N0	64	N2	132	N3	124	Z	
		N2	132	N3	133	N1	129	N0	89		
	Y	N1	108	N2	126	N2	118	N0	53	X	
		N3	149	N0	70	N3	113	N1	74		
V	Z	N3	144	N1	124	N3	104	N2	86	Y	VI
		N2	121	N0	96	N0	89	N1	82		
	X	N0	61	N3	100	N0	97	N1	99	Z	
		N1	91	N2	97	N2	119	N3	121		

Source: Yates, F. (1935) 'Complex experiments', *Journal of the Royal Statistical Society, Supplement*, **2**, 181–247.
Dataset name: oats.

same level of the whole plot treatment factor (variety of oats in this case), so it does not contain the treatment combinations with other levels of this factor. Thus, thinking of it in this way, a split plot experiment has an incomplete block design.

As with all designed experiments, randomization plays a role here. The data are laid out in Table 8.8 in a way that corresponds to their layout in the field. This makes it clearer that the oats varieties X, Y and Z were allocated to the three whole plots within each block at random, and that the four nitrogen levels were allocated to the four subplots within each whole plot at random.

Exercise 8.13

(a) Load the oats dataset into the spreadsheet. Compare the spreadsheet with Table 8.8, so as to ensure that you know how the data in the spreadsheet correspond to the table. (As you will see below, the spreadsheet for a split plot experiment does not always have to be as complicated as this.)

(b) Click on the **Graphics** menu; click on **Point Plot**. Fill in the **Y Coordinates** field as yield; fill in the **X Coordinates** field as nitrogen; fill in the **Groups** field as variety; click on the **OK** button; click on the **Cancel** button.

This produces a plot with the response variable (the yield) on the vertical axis and the nitrogen level on the horizontal axis, with the three varieties indicated by colour codes. Do you think that changing the nitrogen level changes the mean yield? Can you see any pattern in the plot that shows an effect of variety on yield? (You might want to produce a plot with **variety** on the horizontal axis to clarify this.)

(c) Click on the **Stats** menu; click on **Analysis of Variance**. Choose **Split-Plot Design** as the **Design**; fill in the **Y-Variate** field as yield; fill in the **Treatment Structure** field as variety*nitrogen (because of the factorial treatment structure); fill in the **Blocks** field as block; fill in the **Whole Plots** field as wholplot; fill in the **Sub-plots** field as subplot; click on the **OK** button.

Find the ANOVA table in the **Output** window. How does it differ from the ANOVA tables you have seen so far in this chapter? What does it tell you about the effects of the treatment factors? Produce a plot of means to investigate the pattern of responses, as you did for the factorial experiments in Chapter 7. What do you conclude? Produce the standard ANOVA residual plots. Does the model seem to be appropriate?

(d) Scroll up in the **Input Log** window until you find the BLOCK command that tells GENSTAT how the blocks are set up in this experiment. Note what it says, and compare it with the BLOCK command in Exercise 8.10 (which also had a relatively complicated blocking structure).

The oats spreadsheet was quite complicated in that it had factors to define the whole plot and the subplot that each experimental unit corresponded to. In fact, these factors are not necessary in a case like this where there is only one whole plot within each block for each of the levels of the whole plot treatment factor (**variety** in this case). All the wholplot factor does is to distinguish the three whole plots within each block. But they are also distinguished from one another by the factor **variety**. The values of wholplot and **variety** do not match – but their exact values do not matter here. Their only purpose as far as the blocking is concerned is to indicate which subplots are grouped together into which whole plots, and both wholplot and **variety** do this equally well. Furthermore, the factor subplot is not necessary either! In this kind of experiment, it merely tells GENSTAT what the experimental units are; but GENSTAT always assumes that each row of the spreadsheet corresponds to an experimental unit.

(e) Check all this by returning to the **Analysis of Variance** dialogue box, changing the **Whole Plots** field to variety, and deleting the contents of the **Sub-plots** field so that it is blank. Click on **OK**. Check that the output corresponds to what you obtained previously. What is the form of the BLOCK command this time?

Although doing the analysis this way could have saved space in the spreadsheet, you might think it unsatisfactory because it makes it less clear which aspects of the design are to do with blocks and which are to do with treatments. You would be right. But you should be aware that some statistical software other than GENSTAT will simply not cope with a description of an experiment with a split plot design that is set out in the way we originally used.

■

It is normally relatively easy to tell that an agricultural field experiment has a split plot design, because of the way that the experimental units are laid out in the field. Split plot experiments are reasonably common in this context, because some agricultural treatments (often those involving machinery, such as ploughing methods or irrigation methods) cannot be applied to small areas of land. The split plot structure of other types of experiment can be harder to envisage sometimes. You should remember the key characteristic: that there are blocks of experimental units in which one of the treatment factors is applied to the block as a whole. (The terminology – that these blocks are called 'whole plots' – does not help; and you should bear in mind that, as in the oats experiment, the whole plots may themselves be grouped into larger blocks.) Some statisticians refer to 'split unit designs' rather than 'split plot designs' when the units are not plots in a field.

8.3.2 Confounding

In this section we look at a particular form of incomplete block design, used in factorial experiments, in which one arranges things so that one (or more) of the interactions is tested only in the blocks stratum.[2] This is similar to what happened in a split plot experiment where the main effect of one of the treatment factors was tested only in the whole plots stratum. (Remember that, in a split plot design, whole plots are a sort of block.)

The advantage of designing the experiment this way, as with balanced incomplete block designs, is that smaller blocks can be used. Often the practical circumstances of the experiment limit the size of blocks available, as they did in the dishwashing experiment (Example 8.4). The disadvantage is that one can usually get only a very imprecise estimate of any interactions that are tested in the blocks stratum, because differences between blocks are likely to be much bigger than differences between experimental units within the same block. But in many situations this does not matter. Often, for example, as in Example 7.7, the researchers might be happy to assume that some higher-order interaction is zero. If such an interaction is tested in the blocks stratum, then it will not be estimated very accurately, but this will not matter.

The interactions that are estimated in the blocks stratum are said to be *confounded* with blocks (because their estimates are inextricably mixed up with differences between blocks). To see how this confounding can be achieved, let us look at an example.

Example 8.7
The data shown in Table 8.9 were generated in an experiment carried out in a chemical plant. The researchers were interested in the effect of three factors on the yield of an organic chemical produced in a particular process. The process involved the use of ammonium chloride as a raw material. The factors of interest, each of which had two levels, were:

- the quality of the ammonium chloride involved, in terms of how finely it was ground (coarse or fine);

- the amount of ammonium chloride used (usual amount, or with an extra 10% added);

---

[2] Actually we do not deal in detail here with the situation where more than one interaction is tested in the blocks stratum. The principles are the same as with one interaction, but the details can get complicated.

- which of two apparently identical production units (Unit 1 or Unit 2) the process took place in.

The treatment structure is thus a 2^3 factorial structure. There are $2^3 = 8$ different treatments (labelled 1 to 8 in Table 8.9 for convenience).

Table 8.9

Production unit	Ammonium chloride amount	Ammonium chloride quality	Treatment	Block 1	Block 2	Block 3	Block 4
1	usual	coarse	1	155	—	—	164
		fine	2	—	162	171	—
	extra	coarse	3	—	168	175	—
		fine	4	157	—	—	171
2	usual	coarse	5	—	156	161	—
		fine	6	150	—	—	153
	extra	coarse	7	152	—	—	162
		fine	8	—	161	173	—

Source: Davies, O. L. (ed.) (1956) *The Design and Analysis of Industrial Experiments*, 2nd edition, London, Longman.
Dataset name: `chloride`.

The process involved blending together batches of base material, and it was known from previous experience that yields could differ considerably between different batches of base material even if the treatment factors did not change. Thus it would seem appropriate to treat all the runs of the process produced from the same batch of base material as a block. However, because of the way the blending process was performed, each batch of base material was sufficient for only four production runs. In other words, each block could only contain four experimental units, not the eight required to use complete blocks.

The pattern of allocation of experimental units to blocks is shown in Table 8.9. Why was the allocation done this way? We see that there are two different types of block in the experiment. What we shall call 'Block Type I' (Blocks 1 and 4) contains Treatments 1, 4, 6 and 7. The other type, 'Block Type II' (Blocks 2 and 3), contains the other treatments, 2, 3, 5 and 8. In Example 7.4, you saw that the contrast for the three-factor interaction in a 2^3 factorial experiment is

$$(\mu_8 - \mu_7) - (\mu_6 - \mu_5) - (\mu_4 - \mu_3) + (\mu_2 - \mu_1)$$
$$= \mu_8 + \mu_5 + \mu_3 + \mu_2 - \mu_7 - \mu_6 - \mu_4 - \mu_1$$

(where the μ_js represent the treatment means, with the treatments numbered as in Table 8.9). In this contrast, all the treatment means with a + sign appear in blocks of Block Type II, while all those with a − sign appear in blocks of Block Type I. Thus, in order to estimate the three-factor interaction, one would have to work out the average block total for blocks of Block Type II and subtract the average block total for blocks of Block Type I. This comparison is between different blocks; it is thus not estimated very accurately (in the sense that its variance will be large). GENSTAT performs the relevant testing in the blocks stratum.

What about the main effects and two-factor interactions? We shall look at one example of each (namely those that correspond to the contrasts explicitly written out in Example 7.4).

The contrast for the main effect of ammonium chloride quality is

$$\tfrac{1}{4}[(\mu_2 - \mu_1) + (\mu_4 - \mu_3) + (\mu_6 - \mu_5) + (\mu_8 - \mu_7)]$$
$$= \tfrac{1}{4}[(\mu_4 - \mu_1) + (\mu_6 - \mu_7) + (\mu_2 - \mu_3) + (\mu_8 - \mu_5)].$$

The point of writing this in the second way is to show that each of the differences between a pair of μ_js in round brackets involves subtracting two treatments that appear in the same block type. The first two comparisons appear in Block Type I and the other two in Block Type II. Thus the contrast can be estimated by looking at differences between experimental units within blocks. It can be estimated relatively accurately (smaller variance) and is tested in the units stratum.

The contrast for the interaction between ammonium chloride quality and ammonium chloride amount is

$$\tfrac{1}{2}[(\mu_4 + \mu_8 - \mu_3 - \mu_7) - (\mu_2 + \mu_6 - \mu_1 - \mu_5)]$$
$$= \tfrac{1}{2}[(\mu_4 - \mu_6) + (\mu_1 - \mu_7) + (\mu_5 - \mu_3) + (\mu_8 - \mu_2)].$$

Again the contrast can be expressed in terms of differences between experimental units within blocks.

Indeed, the same goes for all the other main effects and all the other two-factor interactions. GENSTAT will check that everything comes out in the appropriate stratum when you carry out the ANOVA in the next exercise. ∎

You have seen just one example of a factorial experiment where an interaction is confounded with the blocks, but the idea can be applied in many ways. Any interaction (or even any main effect) can be confounded. To work out how to do the confounding in a factorial experiment where all the factors have two levels and each block can only accommodate half of the treatment combinations, you simply write down the contrast for the interaction you wish to confound with the blocks. There will be two block types. All the treatments whose mean has a + sign in this contrast should be put together in one block type, and all the treatments whose mean has a − sign in the contrast should be put together in the other block type.

Exercise 8.14

(a) Load the dataset stored in chloride.gsh into the spreadsheet. Compare the spread-sheet with Table 8.9.

(b) Click on the Graphics|Point Plot menu item. Fill in the Y Coordinates field as yield; fill in the X Coordinates field as treat; fill in the Groups field as block; click on the OK button; click on the Cancel button.

What does this plot tell you about differences between blocks? Why is this plot difficult to use for comparing treatments?

(c) Since the graph produced in part (b) is difficult to interpret, we shall look at a better one. We shall look at the yield adjusted for the blocks. That is, we shall regress the yield on the factor that defines the blocks, and look at the residuals from this regression. (Actually the regression will be done using the GENSTAT ANOVA command, but we could use the regression commands instead. As you know, the model is the same.) These residuals give the *extra* differences in yield after the effect of the blocking has been taken into account, and are thus likely to produce a clearer picture.

You can calculate the yields adjusted for blocks using the following method. Click on the **Stats** menu; click on **Analysis of Variance**. Enter **General Analysis of Variance** in the **Design** field; fill in the **Y-Variate** field as **yield**; fill in the **Block Structure** field as **block**; click on the **OK** button. (Leave **Treatment Structure** blank.)

Now, click on the **Save** button. Click on **Residuals**; fill in the **In** field as **adjyield**; click on the **OK** button. The adjusted yields will be stored in the variate **adjyield** and you should be back at the **Analysis of Variance** window; click on the **Cancel** button.

Now, plot the adjusted yields, using the same method as in part (b), but fill in the **Y-variate** field as **adjyield** instead of **yield**. What does this graph tell you about differences between yields for different treatments?

(d) Click on the **Stats** menu; click on **Analysis of Variance**. Choose **General Analysis of Variance** in the **Design** field; fill in the **Y-Variate** field as **yield**; fill in the **Treatment Structure** field as **clamount*clqualty*produnit**; fill in the **Block Structure** field as **block**; click on the **OK** button.

In the **Output** window, you should be able to find an ANOVA table. What does it tell you about differences in yields for the different treatment combinations?

(e) Look at the Tables of means in the **Output** window, in the light of the results of the ANOVA. What would you tell the manager of the production process about the size of differences in yield for different treatments? (You should calculate appropriate confidence intervals to give ranges of plausible values for the differences.)

8.4 Designing experiments

You will be aware from what you have read in this chapter so far that the process of designing experiments can be complicated in some cases. Previous sections of this chapter, and this book in general, concentrate on statistical modelling and data analysis rather than on designing data collection methods, so that a detailed study of how to design experiments is beyond the scope of the book. Nevertheless, this section draws your attention very briefly to some of the main ideas relevant to the practical process of designing experiments.

Roughly speaking, the process of experimental design is one of coming to an appropriate compromise between the limited resources available for the experiment and the necessity to produce sufficiently accurate estimates of the quantities of interest to the experimenter.

In order to come to a satisfactory compromise, the experimenter must choose an appropriate blocking scheme. As you have seen, blocking is a way of arranging an experiment so that the most important comparisons are made between treatments that are applied to a reasonably homogeneous set of experimental units – or, in other words, so that the important comparisons are made within blocks, where the blocks have been chosen sensibly. It is important to think carefully about how to set up the blocks. One must consider just which groupings of experimental units are going to be similar to one another in ways that might affect the responses. Very often, a simple scheme with just one type of block is perfectly adequate, but sometimes a more complicated blocking structure (such as a latin square design, or a split plot design where the whole plots are themselves grouped into blocks) is necessary. As you have seen, there may be physical constraints on the size of the blocks, or on the way that certain treatment factors can be applied to experimental units, but you have seen some ways of dealing with such constraints (and there are others, beyond the scope of the book).

Another important consideration is the amount of replication in the experiment. How many experimental units should each treatment be applied to? Generally speaking, the more experimental units there are for each treatment, the more accurately the treatment parameters can be estimated; but almost always it costs more in time and money to run the experiment on more experimental units. It is necessary to come to an appropriate trade-off. Suppose you were particularly interested in a certain contrast between the treatment means. The variance of that contrast will depend on the replication, as well as on the error variance between experimental units. If one has some idea of the size of the error variance before doing the experiment (say from previous experiments in similar circumstances), one can choose the replication to obtain an appropriate variance for the contrast(s) of interest. There are some quite sophisticated methods for doing this, but again they are beyond the scope of the book. One thing to bear in mind, however, is that there must be *some* replication, or there will be no estimate of the error variance to work with and it will not be possible to analyse the data. At any rate, it will not be possible unless one makes assumptions, such as those in Section 7.4, about the lack of certain interactions; and, as you saw there, such assumptions are not always warranted.

Throughout earlier sections of this chapter, a lot of emphasis has been placed on randomizing experiments properly. Unless an experiment is adequately randomized, it can be very difficult to draw conclusions about the causes of differences between experimental units, as you saw in the grapefruit example in Section 8.1. In complex experiments, randomization can be a time-consuming process, but that is no reason to miss it out! There is a GENSTAT menu item, **Design** under **Stats**, that can help with this process.

Finally, you should always distinguish clearly between the blocking structure of an experiment and its treatment structure. Comparisons between treatments are the quantities you are interested in – the reasons for doing the experiment – while the blocks are there simply to improve the accuracy of these treatment comparisons. In GENSTAT, it is not easy to get blocking factors and treatment factors confused, because you have to enter them in quite different places in the dialogue boxes and commands. But you should be aware that some other software does not make this so straightforward – you may have to keep track yourself of which factors relate to blocks and which to treatments, and unless you are quite sure what you are doing it is easy to go wrong.

Summary of methodology

1 Blocks are groups of experimental units that are relatively homogeneous in that the responses on two units within the same block are likely to be more similar than

the responses on two units in different blocks. Dividing the experimental units into blocks is a way of improving the accuracy of comparisons between treatments.

2 An experiment where each treatment appears the same number of times (usually once) in each block is said to have a randomized block design (or, to be more precise, a randomized complete block design). Data from such designs can be analysed using ANOVA.

3 The model for a randomized block design with each treatment appearing once per block is

$$Y_{jl} = \mu + \gamma_j + \tau_l + \epsilon_{jl}$$

where Y_{jl} is the response for the unit in block j with treatment l, the γ_js are parameters for the blocks, the τ_ls are parameters for the treatments, each $\epsilon_{jl} \sim N(0, \sigma^2)$, and all the ϵ_{jl}s are independent of one another. Thus the blocks are assumed not to interact with the treatments.

4 In some experiments, each block does not contain every treatment combination, perhaps because there are physical restrictions on the size of the blocks. Such experiments have an incomplete block design. These experiments can be analysed using ANOVA, and (if they possess a property known as first-order balance) GENSTAT can work out which treatment comparisons are made within blocks and which are made between blocks, by putting the corresponding sums of squares into appropriate strata.

5 In a balanced incomplete block design, each pair of treatments occurs together in the same number of blocks. Thus all treatment comparisons have the same accuracy.

6 A split plot design is a type of incomplete block design with two (or more) treatment factors. Its experimental units are known as subplots, and they are grouped into blocks known as whole plots. The whole plots may themselves be grouped into larger blocks. Because of physical or other constraints, one (or more) of the treatment factors can be applied only to a whole plot at a time. The effects of this factor are compared in the whole plots stratum.

7 Another way of producing an incomplete block design in a factorial experiment is to confound a treatment interaction with the blocks. The confounded interaction will be estimated less precisely than the other interactions and main effects.

8 Some experiments have more than one system of blocks. One example is the latin square design. This has two systems of blocks, which can be thought of as 'rows' and 'columns'. The number of rows, the number of columns and the number of treatments are all equal in this case, and the treatments are arranged so that each appears in exactly one row and exactly one column.

9 It is important to randomize experiments properly, by allocating treatments to experimental units at random. Exactly how this is done depends on the experimental design being used.

9

Binary regression

Previous chapters have covered a wide variety of linear models, but in all of them the response data were quantitative and were assumed to be random samples from normal distributions. This chapter will make a first departure from these assumptions about the response data, and will consider how to develop models for data where the response variable is binary. In so doing, we stray outside the general linear model that you have (implicitly) been studying so far, and meet models that fall into the wider framework of the generalized linear model, which will be the subject of the remainder of the book.

Binary response variables are encountered in many areas of application of statistical modelling. For example, a patient may be cured of a disease or not; a species may be present or absent in an area; a subject may give a wrong or correct reply; a component may be defective or non-defective; a defendant may receive a custodial or non-custodial sentence. Even when a continuous response variable is available, it may sometimes be replaced by a binary one if the main interest is in a particular threshold value. For example, when the police measure a driver's blood alcohol level, the outcome of interest is whether it is above or below the legal limit. Without loss of generality, we shall always code the two possible outcomes by 0 and 1. It is arbitrary which outcome is coded 1, but it can often help interpretation of the results if 1 is associated with 'success' (e.g. a patient is cured, a species is present, a subject gives a correct answer) and 0 with 'failure'. (The GvHD example of Chapter 1 is, however, a case in which 1 is associated with the unhappy event of developing a certain condition.)

As in previous chapters, the response for the ith individual will be represented by the random variable Y_i, but now Y_i is discrete taking only the values 0 and 1. We also assume independence between responses for different individuals. Let the probability of success $P(Y_i = 1)$ be p_i. In other words, Y_i follows a Bernoulli distribution with parameter p_i (Section 2.5.1). We know that the mean of this distribution is $E(Y_i) = p_i$ and its variance is $V(Y_i) = p_i(1 - p_i)$. Note that the success probability $P(Y_i = 1)$ is the expected value of the response variable – and this mean is the quantity that we hope to explain, estimate or predict by building a linear model. So, although the response data are discrete, our model will estimate probabilities, p_i, and these are quantities taking values between 0 and 1. We may concentrate on $P(Y_i = 1)$ since $P(Y_i = 0) = 1 - P(Y_i = 1)$.

Exercise 9.1
Suppose we have a set of data consisting of n independent observations on a binary response variable Y, coded 0 or 1, and continuous explanatory variables x_j, $j = 1, 2, \ldots, k$. What

difficulties can you see in trying to fit and interpret a linear model of the form

$$Y_i = \alpha + \sum_{j=1}^{k} \beta_j x_{i,j} + \epsilon_i, \quad i = 1, 2, \ldots, n,$$

to the data using the method of least squares? Would the estimated model be appropriate for predicting Y_i? Would the theoretical assumptions of the multiple linear regression model in Chapter 6 still be tenable? Would it be appropriate to apply F and t tests in this situation?

The solution to Exercise 9.1 indicates that we need to develop new methods and theory in order to fit realistic regression models when the response variable is binary. As far as the explanatory variables (which may be quantitative or categorical) are concerned, this chapter covers similar ground to Chapter 6 on multiple linear regression. However, you have just seen that the methods of that chapter are not appropriate when the response variable is binary. All is not lost! Once you have been shown how binary regression can best be formulated, it will turn out that the ideas and methods of Chapter 6 have direct analogues for binary regression and so the expertise, experience and particularly the use of GENSTAT that you have already gained will hold you in very good stead in this chapter too.

9.1   The logistic function

Example 9.1

Table 9.1 shows the first five datapoints from a file containing data on 435 adults who were treated by the University of Southern California General Hospital Burn Center. The explanatory variable, **larea**, is log(area + 1), where **area** is the body area covered by third-degree burns. The binary response variable **survival** records whether or not the patient survived: $Y_i = 1$ for survival, 0 for death. For prediction purposes, it would be very useful to be able to estimate the survival probability $p = E(Y)$ as a function of $x = $ **larea**.

Table 9.1

larea	survival
2.301	0
1.903	1
2.039	1
2.221	0
1.725	1
⋮	⋮

Source: Fan, J., Heckman, N. E. and Wand, M. P. (1995) 'Local polynomial kernel regression for generalized linear models and quasi-likelihood functions', *Journal of the American Statistical Association*, **90**, 141–50. Dataset name: burns.

Figure 9.1(a) is a scatterplot of **survival** against **larea** – there is much overplotting of symbols here. Figure 9.1(b) displays boxplots of the **larea** values of the two groups defined by the responses. The plots certainly suggest that **larea** has some effect on survival, with,

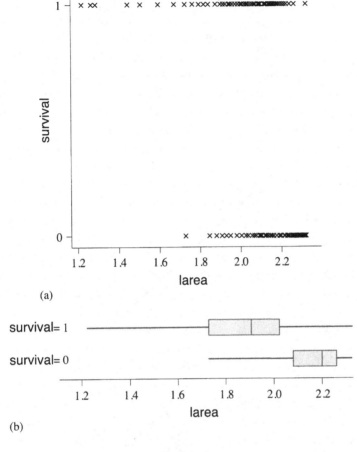

(a)

(b)

Fig. 9.1

unsurprisingly, those with greater burn areas having less chance of survival. The plots, however, are not much help in developing a model to estimate the probabilities p_i.

If we were dealing with a categorical explanatory variable, an obvious way to estimate the probabilities would be to calculate the proportion of patients in each category who survived. As a first approximation, we can use the same method for a quantitative explanatory variable by grouping the data into intervals. Table 9.2 shows the results of grouping larea into intervals with midpoints as given. This grouping method is a rough illustrative method and *not a technique that you will have to learn and apply.*

Figure 9.2 shows the proportions surviving plotted against the midpoints of the larea intervals. Also drawn on Figure 9.2 is the simple linear regression line, with equation

$$\text{probability of survival} = 2.59 - 0.988 \, \text{larea}.$$

There are obvious and considerable difficulties with the straight line fit. As was suggested in Exercise 9.1, the straight line goes outside the allowed values for probabilities, which must

Table 9.2

Group midpoint	Number surviving	Total number	Proportion surviving
1.35	13	13	1.000
1.60	19	19	1.000
1.75	67	69	0.971
1.85	45	50	0.900
1.95	71	79	0.899
2.05	50	70	0.714
2.15	35	66	0.530
2.25	7	56	0.125
2.35	1	13	0.077

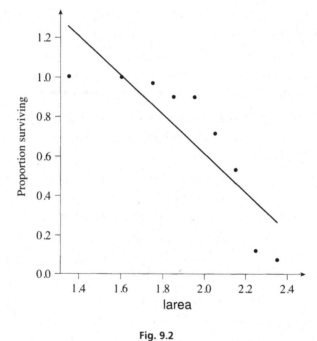

Fig. 9.2

be between 0 and 1. In particular, for patients with larea < 1.6, the predicted probability of survival is more than one. In any case, the straight line does not fit the data well anywhere!

A straight line (but not this one!) may actually be an acceptable description in the range of values of larea (about 1.9–2.3) for which the survival proportion plunges from around 1 to close to 0. But it needs to level off to 1 at lower values of larea, and also (though less obviously for this dataset) to 0 at higher values. This suggests that instead of fitting a straight line, something more like an elongated backwards S-shaped curve would be more appropriate. ∎

Exercise 9.2

Ten samples of alloy fasteners used in aircraft were tested, each sample at a different pressure load. The results are summarized in Table 9.3.

Table 9.3

Load (psi)	Number failed	Sample size
2500	10	50
2700	17	70
2900	30	100
3100	21	60
3300	18	40
3500	43	85
3700	54	90
3900	33	50
4100	60	80
4300	51	65

Source: Montgomery, D. C. and Peck, E. A. (1982) *Introduction to Linear Regression Analysis*, John Wiley, New York. Dataset name: `fasten`.

Make a scatterplot of the *proportions* of fasteners failing at each load against load. You can calculate the proportions in GENSTAT from the **failed** and **samsize** variates. See the solution if you can't work out how. What shape of curve is suggested to model the probability of failure as a function of load?

In fact, the `fasten` data is fairly typical of a wide variety of experimental situations where a binary response variable is observed at increasing levels of some load, dose, stimulus, etc. In this chapter, we shall only consider cases where there is just *one* binary observation for each individual treated separately; more complicated cases will be considered later. A hypothetical S-shaped curve for such binary response data is drawn in Figure 9.3. A curve like that in Figure 9.3 is generally called a *sigmoid*. There are many different mathematical

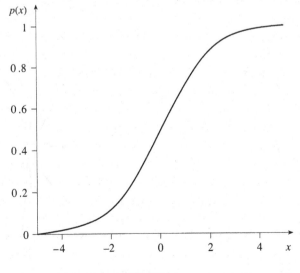

Fig. 9.3

functions $p(x)$ that produce sigmoid curves, but the function plotted in Figure 9.3 is actually

$$p(x) = \frac{1}{1 + \exp(-x)}. \tag{9.1}$$

Such a curve has the kinds of properties needed to model the sort of relationship between a probability and a continuous explanatory variable as exemplified by the fasten data:

- the probability p is constrained to lie between 0 and 1 for all x;

- the relationship between p and x is almost linear from about $p = 0.2$ to about $p = 0.8$;

- p tends to 0 as x becomes small;

- p tends to 1 as x becomes large.

Obvious drawbacks of function (9.1), however, are that it is centred on $x = 0$ and has a fixed range (from about $x = -5$ to about $x = +5$) on which its values differ markedly from 0 or 1. Also, it will not model situations like the burns data where the relationship is decreasing.

A more general sigmoid function is

$$p(x) = \frac{1}{1 + \exp[-(\alpha + \beta x)]}. \tag{9.2}$$

This function has the same properties as (9.1), but is more flexible with respect to the scale and range of x. It also allows both increasing and decreasing relationships to be modelled. Figures 9.4(a) and (b), which show (9.2) drawn for different values of α and β, demonstrate how versatile the sigmoid function (9.2) can be: with α and β determining the location, slope and spread, this curve can be made to suit any scale and range of x and to suit an increasing or decreasing relationship. The curve is actually symmetric about the point $x = -\alpha/\beta$, $p = 0.5$ where its slope is $\beta/4$.

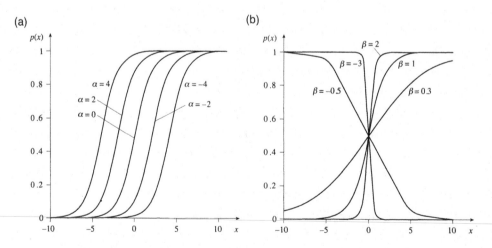

Fig. 9.4 (a) $\beta = 1$ and various values of α (b) $\alpha = 0$ and various values of β.

This is all very well, you may be thinking, but this is a book about linear models, and (9.2) looks distinctly non-linear. With a little manipulation, however, we can produce an equation that looks more like a linear model. Solving (9.2) for $\alpha + \beta x$ yields

$$\log \left(\frac{p(x)}{1 - p(x)} \right) = \alpha + \beta x. \tag{9.3}$$

The right-hand side of (9.3) is now the same as the simple linear regression model you met in Chapter 4; but, on the left-hand side, instead of $E(Y)$, we now have a function of $p = E(Y)$. The function $\log[p/(1 - p)]$ is known as the *logistic* transformation of the probability p. It is also known as the *logit* transformation and written logit(p). (The term 'logit' is short for 'logistic unit'.) Note that this function *links* the expected value of the response variable, $E(Y)$, to the linear component of the model, $\alpha + \beta x$. For this reason it is known as a *link function*.

Box 9.1

The logistic (logit) link function of a probability p is $\log[p/(1 - p)]$.

Up to this point in the book we have always had $E(Y)$ on the left-hand side of our models and have not used a link function – or we could say we have implicitly been using the identity function $f(y) = y$ as a link function. The link function is a feature of generalized linear models, and you will see more use of different link functions in subsequent chapters. There are other link functions for binary response variables, corresponding to sigmoid curves similar in shape to Figure 9.3, but here we shall use only the logistic link function since it is the one most commonly used; it gives rise to the term *logistic regression*.

Why is there a preference for the logistic link function over other link functions in binary regression? There are good reasons, to do with interpreting the model when we come to estimate it and use it for prediction. This will become clearer when we apply logistic regression to some datasets later in the chapter, but for now we note the following properties of the logistic function.

The ratio $p/(1 - p)$ is the ratio of the probability of success ($Y = 1$) to the probability of failure ($Y = 0$). This ratio is known as the *odds* of success. The left-hand side of the model (9.3) is therefore the logarithm of the odds (the *log odds*) of success.

As well as calculating the odds o from the probability p, as $o = p/(1 - p)$, we can calculate the probability from the odds, as $p = o/(1 + o)$.

When comparing probabilities, a most useful statistic is the ratio of odds (the *odds ratio*). The model (9.3) relates differences in an explanatory variable x to odds ratios through differences in the log odds. For example, if x is an indicator variable taking values 0 or 1 representing two groups, and if (temporarily) p_0 and p_1 refer to success probabilities when $x = 0$ and 1 respectively, then

$$\log \left(\frac{p_1}{1 - p_1} \right) - \log \left(\frac{p_0}{1 - p_0} \right) = (\alpha + \beta) - \alpha = \beta.$$

That is, the coefficient β represents the difference between the log odds for the two groups. This difference in logs is the same as the log of the odds ratio (i.e. the *log odds ratio*). Equivalently, $\exp \beta$ is the odds ratio for the two groups; this is sometimes called the *odds multiplier* because it is the number the odds are multiplied by when changing from the $x = 0$ group to the $x = 1$ group.

For quantitative x, let p_x and p_{x+1} (temporarily) denote the success probabilities when the explanatory variable takes values x and $x + 1$, respectively. Then,

$$\log\left(\frac{p_{x+1}}{1 - p_{x+1}}\right) - \log\left(\frac{p_x}{1 - p_x}\right) = (\alpha + \beta(x + 1)) - (\alpha + \beta x) = \beta. \qquad (9.4)$$

So, β represents the difference in log odds (or the log odds ratio) given a unit increase in the explanatory variable. This is the analogue of the interpretation of β in the simple linear regression model as the effect on $E(Y)$ of a unit increase in x (Section 4.1). Also, $\exp \beta$, the odds multiplier, is the odds ratio for a unit increase in the explanatory variable.

More generally, it is the way in which the coefficients of logistic regression models can be interpreted as increasing or decreasing the log odds of 'success' that makes the logistic link function popular.

There are other reasons for preferring the logistic link function too. Computationally and mathematically, it can be easier to deal with than some of the alternative link functions; but in these days of personal computers and packages such as GENSTAT that readily perform the computations for us, these considerations are not so important. And, of course, the logistic regression model is often seen to fit the data well!

Exercise 9.3

Load the burns data into GENSTAT. (There are 435 rows in the spreadsheet, so you will need to adjust the Student Limits if you are using the student version of GENSTAT.) Choose the Stats|Regression Analysis| Generalized Linear menu item. For the first time, we see the Generalized Linear Models dialogue box. It is not a great deal different from the Linear Regression dialogue box, although it has a few extra features; do not worry about these for now.

Change the Analysis field to Modelling of binomial proportions (e.g. by logits) using the pull-down menu; this alters the option fields a little. Enter the single digit 1 in the Number(s) of Subjects field. (The 1 signifies that we are dealing with one individual at a time.) Enter survival in the Numbers of Successes field (survival takes values 0 for death and 1 for survival). Ignore the Maximal Model field and enter larea in the Model to be Fitted field. Notice that the Transformation (link) field already has Logit in it and this does not need to be changed. Click on OK to run the logistic regression, but do not concern yourself with the output for now.

Now click on the Further Output button, and then on the Fitted Model button in the Generalized Linear Models Further Output dialogue box. Put larea in the Explanatory Variable field in the Graph of Fitted Model dialogue box, and click on OK.

What does the diagram you obtain show? Does it appear that the model might be a reasonable one?

9.2 The logistic regression model

Up to this point we have used proportions as estimates of unknown probabilities p at different values or groups of values of some quantitative explanatory variable x. In Example 9.1 we arbitrarily formed groups of patients with similar values of larea. In Exercise 9.2

samples of fasteners were tested each at a different load. In other situations x may be a categorical variable. In all these cases it is possible to observe a proportion – the ratio of the number of successes to the total number in the group. In defining the logistic regression model, however, we are not going to assume observed proportions are available. Unplanned observational data, with which this chapter is concerned, do not always present themselves in such a way that proportions can be satisfactorily calculated; also, grouping a quantitative explanatory variable obviously loses information. Furthermore, we need a logistic regression model for any number of explanatory variables, and proportions will generally not be available for all possible combinations of values of the explanatory variables.

The random component of the model, therefore, is expressed in terms of the probability of the binary response variable taking only the values 0 or 1, i.e. the ith response Y_i follows a Bernoulli distribution with parameter p_i and having probability mass function

$$p(Y_i = y_i) = p_i^{y_i}(1 - p_i)^{1-y_i}.$$

(See Sections 2.5.1 and 2.5.2.) The linear component of the model on the right-hand side of (9.3) can be extended in the same way as it was for multiple linear regression in Chapter 6 to take account of any number of explanatory variables, $x_{i,j}$, $i = 1, 2, \ldots, n$, $j = 1, 2, \ldots, k$. Notice that the number of explanatory variables is k in this chapter, to avoid any confusion as would be caused by calling it p (as in Chapter 6).

Putting this all together, we define the linear logistic regression model (or simply the logistic regression model) as follows.

Box 9.2

Given n independent Bernoulli observations y_i, $i = 1, 2, \ldots, n$, where $E(Y_i) = p_i$, the **linear logistic regression model** for the dependence of p_i on the corresponding values of k explanatory variables, $x_{i,1}, x_{i,2}, \ldots, x_{i,k}$, is

$$\text{logit}(p_i) = \log\left(\frac{p_i}{1 - p_i}\right) = \alpha + \sum_{j=1}^{k} \beta_j x_{i,j}. \qquad (9.5)$$

Note that, unlike the normal multiple linear regression model of Chapter 6, the logistic regression model has to be expressed directly in terms of the distribution of Y given a value of x rather than by introducing an additive random error term, ϵ. (In Chapter 10 you will see how a normal regression model can be written in a form analogous to that in Box 9.2.)

We are now faced with the problem of estimating the $k + 1$ parameters, α and the k β_js, from the n observations. Chapter 2 emphasized the central role of maximum likelihood estimation in this book, and this is what will be used. In Chapter 6, where Y was normally distributed, maximum likelihood estimation was equivalent to least squares estimation; but here Y has the Bernoulli distribution, and least squares estimation is neither the same nor adequate.

The problem of finding estimates of α and the β_js that maximize the likelihood is essentially a computational problem, the details of which need not concern us since we shall be using GENSTAT to obtain the estimates. And GENSTAT uses tried and tested algorithms for performing such calculations.

The estimated regression coefficients of the linear logistic regression model are again interpreted as increasing or decreasing the log odds of 'success' but now, as in the multiple linear regression model, as *partial* regression coefficients showing what happens to the log odds if all the other explanatory variables are held constant. Odds multipliers, $\exp \beta_j$, that are greater than 1 increase the odds of success, while odds multipliers less than 1 decrease the odds of success; these correspond precisely to $\beta_j > 0$ and $\beta_j < 0$, respectively.

Note that the explanatory variables in model (9.5) could include sets of indicator variables representing categorical explanatory variables (factors in GENSTAT terminology) – just as for the multiple linear regression model of Chapter 6. When fitting any kind of regression model to a factor with m levels, GENSTAT automatically generates $m - 1$ indicator variables taking the first level as the reference level. In the case of the logistic regression model, the interpretation of the regression coefficients of the $m - 1$ indicator variables must take this into account. Each indicator coefficient estimates the difference in log odds between its level and the reference level of the corresponding explanatory factor, with all other explanatory variables held fixed. Therefore, in terms of affecting the odds of success, the odds multiplier of the coefficient must be interpreted as the odds ratio for its level relative to the reference level. There are examples of this kind of interpretation later in the chapter.

9.3   Using the logistic regression model

As you have seen in previous chapters, fitting a regression model is not just a question of estimating its coefficients for the set of explanatory variables presented to us. We need to know how well the model fits the data and hence find which explanatory variables, or transformations of them, are worth including in the model. The aim of model building is to arrive at a meaningful and parsimonious model that 'explains' the data. You have already seen how to do this in the normal multiple linear regression context in Chapter 6.

The main point of the rest of this chapter is to see how essentially the same ideas carry over to logistic regression. By using GENSTAT's **Generalized Linear** commands, instead of its **Linear** commands, you will be able to follow almost identical routes in logistic regression as you did in normal regression (as already seen in Exercise 9.3). There will just be a few changed details and some different terminology in places. This section will be concerned with pointing out what new terminology corresponds to what familiar terminology; explaining what the new terms actually mean is deferred until Chapter 10. By the end of this chapter, you should be just as comfortable with *using* logistic regression (in GENSTAT) as you are with using normal multiple linear regression (in GENSTAT).

Example 9.2

Let us continue with the case of logistic regression with a single explanatory variable as needed for the burns data. Let us look further into the regression of the binary response variable survival on the transformed explanatory variable larea. In Exercise 9.3, you looked at the model GENSTAT fitted to these data only in pictorial form. Here, first, is the output that GENSTAT provided in the **Output** window at the same time; following it is reproduced the GENSTAT output from normal linear regression of the quantitative response variable loss on the single explanatory variable hardness from Exercise 4.5 (dataset: rubber).

Logistic regression output

```
***** Regression Analysis *****
```

```
Response variate: survival
  Binomial totals: 1
    Distribution: Binomial
    Link function: Logit
    Fitted terms: Constant, larea
```

```
*** Summary of analysis ***
```

	d.f.	deviance	mean deviance	deviance approx ratio chi pr
Regression	1	188.4	188.4147	188.41 <.001
Residual	433	337.0	0.7782	
Total	434	525.4	1.2106	

```
* MESSAGE: ratios are based on dispersion parameter with value 1
```

```
* MESSAGE: The residuals do not appear to be random;
           for example, fitted values in the range 0.95 to 0.99
           are consistently smaller than observed values
           and fitted values in the range 0.12 to 0.18
           are consistently larger  than observed values
* MESSAGE: The error variance does not appear to be constant:
           large responses are less variable than small responses
```

```
*** Estimates of parameters ***
```

	estimate	s.e.	t(*)	t pr.	antilog of estimate
Constant	22.22	2.20	10.12	<.001	4.479E+09
larea	-10.45	1.05	-9.94	<.001	0.00002887

```
* MESSAGE: s.e.s are based on dispersion parameter with value 1
```

Simple regression output

```
***** Regression Analysis *****
```

```
Response variate: loss
    Fitted terms: Constant, hardness
```

```
*** Summary of analysis ***
```

	d.f.	s.s.	m.s.	v.r.	F pr.
Regression	1	122455.	122455.	33.43	<.001
Residual	28	102556.	3663.		
Total	29	225011.	7759.		

```
Percentage variance accounted for 52.8
Standard error of observations is estimated to be 60.5
* MESSAGE: The following units have high leverage:
        Unit     Response     Leverage
          1        372.0       0.182

*** Estimates of parameters ***

                 estimate      s.e.     t(28)   t pr.
Constant           550.4       65.8      8.37   <.001
hardness          -5.337       0.923    -5.78   <.001
```

The two sets of output clearly have strong similarities: indeed each is simply headed Regression Analysis. In each case, there is a Summary of analysis table, some messages and a table of Estimates of parameters. Where the normal output just reminds you (at the top) of the names of the variables and which is the response variable and which are the explanatory variables (the Fitted terms, including Constant), the binary output also reminds you that the Bernoulli distribution – the Binomial distribution with $n = 1$ – is being used along with the Logit link function, i.e. logistic regression.

Exercise 9.4
Apart from the different variables and the different numbers, what other differences are there between the two outputs?

In the normal case, the theory leading to SPs based on t and F distributions is exact. But in binary regression, no such exact theory exists. There are, however, useful approximations. GENSTAT offers such approximate SPs, but you should remember that their accuracy cannot be guaranteed, and that SP results should always be treated with a little extra caution in binary regression (and, in fact, in all generalized linear models). However, we shall still use some appropriate SPs!

For the individual regression coefficients, approximate SPs are given, but it is rec-ommended to avoid using such SPs if possible. The SPs that will mainly be used with generalized linear models in this book concern values of the *deviance* (or, more usually, of differences between deviances). The precise meaning of 'deviance' is discussed in Section 10.3. A suitable approximation here is *not* the F distribution of normal theory (after all, we saw in Solution 9.4 that no ratio of things corresponding to mean squares was taken) but a χ^2 distribution on an appropriate number of degrees of freedom. Do not worry about the general form of such tests for now, but do Exercise 9.5 to make a first deviance-based test.

Exercise 9.5
The GENSTAT command for calculating χ^2 probabilities is cuchisquare, which can be used in the same way as cut. Obtain an approximate SP for testing the null hypothesis of no effect (on the mean survival probability) of the value of larea. To do this, the appropriate thing to do is to compute the probability of a value greater than or equal to the deviance(Regression) under a χ^2 distribution on 1 degree of freedom. What do you conclude? How does this match up with something you are given in the GENSTAT output?

Exercise 9.6

What is the formula for the fitted regression line? What is the fitted probability of survival when larea = 2? What is the odds multiplier associated with an increase of 0.1 in larea?

Hint   By an extension of (9.4), if x increases by c, the odds multiplier becomes $\exp(c\beta)$.

Residual plots can be obtained for binary regression by the same procedure as for normal regression, but they are not so useful or interpretable. If you are using the student version of GENSTAT, you may wish not to produce the residual plots for the burns data yourself at this stage, since a memory problem arises. If you do want to proceed, the method is as follows. From the **Generalized Linear Models** dialogue box, follow the familiar trail of clicking on **Further Output**, then on **Model Checking**, and then on **OK** in the **Model Checking** dialogue box. This will give GENSTAT's default composite set of

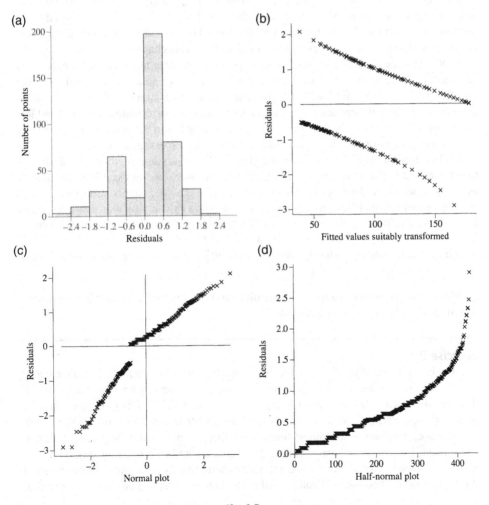

Fig. 9.5

residual plots, as in Figure 9.5. (In the student version of GENSTAT, a fault message occurs, because GENSTAT has insufficient memory space to calculate the smooth curved line on the fitted value plot. However, if you click on **OK** to close the fault dialogue box, the rest of the residual plots are correctly drawn.) Note that the default **Type of Residual** is now **Deviance** rather than **Pearson**. These are another kind of standardized residual, defined in a slightly different way, but with basically the same meaning as the Pearson residuals you used in normal linear regression. We come back to their details in Chapter 11.

There is certainly structure in these plots! However, almost all of it – particularly the obvious two 'lines' of points in the plot of residuals against fitted values – is due to the fact that the response variable is binary. The upper line of (necessarily) positive values corresponds to data with $Y = 1$; the lower line of (necessarily) negative values to $Y = 0$. The reason for these two 'lines' being necessarily positive and negative is that the fitted logistic curve runs between 1 and 0 whereas the Y values are always exactly 1 or 0. (See the diagram in Solution 9.3.) The points follow on from each other so smoothly because they are ordered by values of estimated p. You are also unlikely to be able to work out what exactly GENSTAT is using for 'fitted values' in this plot: it has in fact applied some mysterious transformation to them. You need not bother with such plots at all when performing binary regression, and hence in particular in this chapter.

GENSTAT's messages on residuals (see page 235) are often not very useful either, but are worth keeping an eye on as they are, at this stage, the one source of residual checking we shall use.[1] Fitted values will be consistently smaller or larger than observed values towards the ends of the ranges of the explanatory variable (corresponding to high and low fitted values for the probabilities as in the diagram in Solution 9.3) even if the model fits well, so no extra information is forthcoming from the first message about residuals in this case. (This type of thing is only worth flagging if it happens somewhere in the middle, for then the model could be inappropriate.) The error variances are necessarily non-constant, as the variance of the Bernoulli response variable is $p(1 - p)$; so this part of the second message about residuals tells us nothing also. The extra explanation that 'large responses are less variable than small responses' seems to refer to the lack of a clear (right-hand) 'tail' of zeros relative to the existence of such a tail of ones.

All told, and referring to the figure in Solution 9.3, the logistic regression model seems to be a reasonable one for these data.   ∎

Multiple logistic regression is not difficult to do now that you have seen logistic regression with a single explanatory variable.

Exercise 9.7

Let us now start to analyse the bone marrow transplant data that we looked at in Chapter 1.

This dataset, gvhd, consists of a binary response variable **gvhd** (1 if the individual develops the disease, 0 if not) and six explanatory variables. The latter are: the transplant recipient's age, **recage**; the recipient's sex, **recsex** (1 for female, 0 for male); the donor's age, **donage**; the donor's sex and, if female, whether she had ever been pregnant, **donmfp** (0 for male, 1 for never pregnant female, 2 for female who has been pregnant); the type of leukaemia, **type** (coded 1, 2 or 3); and a further quantitative variable, **indx**. These constitute three quantitative, two nominal categorical (with three classes) and one binary explanatory

---

[1] See Chapter 11 for more on residuals in generalized linear modelling.

variables. In gvhd.gsh, the categorical explanatory variables are factors (but note that the response variable gvhd is a variate).

Load gvhd into GENSTAT. Transform indx by taking logs to give a replacement explanatory variable, lindx say. (In Chapter 4, such a transformation was considered for making a relationship more linear, but this is not so here. The reason for such a transformation in this case is related to the fact that, as GENSTAT warns us on reading in the data, the index variable has a skewed distribution. This will be discussed further in Chapter 11. In fact, for this dataset, the transformation is not crucial, although it does produce slightly better results. Some analysts make such a transformation in such a situation as a matter of course; others are more reluctant to do so.)

Now perform the multiple logistic regression of gvhd on all six explanatory variables (recage, recsex, donage, donmfp, type and lindx). You should be able to work out how to make GENSTAT do this, given your experience so far.

In the resulting output, why is the d.f.(Regression) not equal to six, the original number of explanatory variables? Use the GENSTAT output to make an appropriate approximate test of the null hypothesis that all the β_js are zero. What do you conclude? What has GENSTAT used as the test statistic, and what (approximate) null distribution has it referred the test statistic to? On the basis of the estimated regression coefficients, which explanatory variables appear to be the most important predictors of gvhd?

So, the 'full' six-explanatory-variable model is quite interesting. But might a model with fewer explanatory variables be even more useful? To find out, we can follow through steps analogous to those taken in the multiple normal regression case in Chapter 6.

Example 9.3
We have looked at the overall model for the gvhd data, which suggested that only lindx is a potentially good predictor of gvhd.

We *could* go on to produce a scatterplot matrix (though we would have to redesignate recsex, donmfp and type temporarily as variates (in the spreadsheet) for DSCATTER to work). However the binary nature of the response variable means that, when the individual regressions of gvhd on the six explanatory variables are plotted, only those against quantitative explanatory variables would tell us anything at all, and in fact they tell us little.

Instead, one possibility is to perform all six individual regressions in turn. The regression deviances, d.f.s and SPs that result (from the appropriate χ^2 distributions) are shown in Table 9.4.

Table 9.4

	d.f.	Regression deviance	SP
recage	1	7.03	0.008
recsex	1	0.96	0.327
donage	1	5.57	0.018
donmfp	2	8.68	0.013
type	2	7.89	0.019
lindx	1	13.31	<0.001

Individually, then, it seems that gvhd may be strongly related to lindx (which also showed up best in the multiple regression), and quite strongly related to all the other explanatory variables with the exception of **recsex**. By the way, this confirms the assertion alluded to in Chapter 1 that **indx** is, individually, the most important predictor of **gvhd**. It is the fitted model involving only **lindx** that is given at (1.2).

The correlation matrix can be obtained in just the same way as in Chapter 6 using the Stats|Summary Statistics|Correlations menu item after **recsex**, **donmfp** and **type** have, temporarily, been defined as variates rather than factors. The relevant part of the correlation matrix is as follows.

```
recsex   -0.010
donage    0.653   -0.147
donmfp    0.271   -0.320    0.059
  type    0.192    0.102    0.134   -0.182
 lindx    0.267    0.116    0.090    0.001    0.127

          recage   recsex   donage   donmfp    type
```

Interestingly, most correlations between explanatory variables are very small. The only one worthy of note is the 0.653 correlation between **donage** and **recage**, but this does not necessarily disallow the two from appearing together.

With this background, we can let GENSTAT's stepwise regression routine loose on the data. Start (for instance) from fitting the full six-variable model. The GENSTAT command is exactly the same as in the multiple normal regression case; so, for this example, we enter the following.

```
STEP [MAX=12;IN=4;OUT=4] recage,recsex,donage,donmfp,type,lindx
```

Details of the changes to the method that GENSTAT uses need not concern us now. The output looks much as you should expect, with Residual mean deviances replacing Residual mean squares. The model chosen has two remaining explanatory variables, **lindx** and **donage**. The regression deviance for this regression is 19.98 on 2 d.f. (You would have to calculate the corresponding SP for yourself: PRINT cuchisquare(19.98;2) gives 0.00005. GENSTAT doesn't calculate an SP for you in this case as a reminder of the inappropriateness of standard SPs after such a model-fitting process.) The t statistics of the constant term and of the two retained explanatory variables are each between 2 and 3 in absolute value. The same model is found if you start from the null model. You can run the null model simply by leaving the Model to be Fitted field blank. (You do not have to type in a space!)

Comfortingly, **lindx** and **donage** are two of the five main explanatory variables suggested by the individual regressions (Table 9.4). Another is **recage**, and this, as we noted, is moderately correlated with **donage**; so leaving it out seems reasonable. However, we could check this by looking at the model containing each of **lindx**, **donage** and **recage**. (Intriguingly, in its stepwise procedure, GENSTAT chose to drop **recage** at the very first step.) Fitting the model of **gvhd** regressed on the three variables **lindx**, **donage** and **recage** gives a regression deviance of 20.38 (on 3 d.f.). The regression deviance of the model containing **lindx** and **donage** only is 19.98 (on 2 d.f.). The increase in regression deviance is just $20.38 - 19.98 = 0.4$, and to achieve this increase we have 'used up' one degree of freedom. (It would be exactly the same to take the difference between the two residual deviances.) The appropriate way to calibrate this deviance change is to compare it with the $\chi^2(1)$ distribution. This gives an SP of 0.527, and so there is no evidence in favour of adding **recage** to the model. There turns out to be no evidence in favour of adding **type** either.

You can make GENSTAT perform this test for you in the following way. First fit again the logistic regression model using just **lindx** and **donage** as explanatory variables. Next click on **Change Model**; enter **recage** in the **Terms** field; click on the **Add** button. This tells GENSTAT to fit the logistic regression model with **recage** added to **lindx** and **donage** as explanatory variables. Scroll up the GENSTAT output until you reach the Summary of analysis table. Beneath this table there is an extra line headed Change. This gives the information for the test we have just performed: the d.f. column contains 1 (actually written as −1), the deviance column contains 0.40 (actually the deviance difference, and written −0.40), and the χ^2 probability is given as 0.527.

Exercise 9.8

Fit the model containing **lindx**, **donage** and **recage** to the gvhd data. Use the **Change Model** dialogue box in a slightly different way to remove the **recage** term from the model. Do you get the same information in the GENSTAT output?

The fitted model remains

$$\text{logit}(p) = -5.45 + 2.178 \text{ lindx} + 0.1459 \text{ donage}. \qquad (9.6)$$

Exercise 9.9

What regression fit to these data was given at (1.1) in Chapter 1? Get GENSTAT to recreate this fit, and suggest, on the basis of this fit, what improvement (if any) you would now make to the model (9.6).

Yet further investigation of this dataset could involve introducing interactions between explanatory variables, but we shall not do so here. ■

9.4 Exercises in logistic regression

This section consists of two large exercises in which you will use GENSTAT to analyse further real datasets.

Exercise 9.10

Wetland areas are well protected in the Ontario region of Canada, but lands around them are not so well regulated. A study was carried out to see if the species richness of the wetlands is affected by the characteristics of the surrounding areas. The occurrence or non-occurrence of many species of plants and animals was recorded for a number of wetlands, the 27 of which that had complete records being analysed here. Also recorded were the area of the wetland (ha), and the road densities (m/ha) and percentage of forest cover in the adjacent land. The dataset **wetbirds** records the data for two species of birds, the green-backed heron and the wood duck, for each of the 27 wetlands. The data are coded as follows.

gbheron	1 = green-backed heron present, 0 = green-backed heron not present
woodduck	1 = wood duck present, 0 = wood duck not present
area	area of wetland (ha)
tht1	total hardtop road density within 250 m of wetland edge (m/ha)
tht2	total hardtop road density within 500 m of wetland edge (m/ha)
tht3	total hardtop road density within 1000 m of wetland edge (m/ha)
tht4	total hardtop road density within 2000 m of wetland edge (m/ha)
cfor1	percentage of forest cover within 250 m of wetland edge
cfor2	percentage of forest cover within 500 m of wetland edge
cfor3	percentage of forest cover within 1000 m of wetland edge
cfor4	percentage of forest cover within 2000 m of wetland edge
grav1	gravel road density within 250 m of wetland edge (m/ha)
grav2	gravel road density within 500 m of wetland edge (m/ha)
grav3	gravel road density within 1000 m of wetland edge (m/ha)
grav4	gravel road density within 2000 m of wetland edge (m/ha)

Source: data kindly provided by J. Houlahan, via A. J. Bertie.
Dataset name: `wetbirds`.

There are two binary response variables, **gbheron** and **woodduck**, and 13 possible explanatory variables. The various wetlands differ considerably in area, and so again it proves useful to take logs of this explanatory variable; do this first after loading the data, and replace **area** (throughout this question) by **larea** = log(area). (You should not try to take logs of other variables that GENSTAT flags as skew.)

(a) Perform an exploratory analysis of the data in the following way. Calculate the matrix of correlation coefficients for all the explanatory variables. Identify any blocks of closely related variables: does this make intuitive sense? Choose a 'representative' set of four explanatory variables that seem to be relatively unrelated. Make a scatterplot matrix of these four explanatory variables and both response variables. Which of the representative explanatory variables seem to have most effect on **woodduck** and on **gbheron**?

(b) Take **woodduck** as the response variable. Start by fitting single-variable logistic regression models for **woodduck** based on each of **larea**, **tht4**, **cfor4** and **grav4** in turn. Which of these variables seems most important in modelling **woodduck**? Next, fit a multiple logistic regression model for **woodduck** in terms of *all thirteen* explanatory variables (with **larea** replacing **area**). Which variables does this suggest as most important?

(c) Perform a (book default) stepwise regression (starting from all 13 variables). Does the resulting model seem a useful one? If not, perform an alternative stepwise regression (still with the book default parameters) starting from the null model. Does this give the same result? If not, does the alternative result seem more useful? (Do not attempt to compare the two results formally in terms of an SP.)

(d) Regress **woodduck** on the single explanatory variable **grav2** and make a picture of the fitted model. What do you see?

(e) Remove wetland 9 temporarily from the dataset and repeat parts (c) and (d). What results do you get now?

(f) Interpret the meaning of your preferred model from part (c) in terms of the probability of the presence of wood ducks in the wetlands.

(g) Repeat the analysis of parts (b) to (d) with **gbheron** as the response variable (and with wetland 9 included in the dataset). What do you find in this case?

Note that, because in part (c) of Exercise 9.10, grav2 was not one of the six variables chosen by the backward stepwise regression, it is not legitimate to compare the difference between the regression deviances of the two models with a χ^2 distribution. To do so, one of the two models to be compared would have to be a subset of the other, i.e. one model would have to be *nested* within the other (as in Example 9.3).

Exercise 9.11

The data for this exercise comes from a study by D. Baldus to assess the effect of race in the decision to impose the death penalty on those convicted of murder in the US state of Georgia. Baldus's statistical analysis was used as grounds for a writ of habeas corpus to the District Court in the case of a black defendant sentenced to death (McClesky v. Zant, 481 US 279, 1987). The District Court rejected Baldus's statistical analysis, but this was overturned on appeal. The data in the file death.gsh is a subset of the variables and observations from this study. There are nine variables as follows.

death	1 = death sentence, 0 = life sentence
BD	2 = black defendant, 1 = white defendant
WV	2 = one or more white victims, 1 = no white victims
AC	number of statutory aggravating circumstances
FV	2 = female victim, 1 = male victim
VS	2 = victim was a stranger, 1 = victim was not a stranger
V2	2 = two or more victims, 1 = one victim
MS	2 = multiple stabs, 1 = no multiple stabs
YV	2 = victim 12 years of age or younger, 1 = victim over 12

Source: Finkelstein, M. O. and Levin, B. (1990) *Statistics for Lawyers*, New York, Springer-Verlag. Dataset name: death.

The binary response variable is **death**. With the exception of AC, all the explanatory variables are binary (and in death are designated as factors). AC is a quantitative variable, taking the values 1, 2, ... , 7 (i.e. AC is a count variable). The data in death are a sample of 100 cases from the 'middle range' with respect to aggravation of the crime.

Since most of the explanatory variables are binary, an exploratory analysis of the data requires different tools from earlier in the chapter. The main tool should be contingency tables of **death** with each of the explanatory variables. You can do this for AC as well as for all the binary variables by ignoring the fact that AC's counts are meaningful and pretending that it is a nominal variable.

(a) Load the death data into the GENSTAT spreadsheet. Update the server. Choose the Data|Form Groups menu item. In the resulting dialogue box, enter AC in the Data field and fAC (say) in the Save In field. Leave the Method as Mapping and click on OK. The new factor fAC now contains the same data as the variate AC. Use the Form Groups dialogue box in a similar way to produce a new factor fdeath which contains the same data as the variate death.

To obtain the desired contingency tables, choose the Stats|Summary Statistics|Summaries of Groups (Tabulation) menu item. In the Summary by Groups dialogue box, under Type of Summary, click on Means, to switch this option off, and on No. of Observations, to switch this option on. Enter fdeath and, say, BD into the Groups field. To get fdeath as the variable corresponding to the rows of the table (as befits a response variable), ensure that fdeath is in the Groups field above the other variable. GENSTAT insists that you enter something in the Variate field, though in this case it does not use the information, so enter death in this field. Click on OK to produce the contingency table of counts in cells formed by cross-classifying death and BD.

Repeat the above to obtain the contingency tables of death against all eight explanatory variables.

In which cases does the response pattern appear to differ depending on the value of the explanatory variable? (These are the explanatory variables that appear to exert an important influence on the response variable individually.)

(b) The important part of the correlation matrix for the explanatory variables is given below. (To obtain this matrix, you could use the usual Correlations menu item, but only after converting all the explanatory variables to variates.) What does it tell you?

WV	-0.467					
AC	0.106	0.087				
FV	-0.062	-0.046	-0.324			
VS	0.239	-0.034	0.110	-0.122		
V2	-0.140	-0.024	-0.100	-0.021	-0.181	
MS	-0.022	0.008	-0.193	0.152	-0.254	-0.060
YV	0.056	-0.077	-0.026	0.041	-0.153	0.200	-0.109
	BD	WV	AC	FV	VS	V2	MS

(c) Return all the variables to their original designations: death and AC as variates, the remainder as factors. Then fit, in turn, each of the eight logistic regression models having just a single explanatory variable. By looking at the regression deviances and their χ^2 SPs, which variables seem to be most important individually?

Perform a multiple logistic regression of death on all eight explanatory variables. Which seem important here?

(d) Perform a stepwise regression to select a model based on a subset of the explanatory variables. Does the model you obtain surprise you? Interpret what the model is telling you in qualitative terms.

(e) The main purpose of this study was to quantify the influence of the colour of the defendant (BD). Investigate whether a model containing WV, VS and BD is also a reasonable one in the light of these data.

(f) Perhaps **BD** influences **death** in more subtle ways. Starting from the **WV, VS** and **BD** model, investigate whether it is worthwhile adding any of the interactions between these three factors into the model. Refit the **WV, VS, BD** model, if it was not the last model you fitted. Then click on the **Change Model** button, to enter, say, the interaction **WV.VS** first (in the **Terms** field) and then click on the **Try** button. (**Try** is equivalent in this context to **Add** and then **Drop**; if you were to use **Add** instead, you would need to remember to **Drop** the term you have just **Added** before **Adding** the next one.) Then do the same for each of the other two interaction terms, until all three interactions have been (individually) added. Are any of these interactions worth adding?

(g) It has become clear that **BD** does not seem to have any effect on the probability of a death sentence. There is clearly a strong preference for the regression model involving just **WV** and **VS**. Make a quantitative interpretation of the fitted model involving just these two explanatory variables, i.e. how are the odds ratios in favour of the death sentence affected by these two explanatory variables?

Summary of methodology

1 Binary responses, Y_i, coded 0 for 'failure' and 1 for 'success', follow Bernoulli distributions with unknown probabilities of success p_i. Given corresponding observations on a set of explanatory variables, the purpose of binary regression analysis is to use these to estimate p_i and to assess the effects of the explanatory variables on the response variable.

2 The linear regression model and analysis for a continuous response variable, based on the normal distribution, are not appropriate for a binary response variable.

3 The linear logistic regression model for the dependence of p_i on the corresponding values of k explanatory variables, $x_{i,1}, x_{i,2}, \ldots, x_{i,k}$, is

$$\text{logit}(p_i) = \alpha + \sum_{j=1}^{k} \beta_j x_{i,j}.$$

The regression coefficients of this model are estimated by the method of maximum likelihood. The logit function, $\text{logit}(p) = \log[p/(1-p)]$, is a link function, and the linear logistic regression model is an example of a generalized linear model.

4 Logistic regression model building with respect to explanatory variables uses methods analogous to the case of normal regression model building. Exploratory analysis and variable selection techniques are used to find a plausible model with a good overall fit.

10

What are generalized linear models?

You have already met two particular cases of generalized linear models. The first of these is the normal linear regression model, manifest in many different versions in Chapters 4–8, in which dependence on the explanatory variable(s) is through a linear function and the conditional distribution of the response variable given the explanatory variable(s) is normal. The second special case is the binary regression model explored in Chapter 9. Binary regression was dealt with separately from the normal linear regression, but many similarities in the treatments were stressed. A third type of generalized linear model, the Poisson regression model, is introduced in Section 10.1.

The general framework of the *generalized linear model* – henceforth to be referred to as the GLM for short – affords a single approach to the analysis of a whole host of useful regression models: the normal linear regression model, binary and Poisson regression models, and others, some of which appear in the remainder of the book. The things that vary between different GLMs are: (i) the distribution of the response variable (conditional on the explanatory variables); (ii) the way in which (some parameter such as the mean of) that distribution is linked to the explanatory variables. An important common attribute under (ii) is that a linear function of the explanatory variables is always involved.

Sections 10.2–10.4 develop the general formulation of the GLM and exhibit the main aspects of this class of models. In these sections, you will have a rest from actually analysing data. Indeed, for a change, you could switch your computer off while studying these sections.

The final section, Section 10.5, comprises several data analyses that can be approached using GLMs. From here on, you will find it useful to think of your modelling approach as a generalization to non-normal data of techniques learned earlier in the book for normal models.

10.1 Poisson regression

Exercise 10.1

The data in Table 10.1 were collected to investigate the hypothesis that media publicity of murder or suicides by the deliberate crashing of private aircraft triggers further such murders

Table 10.1

Index of newspaper coverage	Number of crashes	Index of newspaper coverage	Number of crashes
376	8	63	2
347	5	44	7
322	8	40	4
104	4	5	3
103	6	5	2
98	4	0	4
96	8	0	3
85	6	0	2
82	4		

Source: Phillips, D. P. (1978) 'Airplane accident fatalities increase just after newspaper stories about murder and suicide', *Science*, **201**, 748–50.
Dataset name: crashes.

or suicides. The table gives the number of fatal crashes during the week immediately following each crash (y), together with an index describing the amount of publicity given to the 'triggering' crash (z), for a total of 17 such crashes. The data are plotted in Figure 10.1.

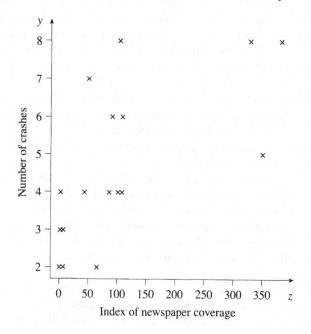

Fig. 10.1

(a) The distribution of the explanatory variable z seems rather skewed and hence 'scrunches up' much of the dataset towards the left of the plot. What transformation of z is most commonly used to make such a relationship between the explanatory variable and the response variable clearer? What problem is there with this transformation in this case? Suggest a way around this problem.

(b) Why are the normal and Bernoulli distributions that have been used so far in the book inappropriate distributions to assume for the responses in this case?

Let us consider what alternative type of model might be contemplated for the responses in Exercise 10.1 (conditional on values of the explanatory variable).

A distribution that provides a probability structure for the non-negative integers, and is therefore often used to model the distribution of data in the form of counts, is the Poisson distribution. It tends to be the first distribution considered for count data, in much the same way that the normal distribution springs to mind for symmetric continuous data.

As a reminder (see Section 2.5.3.), the probability mass function of the Poisson distribution is

$$P(X = x) = \frac{\lambda^x e^{-\lambda}}{x!}, \quad x = 0, 1, \ldots .$$

It seems reasonable, then, to propose that the response variable, here the number of aircraft crashes, given any particular value of the explanatory variable, here $x = \log$ (index of newspaper coverage $+ \frac{1}{2}$), has a Poisson distribution.

If we assume Poisson responses, how do we incorporate the explanatory variables into our model? In the normal response case, what was always proposed was a dependence of the *mean* of the response distribution on some linear function, $\ell(x)$ say, of the explanatory variable(s). We *could* do the same for Poisson responses, but there is a difficulty. The mean, λ, of the Poisson distribution is a positive quantity, since counts are allowed only to be zero or positive (and the case where the mean is zero is trivially uninteresting). The positivity constraint is analogous to the constraint that probabilities lie between 0 and 1 in the binary regression case. However, there is no such restriction on any linear function, e.g. $\ell(x) = \alpha + \beta x$. It seems better to change the dependence on $\ell(x)$ so that the positivity property is respected.

The usual solution to this difficulty is as follows. Because $\lambda > 0$, we can take $\log \lambda$ and obtain something free to vary over both positive and negative values. This is compatible with $\ell(x)$. That is, associate $\log(\lambda(x))$, rather than $\lambda(x)$ itself, with $\ell(x)$ by writing $\log(\lambda(x)) = \ell(x)$ (rather than $\lambda(x) = \ell(x)$). The log function is the analogue of the logit function used to remedy the situation in the binary regression case. (Note that the use of the log function here has nothing to do with the fact that we happened, in Exercise 10.1, to use a log transformation to define x.) Equivalently, you could say that $\lambda(x)$ is directly associated with $\exp(\ell(x))$. The resulting *Poisson regression* model, then, is that

$$Y_i \sim \text{Poisson}(\lambda(x_i))$$

where

$$\log(E(Y_i)) = \log(\lambda(x_i)) = \ell(x_i)$$

and the Y_is are independent given the x_i values.

As you were able to change quickly from using normal regression in GENSTAT to using logistic regression, just such a quick change is possible to using Poisson regression. This is the subject of the next two exercises.

Exercise 10.2

This exercise picks up the analysis of the dataset on murder/suicide air crashes following other publicized crashes of the same type that was introduced in Exercise 10.1.

(a) Load the dataset stored in the file `crashes.gsh` into the spreadsheet. Compare the spreadsheet with Table 10.1, so that you can see that the two are the same. Update the server.

(b) Make the transformation from z (the index of media coverage) to $x = \log\left(z + \frac{1}{2}\right)$. (To do this, click on the **Data** menu; click on **Transformations**. Change the **Transformation** field to **Log (base e)**; fill in the **Data** field as **cover**; change the **Parameters in Equation: c** field to **0.5**; fill the **Save in** field as **lcover** (say). Click on **OK**; click on **Cancel**.)

Choose the **Stats|Regression Analysis|Generalized Linear** menu item. In the dialogue box: make sure the **Analysis** field contains **General Model**; fill in the **Response Variate** as **crashes** and the **Model to be Fitted** field as **lcover**; change the **Distribution** field to **Poisson**; ensure the **Link Function** is set to **Canonical**; click on the **OK** button. (Do not click on the **Cancel** button yet.) Examine the output.

According to Exercise 10.1, the aim of the analysis was 'to investigate the hypothesis that media publicity of murder or suicides by the deliberate crashing of private aircraft triggers further such murders or suicides'. If so, you might expect greater media coverage to result in an increased number of crashes. In the Poisson regression model, what null hypothesis would correspond to no such relationship?

To carry out a test of this hypothesis, the procedure is the same as was used for logistic regression in Chapter 9. Specifically, test the value in the `deviance` column of the `Summary of analysis` table against the χ^2 distribution on `d.f.(Regression)` degrees of freedom. So, using the GENSTAT output, in which this is done for you, test this hypothesis.

(c) Continuing with the Poisson regression analysis of these data, click on the **Further Output** button in the **Generalized Linear Models** dialogue box. In the dialogue box that appears, click on **Fitted Model**, fill in the **Explanatory Variable** field with **lcover**, and click on the **OK** button. This gives a scatterplot of the data with the fitted curve added. Note that the fitted curve is not a straight line; this is the effect of modelling the log of the mean by a straight line. From this plot, do you think that the model fits the data well?

Produce a residual plot by clicking once more on **Further Output** in the **Generalized Linear Models** dialogue box, and clicking this time on the **Model Checking** button. Click on **OK** in the dialogue box that appears. This gives the usual composite set of residual plots (as with logistic regression, the default standardized residuals are deviance residuals, which will be described in Chapter 11). The interpretation of such residual plots is the same as in the normal case. From these plots, do you think that the model fits the data well?

(d) What is the fitted model that you obtained in part (b) of this exercise (in terms of the original variables Y and z)? (You might find it interesting also to manipulate this model into an alternative form by exponentiating each side.)

You will recall that plots of residuals against fitted values can be useful for normal linear regression, but were of little use in the case of binary regression; notice that such plots can once again be useful in the case of Poisson regression. See also Chapter 11.

Exercise 10.3

The dataset `ship1907`, the first five values of which are shown in Table 10.2, is of interest to historians. These data give the tonnage, size of crew and type of power (sail or steam) for 20 British merchant ships in 1907. Interest centres on how crew size depends on tonnage, and whether the relationship is different for different types of power.

Table 10.2

tonnage	crew	power
236	8	sail
739	16	steam
970	15	steam
2371	23	steam
309	5	steam
⋮	⋮	⋮

Source: Floud, R. (1973)
An Introduction to Quantitative Methods for Historians, 2nd edition, London, Methuen.
Dataset name: `ship1907`.

(a) Load the dataset stored as `ship1907.gsh` into GENSTAT. Obtain a scatterplot of the data with **crew** as the response variable, **tonnage** as the explanatory variable, and using the factor **power** to distinguish between sailing and steam ships. Comment on what you see.

(b) Since the response variable comprises counts (of crew members), it is natural to use Poisson regression to analyse these data. To do so, choose the Stats|Regression Analysis|Generalized Linear menu item. In this dialogue box, make sure the Analysis field contains General Model, fill in the Response Variate as **crew** and the Model to be Fitted field as **tonnage,power**, change the Distribution field to Poisson, and click on the OK button. (Do not click on the Cancel button yet.)

Let us not look at the GENSTAT output for now, but concentrate on how well the Poisson regression model fits. To this end, click on the Further Output button, click on Model Checking and then click on OK (so that the plots you obtain are the usual composite set). What are your first impressions from these plots?

Now click on Further Output again, and this time click on Fitted Model, filling in the Explanatory Variable field with **tonnage** and the Grouping Factor field with **power**, and click on the OK button. Describe what you see and discuss any inadequacies there may be with the Poisson regression model.

(c) What transformation of the explanatory variable (**tonnage**) would it be natural to try in order to alleviate the last of the concerns expressed in Solution 10.3(b)?

(d) Make the transformation of Solution 10.3(c). Call the transformed explanatory variable ltonnage. Following the steps of (b) above, fit a Poisson regression model of **crew** regressed on ltonnage and **power**. Produce and look at residual plots and a fitted model plot for the transformed data. Is there any way in which the model might be further improved?

(e) By using the **Change Model** button in the **Generalized Linear Models** dialogue box, perform the Poisson regression of **crew** on **ltonnage** alone. By looking at the resulting output, assess whether the **power** term should be included in the model (i.e., more formally, test whether the regression coefficient associated with **power** is zero). Check the fit of your final model by examining residual and fitted model plots.

(f) What is the final fitted model for the ship1907 data? Use the exponentiation method of Exercise 10.2(d) to manipulate the model for the mean response into an alternative form, and interpret what you get.

10.2 The generalized linear model

Ever since Chapter 3, we have been concerned with models involving a dependence on a linear function of explanatory variables.[1] This linear function can be written in general as

$$\eta_x = \sum_{i=1}^{p+1} \beta_i x_i$$

where $x_1, x_2, \ldots, x_{p+1}$ are the $p + 1$ explanatory variables and $\beta_1, \beta_2, \ldots, \beta_{p+1}$ are the corresponding regression coefficients. The quantity η_x is known as the *linear predictor* and features in all of the earlier models. For instance, in the multiple linear regression model (Chapter 6) there are p different explanatory variables together with an intercept term corresponding to a $((p + 1)$st) constant explanatory 'variable'; in a univariate quadratic regression model (Chapter 6 also; see Exercise 6.3) $p = 2$ and $x_1 = 1, x_2 = x$ and $x_3 = x^2$ where x is the original single explanatory variable; in basic analysis of variance models (as described in Section 6.5) each explanatory variable is associated with a particular group and takes the value 1 or 0 according to whether or not the individual belongs to that particular group; and the more complicated models can be written with explanatory variables that are mixtures of the above.

The normal linear regression model can then be written as

$$Y = \mu(\eta_x) + \epsilon = \eta_x + \epsilon = \sum_{i=1}^{p+1} \beta_i x_i + \epsilon$$

where $\mu(\eta_x) = \eta_x$ (i.e. μ is a function of η_x, which in this case happens to be the identity function) and the random error ϵ is distributed as $N(0, \sigma^2)$. (The reason for this peculiar formulation will become clear very soon.) Alternatively, this is equivalent to writing the model as

$$Y \sim N(\mu(\eta_x), \sigma^2);$$

i.e. given $x_1, x_2, \ldots, x_{p+1}, \mu(\eta_x)$ and σ^2 are the mean and variance, respectively, of a normal distribution for Y. The equivalence is because Y is the sum of a normal random variable ϵ and a constant.

---

[1] GENSTAT will not be needed again until Section 10.5.

For the sake of clarity of the theory in this section and the next, we shall only consider the case where $\sigma^2 = 1$. In the unimportant case of σ^2 known but not equal to one, a rescaling by division by σ would be necessary. In the realistic case where σ^2 is unknown, this extra parameter requires special treatment, as it has already received in earlier chapters. The normal linear regression model of current interest is therefore

$$Y \sim N(\mu(\eta_x), 1). \tag{10.1}$$

The logistic regression model of Chapter 9, in which Y can only take the values 0 or 1, has the form

$$Y \sim \text{Bernoulli}(p(\eta_x)) \tag{10.2}$$

where the mean $p(\eta_x)$ has the property that $\text{logit}(p(\eta_x)) = \eta_x$, where $\text{logit}(w) = \log(w/(1-w))$. (Equivalently $Y \sim B(1, p(\eta_x))$.)

The Poisson regression model of Section 10.1 is

$$Y \sim \text{Poisson}(\lambda(\eta_x)) \tag{10.3}$$

where $\log(\lambda(\eta_x)) = \eta_x$.

Models (10.1), (10.2) and (10.3) have much in common. Each specifies a distribution for an individual's response variable Y conditional on the individual's explanatory variables $x_1, x_2, \ldots, x_{p+1}$. Additionally, each individual's response variable is independent of other individual's response variables. It is the mean of each distribution that carries the dependence of the response variable on $x_1, x_2, \ldots, x_{p+1}$, and that is done through the linear predictor $\eta_x = \sum_i \beta_i x_i$ only. (Recall that the means of $N(\mu, 1)$, Bernoulli(p) and Poisson(λ) are μ, p and λ, respectively.) Indeed, in each case, a simple function of the mean w_x of the distribution, namely $g(w_x) = w_x$ in (10.1), $g(w_x) = \text{logit}(w_x) = \log(w_x/(1-w_x))$ in (10.2) and $g(w_x) = \log(w_x)$ in (10.3), is equal to η_x.

Each of these three models, then, has the same general form: Y follows a distribution, f say, the mean w_x of which is a function of $x_1, x_2, \ldots, x_{p+1}$ through η_x. Let g be the function in question: $g(w_x) = \eta_x$; g is called the *link function* of the generalized linear model. If f is allowed to be a probability density function, if Y is continuous (as in model (10.1)), or a probability mass function, if Y is discrete (as in models (10.2) and (10.3)), we have the following formal definition of a GLM.

Box 10.1

The relationship between a response variable Y and explanatory variables $x_1, x_2, \ldots, x_{p+1}$ follows a **generalized linear model** if

$$Y \sim f(w_x),$$

w_x is the mean of f, and

$$g(w_x) = \eta_x = \sum_{i=1}^{p+1} \beta_i x_i.$$

Here f is a probability density or mass function and g is a link function.

While the words 'generalized linear model' certainly describe any model of this form, the most standard usage is to reserve the term for one particular large class of distributions called the *exponential family* of distributions. The exponential family covers a wide variety of practically useful continuous and discrete distributions, including the normal, Bernoulli/binomial and Poisson distribution among others. The advantage of confining interest to the exponential family is in affording a particularly useful unified theory for estimation, inference and computational algorithms. (The details of this theory are not covered in this book.) All the GLMs available in GENSTAT involve exponential family distributions for the same reasons.

You will not need to be aware of the general formula for the exponential family in this book. You should, though, be aware that not every possible conditional response distribution belongs to the exponential family.

One consequence of exponential family models is that certain link functions, the *canonical link functions*, have a special status. They provide simplifications, which will not be gone into, in the theory and analysis of GLMs. For the three response distributions with which you are now familiar, the canonical link functions are:

$$\text{Normal: identity, } g(w) = w$$
$$\text{Bernoulli: logistic, } g(w) = \log(w/(1 - w))$$
$$\text{Poisson: log, } g(w) = \log(w).$$

These are precisely the link functions that have been used so far. Clearly, the canonical link functions are often the link functions of choice in practice. But this need not always be so.

Sometimes, it is not the link functions g themselves but rather their inverses that are quoted. These *inverse link functions* h arise from solving the equation $g(w) = v$, say, so that $h(v) = w$. The inverses of the canonical link functions above are $h(v) = v$, $h(v) = e^v/(1 + e^v)$ and $h(v) = e^v$, respectively. So, for instance, we could also say that, in the Poisson regression model, $Y \sim \text{Poisson}(w_x)$ and $w_x = e^{\eta_x}$. The link and inverse link formulations are exactly equivalent.

The arguments given in Chapter 9 and Section 10.1 for the logistic and log link functions for binary and count data, respectively, focused on an important property of many link functions: that they transform the mean from whatever scale it is naturally on (from 0 to 1 and from 0 to ∞ in these examples) to the whole real line, so that fitting linear functions makes sense and does not violate any constraints. This is often a desirable feature (but not always: see Section 10.5.3).

Additional requirements of a link function include good fit to the data and ease of interpretation of results (e.g. the odds multiplier properties associated with the logistic link function in Chapter 9).

10.3 Inference for GLMs

In Chapter 9 it was stressed that the GENSTAT output for binary regression had strong parallels with that seen in earlier chapters for normal regression. In Section 10.1 it became clear that the output for Poisson regression is very similar too. Now that it has been established that all three models are GLMs, it becomes reasonable to suppose that there is a general approach to statistical inference for GLMs. This is indeed so, and in this section the key idea behind this unified methodology – that of *deviance* – is explained in general terms.

Inference for GLMs is based, as in much of the rest of statistics, on the likelihood function $L(\theta)$, a reminder of which was given in Section 2.6. The likelihood function is the likelihood of the data given the parameter(s) θ. For GLMs, the likelihood function is a function of the response distribution f, the link function g, and the values of the explanatory variables $x_1, x_2, \ldots, x_{p+1}$ for all of the individuals in the dataset. Since a GLM usually depends on the values of more than one parameter (e.g. the intercept and the slope even in simple linear regression), θ should be thought of as the set of all these parameters.

Parameter estimates $\widehat{\theta}$, the mles (maximum likelihood estimators), are made by maximizing the likelihood. As in Section 2.6, it proves convenient to replace the likelihood function by the log-likelihood function $l(\theta) = \log L(\theta)$. The value $l(\widehat{\theta})$ of the log-likelihood function with θ replaced by $\widehat{\theta}$ measures how well the model describes the observations, once the value of θ has been optimized.

Example 10.1

Suppose $Y \sim N(\eta_x, 1)$. The likelihood function for the data is the product of the n density functions, one for each Y_i, because responses are independent. Equating θ to $\beta = \{\beta_1, \beta_2, \ldots, \beta_{p+1}\}$, this gives

$$L(\beta) = \prod_{i=1}^{n} \frac{1}{\sqrt{2\pi}} \exp\left[-\frac{1}{2}\left(Y_i - \sum_{j=1}^{p+1} \beta_j x_{i,j}\right)^2\right].$$

(Notice that i indexes individuals and j indexes explanatory variables, so that $x_{i,j}$ refers to the jth explanatory variable for the ith individual.)

The log-likelihood function is therefore

$$l(\beta) = C - \frac{1}{2}\sum_{i=1}^{n}\left(Y_i - \sum_{j=1}^{p+1} \beta_j x_{i,j}\right)^2$$

where $C = \log[(1/\sqrt{2\pi})^n]$ is an unimportant constant.

Maximizing this over β gives the mle $\widehat{\beta} = \{\widehat{\beta}_1, \widehat{\beta}_2, \ldots, \widehat{\beta}_{p+1}\}$. The formula for $\widehat{\beta}$ will not be needed explicitly, but remember that it is precisely the same as the least squares estimate that has been used throughout in the normal model. Inserting this into the log-likelihood function shows that, for this normal linear regression model, the value of the log-likelihood function at its maximum is

$$l(\widehat{\beta}) = C - \frac{1}{2}\sum_{i=1}^{n}\left(Y_i - \sum_{j=1}^{p+1} \widehat{\beta}_j x_{i,j}\right)^2.$$

Notice that this quantity is, except for the constant C, minus one half of the residual sum of squares for the fitted model. ∎

Our generalized linear models, however good, estimate the mean of Y conditional on x, and so they typically do not reproduce the values of Y, which are subject to the influence of random sampling from the response distribution, exactly. This, indeed, is their great value: models provide useful simplified descriptions (and predictors) of the data. However, purely on grounds of fit to the data, we would do better by using the *saturated model*, in which a separate parameter is involved for each datapoint: basically, this says model each Y_i by the observed value of Y_i! This rather silly model, which provides no simplification, none the less plays a useful role in calibrating the size of $l(\widehat{\theta})$. The *residual deviance* is defined to be

$$D = 2[l(\text{data}) - l(\widehat{\theta})]$$

where, in an abuse of notation, the parameters of the saturated model have been referred to simply as 'data'. (The multiplier '2' just proves to be convenient.) The residual deviance therefore measures how much fit to the data is lost, in likelihood terms, by modelling; inevitably some will be lost, but not much if the model is good.

D can also be written as $2 \log[L(\text{data})/L(\widehat{\theta})]$. This is twice the log of the *likelihood ratio* (the ratio of the likelihoods of the saturated and fitted models). Likelihood ratio is another important general concept in statistics.

Example 10.1 continued
In the case of $Y \sim N(\eta_x, 1)$, the likelihood function associated with the saturated model is simply

$$l(\text{data}) = C - \tfrac{1}{2} \sum_{i=1}^{n} (Y_i - Y_i)^2 = C.$$

Thus the residual deviance is

$$2[l(\text{data}) - l(\widehat{\beta})] = 2 \left\{ C - \left[C - \tfrac{1}{2} \sum_{i=1}^{n} \left(Y_i - \sum_{j=1}^{p+1} \widehat{\beta}_j x_{i,j} \right)^2 \right] \right\}$$

$$= \sum_{i=1}^{n} \left(Y_i - \sum_{j=1}^{p+1} \widehat{\beta}_j x_{i,j} \right)^2.$$

This is, now, precisely the *RSS* associated with the normal linear regression model. The saturated model has allowed us to calibrate things so that the residual deviance equals the *RSS* in the (unit variance) normal model. ∎

Let us look again at some Summary of analysis tables featured earlier. In the general GLM framework, these are usually called *analysis of deviance* tables. Two of these tables were already compared (in a less general context) in Example 9.2, but it is useful to repeat the tables here. These two tables are for the normal linear regression of loss on hardness for the rubber data and the logistic regression of survival on larea for the burns data. In addition, the table for the Poisson regression of crashes on lcover for the crashes data is repeated from Solution 10.2. The numerical values of the entries need not concern us now.

Normal regression for the rubber **data**

```
***** Regression Analysis *****

Response variate: loss
     Fitted terms: Constant, hardness


*** Summary of analysis ***

                 d.f.          s.s.          m.s.      v.r.  F pr.
Regression        1         122455.      122455.      33.43  <.001
Residual         28         102556.        3663.
Total            29         225011.        7759.
```

Logistic regression for the burns **data**

```
***** Regression Analysis *****

Response variate: survival
  Binomial totals: 1
     Distribution: Binomial
    Link function: Logit
     Fitted terms: Constant, larea


*** Summary of analysis ***

                                    mean   deviance approx
                 d.f.    deviance   deviance    ratio chi pr
Regression        1        188.4   188.4147   188.41  <.001
Residual        433        337.0     0.7782
Total           434        525.4     1.2106
* MESSAGE: ratios are based on dispersion parameter with value 1
```

Poisson regression for the crashes **data**

```
***** Regression Analysis *****

Response variate: crashes
     Distribution: Poisson
    Link function: Log
     Fitted terms: Constant, lcover
```

```
*** Summary of analysis ***

                                            mean   deviance approx
                    d.f.      deviance    deviance   ratio  chi pr
Regression            1         6.614      6.6141    6.61    0.010
Residual             15         8.681      0.5787
Total                16        15.295      0.9560
* MESSAGE: ratios are based on dispersion parameter with value 1
```

The degrees of freedom for the three tables are derived from the same considerations, and take the same values, whichever version of the GLM, including the normal, is being considered: the d.f.(Total) is $n - 1$, the d.f.(Residual) is $n - p - 1$, and the d.f.(Regression) is p. In each of these tables there is just one explanatory variable other than the constant term, so $p = 1$. What changes are the formulae for the deviances and, to some extent, what gets done with them.

The residual deviance, $2[l(\text{data}) - l(\widehat{\theta})]$, appears under the deviance heading in the Residual row in the logistic and Poisson cases, and is equivalent to the s.s.(Residual) in the normal case. In fact, the residual deviance is exactly the s.s.(Residual) in the normal case when $\sigma^2 = 1$. The total deviance is a similar quantity, comparing the fit of the saturated model not to the model of interest but to the other extreme – i.e. to the simplest possible version of the linear model, which disregards any dependence on values of the explanatory variable(s) by modelling the mean of Y by a single constant estimated by the overall mean, $\overline{Y}$, of the responses. The log-likelihood resulting from this oversimplified model is $l(\overline{Y})$, which will, in general, be a large negative number (i.e. the likelihood is small). The corresponding total deviance is $2[l(\text{data}) - l(\overline{Y})]$. By subtracting the residual deviance from the total deviance, the deviance due to the regression is $2[l(\widehat{\theta}) - l(\overline{Y})]$. That is, the regression deviance measures how much better is the model taking into account the explanatory variables than the model ignoring the explanatory variables; again, this measurement is made in terms of log-likelihood.

Example 10.1 continued
The same argument as in our previous visit to the case of $Y \sim N(\eta_x, 1)$, except with $\overline{Y}$ replacing $\sum_{j=1}^{p+1} \widehat{\beta}_j x_{i,j}$, shows that the total deviance reduces to

$$\sum_{i=1}^{n}(Y_i - \overline{Y})^2.$$

This is, as you should have expected, equal to the total sum of squares (*TSS*). (Again, $\sigma^2 \neq 1$ would add a scaling complication.) The regression deviance, which equals the total deviance minus the residual deviance, equals *TSS* – *RSS* in the (unit variance) normal case and so the regression deviance is the explained sum of squares, *ESS*, in the normal model. All three summary tables are therefore produced on the basis of a unified formulation, namely deviance. ∎

Quite generally, deviances and differences between deviances play the role that sums of squares and differences between sums of squares play in the normal linear regression (including ANOVA) models. It will become clear how this works in practice in Section 10.5

and following chapters. (There are some formulae for certain non-normal deviances in Chapter 11.) But, for the moment, a few more words at a more abstract level.

Generally, the most important use of the difference between deviances is in the comparison of the fits of two models where the explanatory variables in one of the models, Model B say, are a subset of the explanatory variables in the other, Model A. In this situation, one says that Model B is *nested* within Model A. One example is the linear regression situation in Example 10.1, in which Model A is the full linear regression model and Model B is the constant model, which is Model A with $\beta_i = 0, i = 2, 3, \ldots, p+1$. Other examples might be: A, quadratic (in x) regression, versus B, linear regression (in which the coefficient of the x^2 term is zero); or A, regression on $p + 1$ variables, versus B, regression on p of those variables (the coefficient of the omitted variable being zero).

In general, suppose Models A and B model the same set of observations of a response variable, y, with the same response distribution f and the same link function g. Model A, however, includes say p_1 explanatory variables (possibly derived ones) additional to those in Model B. The more complicated model, Model A, must provide a better fit to the data than the less complicated model, Model B, because the extra parameters in Model A give the investigator scope for improving the fit over and above that of Model B. The question is how much better, and is that improvement in fit large enough to offset the increased complexity of the model? (Model simplicity is preferred both for interpretation and prediction reasons; see Section 6.2.)

The effect of the reduction from Model A to Model B is measured by the *deviance difference*, the difference between the residual deviances D of Model B (the larger deviance of the two) and Model A:

$$D(B) - D(A) = 2[l(\text{data}) - l(B)] - 2[l(\text{data}) - l(A)]$$
$$= 2[l(A) - l(B)].$$

But this is exactly the same as the difference between the deviances due to the regression for the two models. In fact, it is

$$2[l(A) - l(\overline{Y})] - 2[l(B) - l(\overline{Y})] = 2[l(A) - l(B)].$$

In practice, in this book, such deviance differences are generally calculated in this latter way, using the regression deviances.

The degrees of freedom associated with the comparison between Model A and Model B is equal to p_1, the number of parameters discarded in reducing from Model A to Model B. Approximately, under the null hypothesis that the extra parameters are indeed zero, and hence that Model B is adequate, the deviance difference has a χ^2 distribution on p_1 degrees of freedom. Having such a distributional result allows us to judge when the deviance difference is large enough that we should prefer Model A with its larger number of parameters to Model B. Nifty use of the **Change Model** button can make GENSTAT give you the required information (including the associated SP) in a straightforward way.

In the binary (and Poisson) regression contexts GENSTAT provides further information pertinent to this same question (the 't statistics'). However, it turns out that the χ^2 approximation to the deviance difference is generally the most reliable approximate distributional result, and hence we shall virtually always be using deviance differences from here on to make comparisons between models.

You will have the chance to get to grips with how this works in practice very shortly, in the examples of Section 10.5.

You may be anxious that, in the case of the normal response distribution, the deviance difference, or equivalently the difference between residual sums of squares, was not what you used to compare the fits of two models, one nested within the other. The reason is that the normal distribution, unlike the Bernoulli or Poisson, has that extra variance parameter σ^2 distinct from its mean parameter μ. Any judgement about the magnitude of the difference in RSSs must take into account the magnitude of this variance; if σ^2 is large, then we would require a larger difference in RSSs before adopting the more complicated model than we would if σ^2 were small. The normal variance is also, in practice, unknown and needs to be estimated from the data. It is the estimation of σ^2 which leads to the role of ratios of mean squares and of F distributions rather than χ^2 distributions in normal linear regression (and ANOVA). (See Chapters 4, 5, 6, 7 and 8!) For GLMs in which there is no separate 'variance' or 'dispersion' parameter (such as logistic or Poisson regression), the deviance difference itself is all that need be considered, and compared with the appropriate χ^2 distribution.

Finally, let us return to our examination of the Summary of analysis tables. As discussed above, model comparisons in the logistic and Poisson cases do not require use of the deviance equivalents of mean squares and variance ratios; so you can, for now, safely ignore the mean deviance and deviance ratio columns of the analysis of deviance tables. One reason GENSTAT includes the mean deviance and deviance ratio columns is for consistency with the normal case. Another reason is that these extra columns are used in some non-normal GLMs where there *is* a separate dispersion parameter to be estimated. See Exercise 10.9 and Section 13.4.

10.4   A short history of GLMs

This section is a very short history of the generalized linear model, short partly because GLMs have a rather short history, at least relative to many of the other big ideas of statistics that you have met. While, for example, the normal distribution can be traced back to the eighteenth century, with development and use throughout the nineteenth century, and simple linear regression to developments in the last decades of the nineteenth century, the term 'generalized linear model' first appears in a landmark paper by J. A. Nelder and R. W. M. Wedderburn published in 1972 ('Generalized linear models', *Journal of the Royal Statistical Society*, Series A, **135**, 370–384).

While Nelder and Wedderburn first identified and unified the class of GLMs, they were not working oblivious to other developments in the statistical literature; as with most other scientific breakthroughs, the seminal ideas evolved partly by building on ideas that were already around. Some of these could be found (if you knew what you were looking for!) in the much earlier works of R. A. Fisher. In particular, there had already been quite a lot of interest in models for binary regression (early contributors included Fisher and J. Berkson); the books of D. J. Finney published in 1947 (*Probit Analysis*, Cambridge, Cambridge University Press) and D. R. Cox published in 1970 (*The Analysis of Binary Data*, London, Chapman and Hall) were influential. There were many important contributors to the development of GLM technology after the Nelder and Wedderburn paper, but the methodology was especially popularized by the comprehensive account given by

P. McCullagh and J. A. Nelder's book *Generalized Linear Models* (London, Chapman and Hall, 1st edition 1983, 2nd edition 1989).

Nowadays, GLMs are used almost as much as normal linear models. By the end of the book, you will have seen the vast array of statistical applications in which the data and questions are of a form amenable to analysis by GLMs. So why are GLMs such a relatively recent invention? The answer lies, as with many other statistical techniques, in the development of fast and cheap computer power. Even though the elegant theory of GLMs (barely touched on above) could have been developed without the computer, it would have remained a theoretical curiosity had the ability to implement the ideas not become available; instead, it was the arrival of computing power that gave the stimulus for GLMs, and hence for more realistic models for a variety of situations, to be developed. The normal linear model, available from much earlier, was not held back to such a degree by lack of computers. Because the analysis reduces to working with (lots of) sums of squares, it is possible, if often extremely tedious, to do the necessary calculations by hand. If you read a book on linear modelling from, say, 30 or more years ago, you will find much ingenuity going into organizing such calculations to make them less time consuming. For the normal linear model, the computer takes away the drudgery, making analyses quicker, and thus transferring effort away from calculation towards modelling and interpretation. For the GLM, the computer actually makes possible, on a daily and routine basis, analyses that, if not quite impossible, would be prohibitively expensive in terms of time and effort without, and does them just as quickly and routinely as for normal linear models. This is because the fitting of generalized linear models requires iterative computations.

In addition, starting in the 1970s and in conjunction with the theoretical development of GLMs, special-purpose software was written and made available for use in practice. This first GLM software was – and still is – GLIM, a package produced by Professor Nelder and colleagues. (GLIM = *G*eneralized *L*inear *I*nteractive *M*odelling; the term 'interactive' is not used in quite the modern sense.) By the time of writing, the popularity of GLMs is such that they are also provided in many other more general-purpose statistical software packages, of which GENSTAT – standing simply, as you might have guessed, for *GEN*eral *STAT*istics package – is one. (GENSTAT was in fact one of the first other packages to incorporate GLMs.)

10.5 Some more GLM applications

In this subsection, four different examples of GLM applications, each illustrating different aspects of the general methodology, will be investigated.

10.5.1 Mixing insecticides

The data presented in full in Table 10.3 are the results of an experiment concerning two insecticides, labelled A and B, and their effect on the tobacco budworm *Heliothis virescens*. There were 20 groups of budworms, intended to be of size 30 insects each, but in fact with just 29 insects in three of the groups. Each group of insects was exposed to one of five

Table 10.3

Mixture	Total dose (ppm)	Number of dead insects	Number of insects tested
B	30	26	30
B	15	19	30
B	7.5	7	30
B	3.75	5	30
A25:B75	26	23	30
A25:B75	13	11	30
A25:B75	6.5	3	30
A25:B75	3.25	0	30
A50:B50	26	15	30
A50:B50	13	5	30
A50:B50	6.5	4	29
A50:B50	3.25	0	29
A75:B25	26	20	30
A75:B25	13	13	30
A75:B25	6.5	6	29
A75:B25	3.25	0	30
A	30	23	30
A	15	21	30
A	7.5	13	30
A	3.75	5	30

Source: Giltinan, D., Capizzi, T. P. and Malan, H. (1988)
'Diagnostic tests for similar action of two compounds',
Applied Statistics, **37**, 39–50.
Dataset name: `insect`.

mixtures of insecticides A and B, at one of four total dose levels (measured in parts per million, ppm). The five mixtures were each of A and B alone, each of the 3 : 1 mixtures of A and B, and the fifty–fifty mixture of two. The numbers of insects in each group that were dead 96 hours after treatment were recorded. Aside from wishing to know whether A or B was more effective, interest centred on the joint action of the insecticides when combined: did combinations of the two work better than the single treatments alone or did perhaps combining the two make for less effective treatment? (Treatments are said to be 'synergistic' when they work better in combination than alone, and 'antagonistic' when they work worse in combination than alone.)

What is the response variable here? Well, there are two columns of the table that are relevant. These are both the number of dead insects, ndead, and the number of insects tested, nins. In a sense, it is the proportion of dead insects (ndead/nins) that is the response variable of real interest, although the numbers of insects tested remains relevant. (Here, nins is always 30 or 29. In other cases, there can be a much greater variety in numbers tested; see Section 10.5.2.)

The situation is, of course, just like that in Exercise 9.2. But in Chapter 9 we did not further consider such data, preferring to concentrate on the binary regression case in which the size of each group tested is one.

Exercise 10.4

In what way could you manipulate the `insect` data to make it appropriate for analysis by binary regression? (Do not attempt to perform such a manipulation!)

Proportions dead are, however, naturally modelled by the binomial distribution (see Section 2.5.2). That is, the number dead, say Y, out of the number tested, say n, can be modelled as following a $B(n, p)$ distribution, where p is the (unknown) probability of a random insect succumbing to the treatment and n is known. Now, in general, n varies. But more importantly, so might p. For instance, one might assume that p is near one for a high dose of an effective insecticide, but that it is rather smaller for a much lower dose or for a high dose of a less effective insecticide. What this means is that the binomial parameter p should be allowed to be different for different values of the explanatory variables.

Clearly, we can investigate such *binomial regression* in the GLM framework too. It should be clear to you that what we are arguing for is a model of the form

$$Y_i \sim B(n_i, p_x),$$

where each Y_i is independent, given $x_{i,1}, x_{i,2}, \dots, x_{i,p+1}$, where

$$g(p_x) = \sum_{j=1}^{p+1} \beta_j x_{i,j}.$$

It remains to specify the link function g. But p_x is a probability, lying between 0 and 1; just as in the binary case, a natural link function (with good interpretability) turns out to be the logistic function $g(p) = \log(p/(1 - p))$. In fact, the logit function is the canonical link function for the binomial distribution. (You will thus be relieved to hear that there is no need for the laborious conversion of binomial regression problems to binary regression problems. Indeed, binary regression is the special case of binomial regression where $n_i = 1$ for $i = 1, 2, \dots, n$.)

Exercise 10.5
Load the insect data into GENSTAT.

You may have noticed that the levels of **dose** increase by factors of two within each set of four groups of insects getting the same mixture of A and B. In such a case, it is natural, and common practice, to take logs of dose levels so that the doses are uniformly spaced out on the log scale. Do this now to obtain the new variable **ldose**; we shall employ **ldose** rather than **dose** throughout the exercise.

(a) Use the **Input Log**, in interactive mode, and CALC to obtain an additional variable **prop** = ndead/nins, the proportions of insects that perished. Use this to make a scatterplot of **prop** against **ldose**, with **mix** as the **Groups** variable. Remember that GENSTAT has attached factor levels 1, 2, ... , 5 to the **mix** factor in the order that the factor labels appear in the spreadsheet; so, mixture B is factor level 1, mixture A25:B75 is factor level 2, and so on. GENSTAT identifies factor levels by colours on its plot, listing the colours in order of factor levels, with level 1 at the top. What do you see?

(b) Perform a binomial regression of the insect data, using **mix** and **ldose** as the explanatory variables. The procedure for doing this in GENSTAT is almost exactly the same as for logistic regression. The only difference is that instead of writing 1 in the **Number(s) of Subjects** field, put in **nins**. Explicitly, choose the **Stats|Regression Analysis|Generalized Linear** menu item. In the **Generalized Linear Models** dialogue box, change the **Analysis** field to **Modelling of binomial proportions**. (e.g. by logits), fill in the Number(s) of Subjects field as nins, the Numbers of Successes field as ndead, the Model to be Fitted field as ldose,mix, and keep the Transformation (link) field as Logit. (Notice that prop is not needed.) Then click on the OK button.

   (i) What is the outcome of the test that there is no regression effect?

   (ii) Look at the t statistics. Do you think there is scope for further investigation of the mix factor?

   (iii) Make residual plots and comment on the fit of the model.

(c) To manipulate the individual binary explanatory variables that GENSTAT has created from the factor mix, you need explicitly to create such binary explanatory variables. These binary variables are indicator variables; see Chapter 6. Make the Input Log active and enter interactive mode. Type the following.

```
CALC m13=mix.EQ.2
PRINT m13
```

What has this produced? In a similar way (and by doing each individually), produce variables m22, m31 and ma corresponding to mixtures A50:B50 (which corresponds to factor level 3), A75:B25 (level 4) and A (level 5) respectively. Why is it not necessary to produce a variable corresponding to B?

(d) Use the Generalized Linear Models dialogue box to perform a binomial regression of the response variable on ldose, m13, m22, m31 and ma. How does this correspond to the analysis in part (b)?

Starting from this model as a full model, perform a stepwise regression and comment on the model obtained. Stepwise regression works just as for normal and binary regression. By comparing the respective values of deviance(Regression), check that a reduction from the full model to the model obtained by stepwise regression is justified.

(e) Looking at the Estimates of parameters part of the GENSTAT output for the regression on ldose, m13, m22 and m31 only, comment on the effects of mixed treatments in general. Does any further simplification of the model suggest itself to you?

(f) Are the regression coefficients β_{13} of m13 and β_{31} of m31 equal? If β_{13} were equal to β_{31}, then we could treat m13 and m31 as if they were the same treatment. Therefore, to test $H_0 : \beta_{13} = \beta_{31}$, we can compare the current model (via its regression deviance) with the reduced model obtained by regressing on ldose, m22 and the combined variable m1331, obtained by using the following command.

```
CALC m1331=m13+m31
```

Explain why this CALC command has the right effect. What is the result of the test? Compare the reduced model with the further reduced model in which m22 is also set equal to m1331. What do you conclude?

(g) In qualitative terms, describe the model you have finished up with, and comment on the action of insecticides A and B.

10.5.2 Toxoplasmosis and rainfall

The data stored in toxo.gsh comprise one explanatory variable and two columns describing the response variable. (Source: Efron, B. (1978) 'Regression and ANOVA with zero–one data: measures of residual variation', *Journal of the American Statistical Association*, **73**, 113–21. Dataset name: toxo.) The 34 datapoints correspond to cities in El Salvador. The explanatory variable is rainfall (in mm). The final column contains the numbers of children in each city that were tested for the disease toxoplasmosis (ntest), and the numbers in the middle column show how many of these people tested positive (npos). Interest lay in whether there was any relationship between the amount of rainfall in a city and the incidence of toxoplasmosis.

Exercise 10.6

(a) Load the toxo data into the spreadsheet. Is there anything noteworthy about the numbers in the ntest column?

(b) Update the server. Make a new variable containing the proportions of individuals testing positive for toxoplasmosis. Draw a scatterplot of these proportions against rainfall. What can you see? Does the answer to part (a) impinge on your thoughts?

(c) Fit a binomial GLM to these data following the steps in Exercise 10.5. What is the SP associated with the regression deviance? What do you conclude about the model?

(d) Looking at the output again, observe the warnings about large standardized residuals. (If you like, you could look at a plot of standardized residuals using the **Further Output** button.) What do these suggest about the fit of the model?

(e) Return to the **Generalized Linear Models** dialogue box. Change the **Model to be Fitted** field to pol(rain;3). This tells GENSTAT to fit a cubic model (i.e. a third-degree polynomial model) to the data (on the logistic scale). Test the regression deviance to see if a hypothesis of all zero regression coefficients is tenable in this case. What do you conclude about the fit of the model now?

(f) Ignoring the standardized residuals for the moment, let us pursue the question of which polynomial model is most appropriate. We have discarded the logistic-linear model in favour of a logistic-cubic, but what of a logistic-quadratic? Fit the quadratic model and then a quartic (fourth-degree) model. By subtractions, obtain a table of the extra deviance due to the addition of each extra polynomial term in succession up to the quartic one. Each additional term is on 1 d.f. and each can be tested against the $\chi^2(1)$ distribution. Which degree of polynomial is suggested as the best model by this analysis?

(g) Return to the fitting of the logistic-cubic model as in part (e). Fit this model again. Use the **Fitted Model** option in the **Further Output** dialogue box, and fill in rain as the **Explanatory Variable**, to obtain a diagram on which the logistic-cubic curve is plotted along with the data. Do you think this is a good fit to the data? Does any other curve strike you as being potentially better?

The final lesson of Exercise 10.6 is that, even if we can locate the best fitting model within the class being entertained, perhaps there's something else wrong with the model.

An obvious candidate is the binomial distribution assumption itself. In fact, it has been suggested that these data show a greater degree of variability than is consistent with a binomial model, even one with a mean varying with rainfall. Such a phenomenon, called *overdispersion*, actually happens a lot in practice with data for which Poisson or binomial response distributions are the obvious first stab. However, overdispersion will not be dealt with further in this example. (One way of approaching overdispersion is to introduce an extra variance parameter, rather like σ^2 in the normal linear regression model; an alternative is to trace the cause of the overdispersion and amend the model accordingly. An example with overdispersion is analysed in Section 13.4.) Interestingly, a researcher who has investigated the toxoplasmosis data taking overdispersion into account concluded that there was really nothing usefully modellable here: 'the suitability of rainfall as a predictor of toxoplasmosis incidence is therefore open to question'. This provides a salutary lesson, indeed, to prevent us getting too caught up in expecting useful statistical models of absolutely everything!

10.5.3   Survival of leukaemia patients

The next two examples will illustrate the great generality of the GLM framework by employing yet different response distributions.

The exponential distribution is a simple distribution for positive continuous random variables that is often a useful model for 'survival times', i.e. times until some event occurs. Examples include times until death or until relapse, or perhaps indeed until remission, where human patients suffering from disease are concerned; or lifetimes of mechanical or electrical components, e.g. computer chips, until they fail and need to be replaced. The p.d.f. of the exponential distribution is

$$f(x) = \lambda e^{-\lambda x}, \quad x \ge 0.$$

See Figure 10.2. Notice that the density is a decreasing function, and it has what is often called a 'long tail', allowing occasional large values of x to occur. The parameter $\lambda > 0$ is an event rate, and is related to the mean μ of the distribution by

$$\mu = E(X) = \frac{1}{\lambda}.$$

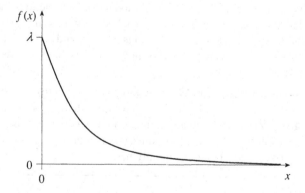

Fig. 10.2   The exponential density function.

Also,

$$V(X) = \frac{1}{\lambda^2}.$$

Exercise 10.7

The exponential distribution is indeed a member of the exponential family of distributions mentioned in Section 10.2. The canonical link function (for μ) is the reciprocal $g(\mu) = 1/\mu$. What does this imply for the relationship between the parameter λ and the linear predictor? Thinking about the allowed range of values for μ, can you envisage any difficulty with the use of this link function? If so, is this necessarily a total bar to its use?

Exercise 10.8

The data in leuksurv.gsh concern the time to death in weeks (time) of 33 patients suffering from acute myelogeneous leukaemia (under the conditions that held sway at the time of the study in the early 1960s). (Source: Feigl, P. and Zelen, M. (1965) 'Estimation of exponential survival probabilities with concomitant information', *Biometrics*, **21**, 826–38. Dataset name: leuksurv.) There are two explanatory variables. One is the white blood count (wbc). The other is a test result for a morphological characteristic of white cells, referred to as 'the AG factor' (ag), made at the time of diagnosis; the factor is either positive or negative.

(a) Load the leuksurv data into GENSTAT. Invoke Graphics|Point Plot and put time as the Y Coordinates, wbc as the X Coordinates and ag as the Groups. Click on OK. Why would a transformation of wbc help? Make the standard transformation of wbc (to give lwbc) and make a new scatterplot with lwbc rather than wbc on the x-axis. Comment on what you see.

(b) As has been said, survival times are often modelled using exponential distributions. Let us therefore analyse these data using an exponential response distribution. Let us also stick with the canonical link function, the reciprocal. (Researchers have considered other link functions for these data but have not obtained major improvements over the use of the canonical link function.) Select the Stats|Regression Analysis|Generalized Linear menu item. Ensure the Analysis field contains General Model. Fill in the Response Variate field as time, the Model to be Fitted field as ag*lwbc, and the distribution field as Gamma (ensuring the Link Function field says Canonical). (Remember that ag*lwbc is an abbreviation for ag+lwbc+ag.lwbc.)

The response distribution is, so far, designated as 'gamma' and not the exponential distribution that you would have expected. In fact, the gamma distribution is a family of distributions for positive continuous random variables that has the exponential distribution as one of its special cases. (You need not concern yourself with details of what the gamma distribution is or of its relationship to the exponential. χ^2 distributions are in fact also special cases of the gamma distribution.) To specify that you want the exponential and not the general gamma distribution, click on the Options button (in the Generalized Linear Models dialogue box). In the Generalized

Linear Model Options dialogue box that appears, make sure that **Fix** is selected under **Dispersion Parameter** and that the **Value** field contains 1. Click on **OK** here and in the other dialogue box.

Just as for other GLMs, a test for whether all the regression coefficients are zero is available by comparing the regression deviance against the χ^2 distribution on d.f.(Regression) degrees of freedom. What is the result of such a test in this case?

(c) By refitting the data with a model lacking the **ag.lwbc** term, test whether the presence of the **ag.lwbc** term is necessary.

(d) By refitting the data with a model lacking not only the **ag.lwbc** term but also each of **ag** and **lwbc** in turn (but not lacking both), decide what seems to be the best model for these data. After running the best fitting model again if necessary, use the **Fitted Model** button in the **Regression Further Output** dialogue box to obtain a graph of the best fitted model.

(e) Write down the finally fitted model in the form of two equations, one for the response mean of individuals for each of the two groups defined by the value of **ag**. Are there any regression coefficients in common? Would there have been any regression coefficients in common in the two equations had **ag** not been in the model? Would there have been any regression coefficients in common in the two equations had **lwbc**, **ag** and **ag.lwbc** all been in the model?

Hint   You might find it helpful to refer back to Example 6.4.

10.5.4   Janka hardness revisited

In Exercise 4.25 of Chapter 4, a dataset, hardness, comprising measurements of hardness and density of 36 Australian eucalypt hardwoods, was analysed. The interest in this dataset was to produce a model that could predict the response variable hardness from values of the single explanatory variable density. In Solution 4.25, an adequate model was obtained using normal linear regression after transformation of both hardness and density by taking logs.

Let us consider an alternative approach. Instead of transforming to normality, it might be at least as good to employ a generalized linear model directly for hardness instead. (An advantage of the latter is that everything is done on the meaningful original scale on which the data were measured.) Let's try this, and compare back with the transformation approach later. The main property of the scatterplot of hardness against density that made a normal linear regression on the original hardness/density scale inappropriate is that the variance increases with the mean. In fact, it seems to do so in a rather smooth way. (See the first figure in Solution 4.25, or you can load hardness into GENSTAT and obtain the scatterplot via **Graphics|Point Plot**.) It happens that the gamma distribution, mentioned in Section 10.5.3, has the property that the variance is the square of the mean, so this could perhaps provide a suitable model. Also, the responses can indeed only be positive. Let us then suppose that a gamma response distribution might be appropriate and see where the GLM path leads us.

Exercise 10.9

Load the hardness data into GENSTAT, if you have not done so already.

(a) Use the Stats|Regression Analysis|Generalized Linear menu item to obtain the Generalized Linear Models dialogue box. Choose **General Model** in the Analysis field, enter hardness as **Response Variate** and density as **Model to be Fitted**, and change the Distribution to **Gamma** and the Link Function to **Identity**. Click on the **Options** button and, under **Dispersion Parameter**, select **Estimate** rather than **Fix**. (This tells GENSTAT to use a general gamma distribution and not to specialize (as in Section 10.5.3) to the exponential distribution.) Click on the **OK** buttons in both dialogue boxes.

What model have you just fitted? Click on the **Further Output** button, and then on the **Fitted Model** button. In the resulting dialogue box, enter density as **Explanatory Variable** and leave the **Grouping Factor** field blank, then click on **OK**. This gives you a picture of the fitted model superimposed on the data. (You will find it useful to keep a copy of this diagram.) What do you think of this fit? How might you go about trying to improve the model?

(b) We shall continue with the gamma response distribution and the identity link function. Is there any reason to take care over the form of linear predictor allowed?

(c) Long experience of researchers using gamma regression has suggested that it is often useful to consider reciprocals of an explanatory variable in such a context. (See the book by McCullagh and Nelder mentioned in Section 10.4.) To this end, create the variate rden via the transformation rden = 1/density. Fit the gamma model, with identity link function, including *both* density and rden by using the **Change Model** button to add rden to the model previously fitted.

A new feature arises when we try to compare the fit of this model with the model of part (a). In the generalized linear models you have fitted so far, there has not been a dispersion parameter to estimate. In gamma regression, there is such a parameter, which plays a similar role to the response variance σ^2 in normal linear regression. (Indeed, this parameter is also similar to that mentioned in the context of overdispersed Poisson or binomial data at the end of Section 10.5.2.) Because of this, the usual approach of comparing deviance differences with a χ^2 distribution is not appropriate. There are two alternative approaches, both of which involve essentially doing the same thing in the GLM output that you have done previously with normal linear regression output (except that deviances replace sums of squares). One approach is to look at the t statistic for the extra explanatory variable (given in the Estimates of parameters) and compare it with a t distribution with the appropriate d.f. (i.e. d.f. (Residual)) for the model containing the extra variable. This approach is not appropriate, however, if the two models you are comparing differ in terms of more than one explanatory variable. In that case, the alternative approach is to calculate an F statistic for the extra explanatory variables by using the regression and residual deviances in exactly the same way as you did with sums of squares in normal linear regression models. The F statistic is compared with an F distribution with the usual d.f. The main snag with both these approaches is that the t and F distributions involved are approximations, and often not very good ones, so the results of such tests should be treated with circumspection.

GENSTAT gives the results of both approaches, but we shall concentrate only on the F test. This is given in the Change line beneath the Summary of analysis table. By looking at this, decide whether the model containing both of these explanatory variables is better than the model containing just **density**. By using the **Change Model** button appropriately, test also whether the model containing both explanatory variables is better than the model containing just **rden**.

(d) Obtain a picture of the fitted model when both **density** and **rden** are in the model (follow the steps in part (a), continuing to put **density** in the **Explanatory Variable** field). Compare this picture with the one you obtained in part (a). Which model do you prefer on these grounds?

(e) In Exercise 4.25, you fitted a normal linear regression model after transforming both hardness and density by taking logs; in this exercise you have fitted a gamma regression model with the dependence on the explanatory variable x being through a combination of x and $1/x$. It would be nice to be able to compare the fits of these two different models on a single picture. Here's one way to do this. First, rerun the gamma regression of **hardness** on **density** and **rden** if it wasn't the last regression that you ran. Then click on the **Save** button and, in the resulting dialogue box, click on **Fitted Values** and give the adjacent **In** field a name, **gamm** say. Click on **OK**. Next, run the normal linear regression of log(**hardness**) on log(**density**), and then follow the same procedure of using the **Save** button to save the fitted values from this model in another variate, **norm** say.

What we are going to do now is to plot each of the sets of fitted values on the same picture, join them up with curves to get an impression of the two fitted models, and do this on a diagram also showing the original datapoints. This is a little fiddly in GENSTAT. In the **Input Log** window in interactive mode, enter the following.

```
CALC enorm=exp(norm)
APPEND [NEWVECTOR=y;GROUPS=g] gamm,enorm,hardness
APPEND [NEWVECTOR=x] density,density,density
PEN 1,2;SYMBOL=0;METHOD=line
PEN 3;SYMBOL=1;METHOD=point
DGRAPH [SCREEN=clear] y;x;PEN=NEWLEVELS(g;!(1 ... 3))
```

The first of these six commands puts the normal linear regression fitted values on the same scale as the others; this is because the normal model is working with log(**hardness**), so we have to take exp(fitted values) to get back to the original **hardness** scale. The two APPEND commands convert the three fitted-value variates (i.e. the gamma variates, the normal variates and the original datapoints) into a single long variate y, make a groups factor g which has three levels corresponding to the three variates that are being combined, and make three copies of density into the long variate x to stay compatible with y. The final three commands are based on what **Graphics|Point Plot** does, but with amendments to join the fitted values (but not the datapoints themselves) by curves.

Be careful to get the commas and semicolons right! If you make an error with the second line, you may have to delete g before trying again. To do this, choose the **Data|Display** menu item and, in the dialogue box that appears, find the row starting g and click on it, then click on the **Delete** button and then on **Cancel**. If you are using the student version of GENSTAT, you may also find you need to adjust the values under **Options|Student Limits**.

Look at the diagram you obtain and comment on the relative goodness of fit of the two models.

Exercise 10.9 demonstrates that apparently rather different models can fit a given dataset just as well as each other. This is fine for most purposes, but can be a problem if, for example, the two fitted models go on to give rather different predictions outside the range of the data; although such extrapolation is to be avoided if possible, sometimes it is inevitably what is of interest.

The similarity in Exercise 10.9 between the results from using a gamma distributed response variable and those from using a normally distributed response variable after a log transformation occurs often when the amount of residual variability is fairly small. The latter is the same as using what is defined to be a *lognormal* distribution on the responses without transformation. The similarity is because the gamma distribution and the lognormal distribution are rather similar for small amounts of variability.

That said, there remains a difference in the way the mean response depends on the explanatory variable in the fitted models of Exercise 10.9. The lognormal one depends on the log(**density**); the gamma one on **density** and 1/**density**. Interestingly, one can get virtually the same results by applying a lognormal model to **density** and 1/**density** – something that was unlikely to occur to us when considering the transformation approach alone in Exercise 4.25. (But gamma regression on log(**density**) is not so good.)

In general, GLM technology often gives an alternative to transformation technology, with the advantage that the GLM retains the original scales of the observations.

Summary of methodology

1 The generalized linear model, or GLM, is a statistical model in which the distribution of the response variable Y given values of the explanatory variables $x_1, x_2, \ldots, x_{p+1}$ has a (continuous) probability density function or (discrete) probability mass function, f, with mean μ depending on $x_1, x_2, \ldots, x_{p+1}$ through

$$g(\mu) = \eta_x = \sum_j \beta_j x_j$$

where η_x is the linear predictor and g is the link function. Normal, binary and Poisson linear models are the most widespread versions of the GLM.

2 The link function is often chosen to be the canonical link function, which is the identity function for normal responses, the logistic function for binary responses, and the log function for Poisson responses.

3 Inference for GLMs is based on the deviance. The deviance is twice the difference between maximized log-likelihoods under various different versions of the model.

To test hypotheses, the regression deviance, or an appropriate difference between regression deviances, is usually referred to the χ^2 distribution, with degrees of freedom calculated as for the normal linear regression model. An alternative procedure is used in GLMs where a dispersion parameter has to be estimated from the data.

4 Further specific examples of GLMs introduced in this chapter include the Poisson GLM, often appropriate for data where the response variable is a count, the binomial GLM, often appropriate for data where the response variable is a proportion, and the exponential and gamma GLMs, useful models for data where the response variable is continuous and non-negative.

Diagnostic checking

In the previous chapters, we have fitted models and then checked the assumptions of the models using residual plots. These plots allow us to identify failures of the model such as the variance not being constant, the fitted relationship being incorrect (e.g. non-linear) and whether the random variation is from a normal distribution. We can also identify observations that do not seem to agree with the fitted model. The residuals are examples of what are known as *diagnostics*, because they help us diagnose what might be wrong with a model or with particular datapoints.

In this chapter, we shall look at ways of identifying individual observations that have, or might have, distorted the model. Sections 11.1 and 11.2 deal with this idea, introducing two new diagnostics (leverage and the Cook statistic) in the context of the normal linear model. In addition, these new ideas and the use of residuals will be extended to generalized linear models in Section 11.3.

11.1 Leverage

Part of the output produced in Exercise 4.5 for the regression of abrasion loss on hardness for the data in rubber.gsh is reproduced below.

```
***** Regression Analysis *****

Response variate: loss
    Fitted terms: Constant, hardness


*** Summary of analysis ***
```

	d.f.	s.s.	m.s.	v.r.	F pr.
Regression	1	122455.	122455.	33.43	<.001
Residual	28	102556.	3663.		
Total	29	225011.	7759.		

```
Percentage variance accounted for 52.8
Standard error of observations is estimated to be 60.5
```

```
* MESSAGE: The following units have high leverage:
    Unit      Response      Leverage
     1         372.0         0.182
```

There is a message at the end of the output; so far in this book you have been asked to ignore such messages. It is a similar-looking sort of message to those generated when there are large residuals. It provides a warning that there might be something odd about this dataset. The oddity that GENSTAT has identified is that the first datapoint has a high *leverage*. This will probably not mean anything to you at the moment. We shall see what leverage is in this section.

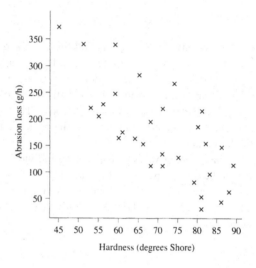

Fig. 11.1

There is a scatterplot of these data in Figure 11.1. This scatterplot originally appeared as Figure 6.1(a), and in Solution 4.1. The first datapoint (unit 1 in GENSTAT's terms) is the one at the far left, with a value of 45 for hardness. It has the smallest x (hardness) value (and, as it happens, the largest y (abrasion loss) value, though that turns out to be irrelevant as far as leverage is concerned). This point has a high leverage value because it is quite a long distance from the rest of the data. The relevant distance for leverage is measured *in terms of the explanatory variable*; in other words, the high leverage is because 45 is an extreme value for hardness. This probably does not convince you that leverage is telling us anything useful!

The relevance of a point with high leverage is that it *might* have a major influence on the regression line. Consider Figure 11.2. Here both plots show fitted models for the regression of abrasion loss on hardness based on the rubber dataset. Purely for illustration, 1000 has been added to the abrasion loss value for one datapoint in each of the scatterplots in Figure 11.2. In scatterplot (a), the point with high leverage (unit 1 with a hardness of 45) is the one which has had its abrasion loss value increased. The point with increased abrasion loss value in scatterplot (b) is unit 5 (hardness 71), which has low leverage, as we shall see later. In each case the fitted regression line for the amended dataset is shown, together with the original regression line fitted before changing the response.

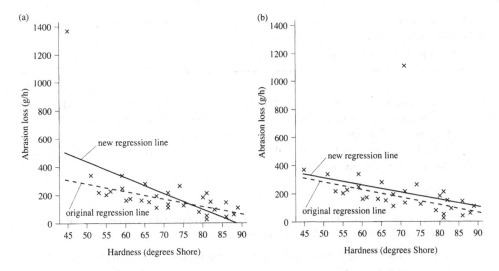

Fig. 11.2   Scatterplots showing the effect of changing the response for a point with high leverage and for a point with low leverage.

(If you want to confirm these results on your own computer, you can do it by loading rubber.gsh into the GENSTAT spreadsheet, typing in the amended response as appropriate, and updating the server. To produce the plots, click on the **Further Output** button from the **Linear Regression** dialogue box (after performing the regression of loss on hardness) and then click on the **Fitted Model** button in the dialogue box that pops up.)

The change made to the high-leverage point has a big effect on the fitted regression line; the fitted line is too steep. Most of the points corresponding to low values of hardness fall below the fitted line; for high hardness values, the points are mostly above the fitted line.

In the case where the change was made to the low-leverage point, the regression line looks much better. The slope appears to be close to the pattern in the bulk of the data. The only effect of adding 1000 to the low-leverage point is that the whole line is moved up slightly.

The general point that is illustrated in Figure 11.2 is that *changes to the response for a high-leverage point have a bigger effect on the estimates of regression coefficients (the $\widehat{\beta}_j s$) than do those for a low-leverage point.*

We use the regression coefficients to decide whether there is a linear relationship between the response variable and the explanatory variables, and to predict the responses for new values of the explanatory variables. Therefore, it would be worrying if a high-leverage point had a response that did not seem to fit the pattern of the rest of the data. The worry would be that the estimated regression coefficients might be very inaccurate; this would mean that our conclusions about which explanatory variables were related to the response variable could be misleading and that the predictions made using the model would be wrong.

Exercise 11.1
Look at Figure 11.1. Do you think the point with high leverage fits in with the pattern of the rest of the data or not?

We shall look at a more systematic way of assessing whether individual points fit the pattern of the rest of the data or not in Section 11.2. In the rest of this section, we shall look at other properties of leverage.

Exercise 11.2

Load the data in rubber.gsh into GENSTAT and fit a simple linear regression model using loss as the response variable and hardness as the explanatory variable. Do not close the Linear Regression dialogue box yet.

Store the leverages as lev by making the Input Log active, entering interactive mode and typing the following command.

    RKEEP leverages=lev

(a) Using Graphics|Point Plot, plot lev (on the y-axis) against hardness (on the x-axis). How is the leverage related to the explanatory variable?

(b) Now, plot lev (on the y-axis) against loss (on the x-axis). How is the leverage related to the response variable?

(c) In the Linear Regression dialogue box, click on the Further Output button, and in the resulting dialogue box click on Model Checking. In the Model Checking dialogue box, make sure that the Fitted Values and Leverage radio buttons are checked, and click on OK to produce a plot of leverages against fitted values. Obtain the Model Checking dialogue box again, and this time produce an index plot of the leverages. (This simply plots the leverages in order of the rows in the spreadsheet.) Comment on how your plots from this part relate to the plots produced in the other parts of this question.

Using the Model Checking dialogue box is probably the most common way in which you will look at plots of leverages. This dialogue box allows you to produce normal and half-normal probablity plots of leverages, as well as a histogram and a composite plot similar to that for residuals, but these plots are far from useful! Since Composite is the default Type of Graph in the Model Checking dialogue box, you will therefore always have to remember to change it to something more useful.

Exercise 11.2 illustrates the fact that leverage depends only on the values of the explanatory variable(s). Points with high leverage have extreme values for the explanatory variable(s). In fact, when there is only one explanatory variable, x, and there are n datapoints, the leverage h_i of the ith datapoint is given by the equation

$$h_i = \frac{1}{n} + \frac{(x_i - \overline{x})^2}{(n-1)s_x^2}$$

where $\overline{x}$ and s_x^2 are the mean and variance of the x values. This is a quadratic function of x_i, and it takes small values for points whose x value is close to the mean of the x_is and higher values for points whose x value is a long way from the mean. Importantly, the general idea of allocating high leverage values to points a long way away from the 'centre' of the pattern of explanatory variables makes sense in the case of more than one explanatory variable, though the actual formula used is a little complicated and will not be given in the book.

There is another situation where the formula for the leverage of a point is very straight-forward. This is when there is just one categorical explanatory variable, as in one-way ANOVA. In this case, if the number of observations at level j of the explanatory factor is n_j, the leverage for those points is simply $1/n_j$. This provides a rationale for why it's sensible for the group sizes to be equal (or at any rate not too different from each other) in one-way ANOVA, because otherwise the points in the small groups have high leverage.

We now turn to an example where there is more than one explanatory variable.

Exercise 11.3

Load the peru dataset into the GENSTAT spreadsheet. In Example 6.2, you saw that the first point in the dataset has a very high value for sbp. This point was excluded from the subsequent analysis. For now, we shall also exclude this point. Do so, using the Spread|Restrictions|Add Selected Rows menu item. Update the server.

The selected model in Example 6.2 had two explanatory variables, years and weight.

(a) Plot years against weight (i.e. plot the two explanatory variables). Which points would you expect to have high leverage? (Pick out no more than three points.) Look in the spreadsheet to find out which rows these points correspond to.

(b) Now fit a multiple linear regression model to sbp, using years and weight as the explanatory variables. Which points does GENSTAT identify as having high leverage?

(c) We can try to find out whether a particular high-leverage point is having a major effect on the regression coefficients (the $\hat{\beta}_j$s). Amend the spreadsheet to exclude row 39 from the dataset and update the server. Now fit the multiple linear regression model again. Have the regression coefficients changed much? (The coefficients for the regression with and without row 39 should be in the Output window. If you accidentally lost the earlier set, you can find them in Example 6.2.)

Look at the values for the 'high' leverages for the two analyses (given in the messages about units with high leverage). Notice that omitting row 39 causes the leverages for other rows to change.

We can see from Exercise 11.3 that it is possible in models with two explanatory variables to find points with high leverage simply by looking at a scatterplot of those explanatory variables. Many problems will involve more than two explanatory variables. In these problems it is far easier to have the leverage values calculated by GENSTAT than to try to identify high-leverage points using graphs.

To end our look at leverage, let us consider briefly how GENSTAT decides which points to identify as having a high leverage.

It is possible to show that the sum of the leverages for all n points in the dataset is q, the number of regression coefficients in our model, including any intercept. (In earlier chapters we denoted by p the number of explanatory variables, so $q = p + 1$.) So, for the data in peru.gsh, we have been fitting a model with three coefficients: one for the constant and one each for years and weight, so the leverages of all the points add up to 3. Thus the mean leverage value is q/n. Further, the leverage of the ith observation lies between $1/n$ and $1/n_i$, where n_i is the number of observations having exactly the same set of values for all the explanatory variables as the ith point. In many observational studies, the pattern of explanatory variable values is unique for many observations, and thus the leverage can be as high as 1. A common rule of thumb, and one employed by GENSTAT, is to label a leverage as high if it exceeds twice the average leverage, i.e. if it is greater than $2q/n$. Note that this is merely a rule of thumb; there is no strong theoretical reason for its use. The rule is used because it seems to work reasonably in practice. Some people use $3q/n$ instead of $2q/n$.

In the rubber example, whose output is shown at the beginning of this section, we fitted the $n = 30$ observations with a linear model with a single explanatory variable. Taking the model intercept into account, for that model we have $q = 2$. Thus GENSTAT would declare any leverage to be large if it exceeded $2q/n = 4/30 = 0.133$. As the output indicated, the leverage for the first data point, 0.182, exceeded this threshold. But, as the plot in Solution 11.2(a) confirms, none of the other leverages exceeds the threshold.

11.2 The Cook statistic

Leverage gives us a way to identify points that have the *potential* to exert too big an influence on the fitted regression model, in terms of changing the coefficient estimates and through them the predictions of the model. *Cook's distance* is a way to identify points that actually *do* exert too big an influence.[1] Those points that do exert a very big influence are called *influential points*. In this section we shall investigate what Cook's distance is, and how it can be used. The discussion of how Cook's distance is defined is mathematically slightly complicated, particularly as GENSTAT actually calculates a modified form of it. The important thing to bear in mind is that the purpose of Cook's distance is to give a measure, for each point, of the change that would occur in the estimated regression coefficients if that point were left out of the analysis altogether. However, we start with a slightly different way of expressing Cook's distance, in terms of fitted values.

Suppose we want to work out Cook's distance for some particular point in the dataset, point i. Suppose further that $\widehat{Y}_j$ denotes the fitted value for point j ($j = 1, 2, \ldots, n$). These values are worked out using the estimated regression coefficients obtained from fitting a model including all n datapoints. We could also estimate the regression coefficients when point i is excluded from the dataset. These coefficients can be combined with the values of the explanatory variables for point j ($j = 1, 2, \ldots, n$) to obtain an alternative fitted value for point j. This alternative fitted value will be written as $\widehat{Y}_{j(i)}$.

Cook's (squared) distance for point i is then

$$D_i^2 = \frac{1}{qs^2} \sum_{j=1}^{n} (\widehat{Y}_j - \widehat{Y}_{j(i)})^2 \tag{11.1}$$

where q is the number of regression coefficients, as in Section 11.1, and s^2 is the usual estimate of the error variance (from the regression including all the points). The formula in (11.1) essentially measures the overall size of the change in the fitted values when point i is excluded from the regression. The inclusion of s^2 in the denominator has the consequence that the value of Cook's distance does not depend on the scaling of the responses – for

---

[1] Readers of a cynical mind might think the authors have now come clean and are explicitly cooking the analysis! Not so: these ideas are named after their originator, Professor R. D. Cook.

example, if all the responses were doubled, the value of D_i^2 would not change. The fitted values are calculated using the estimated regression coefficients, so the changes in fitted values tell us about changes in the estimated regression coefficients.

Omitting each point in turn from a dataset and performing a separate regression each time would be slow if there were lots of datapoints. Fortunately, there is a way to work out Cook's distance directly from things that GENSTAT would normally calculate anyway. It turns out that

$$D_i^2 = \frac{1}{q} r_i'^2 \frac{h_i}{1 - h_i} r \tag{11.2}$$

where r_i' is the standardized residual for point i and h_i is its leverage. This makes intuitive sense in terms of what we want Cook's distance to measure. We know that points with high leverage *can* have a big impact on the regression coefficients, but that in practice they will do so only if their responses are a long way from the pattern of responses for the rest of the points, in which case they are likely to have large standardized residuals. If a point has a standardized residual (r_i') that is large in absolute value, and a large leverage (h_i), then its values of $r_i'^2$ and of $h_i/(1 - h_i)$ will both be large too, so that it will then have a large value of D_i^2. On the other hand, a point with large leverage but a small standardized residual (so that its response is generally in line with those of the other points) will have a smaller Cook's distance, as will a point with a large standardized residual but a small leverage.

To make things slightly more complicated, what GENSTAT calls **Cook** is not Cook's distance, but the *modified Cook statistic*. (The text will often call this simply the *Cook statistic*. The unmodified version will never be calculated in the book.) The main modification in the modified Cook statistic is that the standardized residual is replaced by the *deletion residual*. The simple (unstandardized, or raw) residual for point i is just the difference between the response and the fitted value, $Y_i - \widehat{Y}_i$. To standardize this, we have to divide it by its estimated standard deviation, which is in fact $s\sqrt{1 - h_i}$. Thus the standardized (Pearson) residual is actually

$$r_i' = \frac{Y_i - \widehat{Y}_i}{s\sqrt{1 - h_i}}. \tag{11.3}$$

We shall write the (standardized) deletion residual as r_i^*. The deletion residual can be defined as what you get if you calculate the difference $Y_i - \widehat{Y}_{i(i)}$ between the response for point i and the fitted value for that point found when point i is excluded from the dataset, and then divide by an estimate of the standard deviation of this quantity. It turns out that it can be calculated without the value of $\widehat{Y}_{i(i)}$, using the formula

$$r_i^* = \frac{Y_i - \widehat{Y}_i}{s_{(i)}\sqrt{1 - h_i}} \tag{11.4}$$

where $s_{(i)}$ is the estimate of σ produced by fitting the model excluding point i. This is very similar to the formula for standardized residuals. The difference between (11.4) and (11.3) is that (11.4) is based on the estimate of σ from a regression in which point i has been excluded, or *deleted*.

You might think that (11.4) returns us to the problem of having to fit the regression model many times, but this is not so. An alternative, but not at all obvious, formula for r_i^* is

$$r_i^* = r_i' \sqrt{\frac{n - q - 1}{n - q - r_i'^2}},$$

so that the deletion residuals can easily be worked out from quantities GENSTAT already finds in a regression calculation.

The squared modified Cook statistic is

$$C_i^2 = \frac{n - q}{q} r_i^{*2} \frac{h_i}{1 - h_i}.$$

Apart from the replacement of the standardized residual r_i' by the deletion residual r_i^*, this formula is the same as (11.2) multiplied by $(n - q)$. The modified Cook statistic is

$$C_i = \sqrt{\frac{n - q}{q} r_i^{*2} \frac{h_i}{1 - h_i}} \tag{11.5}$$

and this is what GENSTAT produces as **Cook**. The multiplication by $(n - q)$ has the effect of putting the values of the Cook statistic on the same scale as the deletion residuals. If a point's leverage takes the mean leverage value of q/n, then $C_i^2 = r_i^{*2}$.

In summary, Cook's distance measures how much the fitted values change when point i is missed out of the regression. Cook's distance is calculated using the standardized residuals from the full regression. The modification is essentially to use deletion residuals instead of the standardized residuals. The modification is claimed to be an improvement, so GENSTAT uses it. The modified form still involves the residuals and the leverage for a point, as did the original form in (11.2), so it still picks out points which have had influence on the estimated regression coefficients in the same sort of way.

Exercise 11.4

Load the data in rubber.gsh into GENSTAT. Perform the simple linear regression of loss (response variable) on hardness (explanatory variable). Do not close the Linear Regression dialogue box yet.

(a) Click on the **Further Output** button in the Linear Regression dialogue box; then click on the **Model Checking** button. In the dialogue box that now appears, click on **Index**, then click on **Cook** and finally click on the **OK** button. Does the Cook statistic for the first datapoint (the one identified as having high leverage in the regression output at the beginning of Section 11.1) seem exceptionally large compared with the other Cook statistics? This is a subjective judgement, as there is no standard rule of thumb for when a Cook statistic is exceptionally large.

(b) Having observed a strange pattern (described in Solution 11.4(a)), we should go a little further in our investigation. Use GENSTAT to produce an index plot of the Pearson residuals. All you need do to produce the plot is follow the instructions in part (a), except that you should not click on **Cook**. Describe the structure of this index plot. Looking at the data in the spreadsheet, can you think of any reason for this structure?

(c) Produce an index plot of the deletion residuals. There is a button for choosing deletion residuals in the **Model Checking** dialogue box. How does it compare with the index plot of the Pearson residuals?

The diagrams produced in Exercise 11.4 indicate that the high leverage of point 1 does not result in that point having a major influence on the fitted line. Also, the pattern in the residuals and the Cook statistics is still present after removing point 1. There is something strange about these data; it seems likely that the experimental conditions changed during the data collection process. The only way to establish what had happened would be to discuss it with the experimenters, but we do not have that option.

Now, we shall return to the data in peru.gsh, which we looked at in Exercise 11.3.

Exercise 11.5

Load the data in peru.gsh into GENSTAT. (If this datafile is already open, use the Spread|Restrictions|Remove All menu item to unrestrict all of its rows.)

(a) First, we shall look at what happens if we do not exclude the first point from the analysis. Update the server and perform the multiple linear regression of **sbp** on the explanatory variables **years** and **weight**. Produce an index plot of the Cook statistics. (Use the same method as in Exercise 11.4.) Is the Cook statistic for point 1 large relative to the other Cook statistics?

(b) Now, exclude point 1 from the analysis, perform the multiple linear regression and produce a new index plot of the Cook statistics. Which points have high Cook statistics? Are these points the same as the points with high leverage?

We can see from Exercise 11.5 that a point can have high leverage without changing the fitted regression line much. We have also seen that a point which does not have high leverage can still have a big impact on the regression coefficients.

Having identified influential points, we are then left with the problem of what to do about them. For the peru data, there are (at least) two issues that you might want to discuss with the researchers (if this were possible).

- Is it reasonable to believe that, forty years after someone moves to lower altitude, that person's blood pressure keeps falling at the same rate as, say, five years after moving to lower altitude? Cases 38 and 39 have both been living at low altitude for forty years or more. The other cases have been living at lower altitude for less than thirty-five years, most for less than thirty. If it is not reasonable to believe that cases 38 and 39 have the same rate of decrease in blood pressure as the other cases, then they should not be used as they are outside the range of validity for the model.

- Observation 4 is not a high-leverage point, but is influential. It would be useful to ask the researchers to check the value for **sbp** for this person, by going back to the original records. (There ought to be original records, as it would be usual to measure the blood pressure, write down the reading and enter the readings into a computer later.) It would also be of interest to check precisely how long case 4 had been living at low altitude. The value for **years** is 1: does this mean that this person has been living at low altitude for over a year, or could it have been a much shorter period?

As we do not have access to the researchers, we would have to fall back on either presenting analyses with and without influential points or (if we feel strongly enough) omitting the influential points from the analysis and explaining why we do not think they are valid for inclusion. The fact that they are influential is *not* by itself a valid reason for excluding points.

We finish the section with a reanalysis of another dataset you have already met.

Exercise 11.6

Load the data in `crime.gsh` into GENSTAT. You analysed these data in Exercises 6.5 and 6.6.

(a) In Exercise 6.5, our final fitted model for these data regressed the response variable (**crime**) on five explanatory variables, **malyth**, **school**, **pol60**, **unemid** and **poor**. Explore the fit of this model further by fitting it again and then producing appropriate plots of residuals, leverages and Cook statistics. What do you conclude?

(b) You will have found in the first part of this exercise that a few datapoints have Cook statistics considerably in excess of most of the others. Can you explain why the two points with the highest Cook statistics (points 11 and 29) have such high values, in terms of their leverage and/or residuals?

(c) The high Cook statistics of points 11 and 29 mean that they have a considerable influence on the regression coefficients. Investigate how important this influence might be by excluding the point with the highest Cook statistic (point 29) and going again through the process of choosing a model by stepwise regression. Use the method of Exercise 6.5(c). Do you end up with the same model as in Exercise 6.5? Comment on the output in comparison with that obtained in Exercise 6.5.

11.3   Diagnostics for generalized linear models

So far we have concentrated on diagnostics for normal linear models. A fitted generalized linear model can also be checked to see if it is a valid and appropriate description of the data. The sort of questions we need to investigate are similar to those for normal linear models.

- Are the assumptions of the generalized linear model tenable?

- Is the model a good fit to the data?

- Are particular observations not well fitted by the model?

- Do some observations have undue influence on the model?

(For generalized linear models, the question of whether the model is a good fit to the data includes the question of whether the most appropriate link function has been used, though we shall not pursue this here.)

We shall look at diagnostic tools for generalized linear models in this section. These tools are generalizations of those that you have already seen. In the special case of a generalized linear model that happens to be a normal linear model, the diagnostics that are introduced in this section reduce to the ones you have seen already in the first two sections of this chapter (and earlier in the book). We deal first with residuals for generalized linear models, where the situation is rather more complicated than for normal linear models. Then we consider leverage and influence (in Section 11.3.2).

11.3.1   Residuals for generalized linear models

The first type of residuals that we used in this book were *simple* (raw) residuals, which are merely the difference between the observed and fitted values of the response variable. In Chapter 6, we started to use *(standardized) Pearson* residuals instead of the simple residuals, so that each residual had a standard error of 1. The third type that we have seen, but not used very much, is the *deletion* residual, which was introduced in Section 11.2. There are generalized linear model equivalents of all these types of residuals.

For a generalized linear model, we follow the case of the normal linear model and define the simple residual for the ith datapoint as $Y_i - \widehat{Y}_i$. However, these simple residuals are difficult to interpret for most generalized linear models, and are used very little. For example, if Y_i is binary, the simple residuals will tend to be larger when the success probability, p_i, is about 0.5 than when it is near 0 or 1. If $p_i = 0.5$ then the simple residual is close to either 0.5 or -0.5, because Y_i is 0 or 1 and, assuming that the fitted model is good, because $\widehat{Y}_i = \widehat{p}_i$ will be about 0.5. On the other hand, if $p_i = 1$ then Y_i will be 1 and $\widehat{Y}_i$ should be close to 1, so the simple residual will be small; the same thing happens if $p_i = 0$. The technical reflection of this is that the variance of Y_i is

$$V(Y_i) = p_i(1 - p_i)$$

which is at its greatest when p_i is 0.5 and at its smallest when p_i is 0 or 1.

Since the simple residuals are not satisfactory, a variety of different types of residuals have been developed for generalized linear models.

Pearson residuals

To make the residuals easier to interpret, we could divide each by the estimated standard deviation of Y_i, to give

$$\frac{Y_i - \widehat{Y}_i}{\sqrt{v_i}}. \tag{11.6}$$

Here v_i is an estimate of $V(Y_i)$: for instance, for normal linear regression $v_i = s^2$ for all datapoints; for binary logistic regression $v_i = \widehat{p}_i(1 - \widehat{p}_i)$; for binomial logistic regression $v_i = n_i \widehat{p}_i(1 - \widehat{p}_i)$ (where n_i is the 'number of trials' for datapoint i); for Poisson regression $v_i = \widehat{Y}_i$. The residuals obtained using (11.6) are called *(unstandardized) Pearson residuals* (because they are related to the chi-squared goodness-of-fit test, in a way we shall not go into, and this test was developed by Karl Pearson).

Unfortunately, the Pearson residuals defined by (11.6) still do not have constant variance. This problem is easily overcome, as for normal linear models, by using *standardized Pearson residuals*. The standardized Pearson residual for the ith datapoint is given by the formula

$$r_{Pi} = \frac{Y_i - \widehat{Y}_i}{\sqrt{v_i(1 - h_i)}} \tag{11.7}$$

where h_i is the leverage for point i. (As in Section 11.1, no general formula will be given for h_i. Leverage in the context of generalized linear models will be discussed further in Section 11.3.2.) You can compare equation (11.7) with equation (11.3) and see that the two are virtually identical; they *are* identical for a normal linear model, so in that case the standardized Pearson residuals that we have been using so far in the book match the more general definition of (11.7). In general, the standardized Pearson residuals have mean 0 and variance approximately 1. Unlike standardized residuals for normal linear models, however, their distribution does not usually even approximate to a standard normal one in cases where the distribution of the responses is not normal.

Deviance residuals

In fitting and investigating normal linear models, the residuals appear in several different guises. For instance, the estimate s^2 of the error variance is found from the residual sum of squares, which is the sum of the squares of the natural residuals. The quantity in a generalized linear model that corresponds to the residual sum of squares is the residual deviance, and another form of residual in a generalized linear model is one that is related to the residual deviance in the way that the residuals in a normal linear model are related to the residual sum of squares.

The residual deviance of a generalized linear model can be written as the sum of contributions from each datapoint, so that we get

$$D = \sum_i d_i$$

where D is the residual deviance of the fitted model and d_i is the contribution of point i. As an example, in binomial logistic regression (where n_i is the 'number of trials' for datapoint i), the contribution from point i to the deviance is

$$d_i = 2\left[y_i \log\left(\frac{y_i}{\widehat{y}_i}\right) + (n_i - y_i) \log\left(\frac{n_i - y_i}{n_i - \widehat{y}_i}\right) \right]$$

so that the deviance is

$$D = \sum_{i=1}^{n} d_i = 2 \sum_{i=1}^{n} \left[y_i \log\left(\frac{y_i}{\widehat{y}_i}\right) + (n_i - y_i) \log\left(\frac{n_i - y_i}{n_i - \widehat{y}_i}\right) \right].$$

As another example, in Poisson regression, the contribution from point i to the deviance is

$$d_i = 2\left[y_i \log\left(\frac{y_i}{\widehat{y}_i}\right) - (y_i - \widehat{y}_i) \right]$$

so that the deviance is

$$D = \sum_{i=1}^{n} d_i = 2 \sum_{i=1}^{n} \left[y_i \log \left(\frac{y_i}{\widehat{y}_i} \right) - (y_i - \widehat{y}_i) \right].$$

The *deviance residual* for point i is

$$\text{sgn}(Y_i - \widehat{Y}_i)\sqrt{d_i}.$$

(The function sgn is called the *signum* function and $\text{sgn}(x)$ is 1 if $x > 0$, -1 if $x < 0$ and 0 if $x = 0$. It appears in the definition just to give the deviance residual the same sign as the simple residual would have.)

Like Pearson residuals, the deviance residuals do not have constant variance, so they are usually standardized. The *standardized deviance residuals* are defined by

$$r_{Di} = \text{sgn}(Y_i - \widehat{Y}_i)\sqrt{\frac{d_i}{1 - h_i}}.$$

The standardized deviance residuals have variances of approximately 1. For most types of generalized linear model, the standardized deviance residuals are also approximately normally distributed; an important exception is binary logistic regression (as you have seen in several examples in Chapter 9). Even when they are not normal, the standardized deviance residuals are often good for detecting extreme or outlying observations.

In normal linear models, the standardized deviance residuals are exactly the same thing as the standardized Pearson residuals – and both are exactly the same as the standardized residuals we have been using in Chapters 6–8. For non-normal models, the standardized deviance residuals are not equal (in general) to the standardized Pearson residuals.

Deletion residuals

A third form of residual is based on the change in deviance when the ith observation is excluded from the dataset. In the case of a normal linear model, the process for calculating these residuals can be shown to be equivalent to the process for calculating the deletion residuals that you met briefly in Section 11.2. To calculate the exact values of these residuals would require fitting $n + 1$ regression models (the model including all datapoints, and then in turn the models with each of the n datapoints omitted). This could take quite some time on a desktop computer. This lengthy computation can be avoided by using a good approximation. The following approximation to the *deletion residual* uses the leverage and both the standardized Pearson and standardized deviance residuals, but requires fitting only the model for all n observations:

$$r_{Li} = \text{sgn}(Y_i - \widehat{Y}_i)\sqrt{h_i r_{Pi}^2 + (1 - h_i)r_{Di}^2}. \tag{11.8}$$

These residuals are often called *likelihood residuals*, because of a connection they have with the likelihood function. In this book (and in GENSTAT) they will generally be called deletion residuals, but will be denoted r_{Li} because we have already used r_{Di} for deviance residuals.

All three forms of residuals are available in GENSTAT. Standardized Pearson and standardized deviance residuals are simply called Pearson and deviance residuals in dialogue boxes, statements and on-line help, and both forms are usually described as standardized residuals in output. Deviance residuals are the default form of residuals for generalized linear models (and, as you already know, Pearson residuals are the default for normal regression or ANOVA models fitted via the Stats|Regression Analysis|Linear or Stats|Analysis of Variance menu items). Deletion and Pearson residuals are also options in the Model Checking dialogue box.

Which of the three definitions of residual should you use? The standardized Pearson residuals, measuring the differences between observed and fitted responses, might seem the most natural; but the standardized deviance residuals are more closely related to our fitting criterion of maximizing likelihood (which is why they are the GENSTAT default). In all cases, relatively large residuals indicate observations poorly fitted by the model.

What difference do the three definitions make? You already know that, for normal linear models, the standardized Pearson residuals and the standardized deviance residuals are equal. And you have seen in Section 11.2 that, in some datasets at any rate, the deletion residuals are almost the same as well. Another example is given in Figure 11.3. (You are not expected to be able to reproduce this plot.) Here, the standardized Pearson, standardized deviance and deletion residuals have been calculated for the Poisson regression model that you fitted to the crashes data in Exercise 10.2. The deviance and deletion residuals are

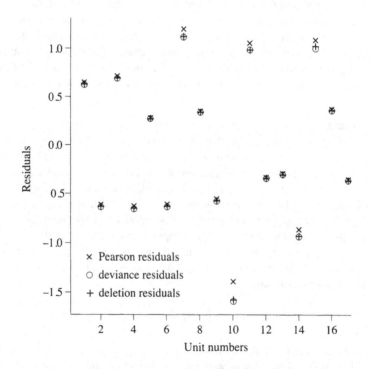

Fig. 11.3 Index plot of (standardized) Pearson, deviance and deletion residuals for a Poisson regression model fitted to the data in crashes.gsh.

almost identical, while the Pearson residuals are in some cases a little different. In fact, it is very often the case, for any generalized linear model, that the standardized deviance residuals are close to the deletion residuals. This is because, typically, the leverages for most points are not very large: looking at (11.8) tells us that if the leverage, h_i, is small, then r_{Li} is based mainly on r_{Di}. In some datasets, the differences between the standardized Pearson residuals and the standardized deviance residuals can be more marked than they are here.

Many statisticians recommend that the standardized deviance residuals are used, because:

- they have approximately a standard normal distribution for most types of generalized linear model, so they are easier to interpret;

- standardized Pearson residuals have been found by experience to be poorer at detecting extreme observations.

In previous chapters, you have seen how various plots of residuals were used to check normal linear models. You might have supposed that it would be useful to produce similar plots for checking generalized linear models. It turns out that the corresponding residual plots for generalized linear models are not always so useful, as you saw in the case of binary logistic regression in Chapter 9. The problem is that, for generalized linear models, strong patterns can be present in the plots of residuals against fitted values, *even when the model is a good one*. This effect makes the usual plot of residuals against fitted values practically useless in the case of binary regression; similarly, plots of residuals against explanatory variables or against transformations of explanatory variables are uninformative because there will generally be a trends in such plots even when the model is good. However, for other sorts of generalized linear model, plots of residuals against fitted values can sometimes be useful. And, in any case, some other kinds of residual plots are useful for checking the adequacy of the generalized linear models (even in the binary case). You have already seen some examples of this in Chapters 9 and 10. Two of these plots are the index plot of residuals and the half-normal plot of residuals, which we shall use in Exercise 11.7.

Exercise 11.7

This exercise concerns the data in `wetbirds.gsh`, which you first met in Exercise 9.10. Load the data into the GENSTAT spreadsheet. We shall consider the binary response variable **woodduck**. There are many explanatory variables, but you saw in Exercise 9.10 that a model containing just one of them, **grav2**, fitted the data well. Fit a logistic regression model with just this one explanatory variable, as you did in Exercise 9.10(d). Do not close the Generalized Linear Models dialogue box yet.

(a) In observational studies like this one, we generally do not expect the data to be listed in any particular order; so, if the model is adequate, you would not expect to see a trend in an index plot of residuals. As you have seen for normal linear models, index plots can also be useful for detecting unusually large residuals (possible outliers).

Produce an index plot of the (standardized) deviance residuals. What do you conclude from it?

(b) You have already met half-normal plots for residuals in the case of normal linear models. For the current model, this type of plot would appear to be examining whether the binary regression residuals are normally distributed, when we know they are not; nevertheless, the half-normal plot is useful for revealing inadequacies of the fit. For binary regression in particular, they tend to show such inadequacies rather better than normal plots do. Possible outliers will appear as isolated points at the top right of the plot. Points systematically deviating from a straight line can in some cases indicate that the model is unsatisfactory, though some deviation from a straight line can be expected even when the model is good.

Produce a half-normal plot of the standardized deviance residuals for the wetbirds model, by obtaining the **Model Checking** dialogue box again and choosing the appropriate options. What do you conclude?

(c) Produce index and half-normal plots of the deletion residuals for this model. How do they compare with the plots of deviance residuals you produced in parts (a) and (b)?

Inspection of index and half-normal plots of deviance or deletion residuals can help to identify observations that are outliers. For generalized linear models other than binary regression models, normal plots may be useful as well. In some cases, e.g. Poisson regression where the mean is not too small, or binomial regression where n is reasonably large, the distributions involved are approximately normal and a normal plot of residuals may show up problems with the model. Outliers in generalized linear models can occur for the same reasons as outliers in normal linear models: they may be genuine extreme values, or they may be due to mistakes in measurement or transcription of the data. In particular, a likely cause of an outlier in binary regression is an observation where the response is erroneously recorded as a 'success' when it is really a 'failure', or vice versa.

Exercise 11.8

In Exercise 10.8, you fitted an exponential regression model to the data in leuksurv.gsh. Load these data into GENSTAT and make a log transformation of the explanatory variable wbc to give lwbc. Fit an exponential model (with canonical, i.e. reciprocal, link function) to these data, as in Exercise 10.8(d), specifying the model as ag+lwbc. Produce residual plots as in Exercise 11.7(a) and (b). Do they provide any evidence of either lack of fit of the model or of the presence of outliers?

11.3.2 Detection of observations with high leverage or influence

An observation is influential if deleting it from the dataset would lead to a substantial change in some feature of the fit of the generalized linear model. It has high leverage if its set of values for the explanatory variables are such that it *potentially could* be influential, whether it is or not depending on the value of the response variable at the point. These concepts in the generalized linear case are therefore very similar to the equivalent concepts in the normal linear case (Sections 11.1 and 11.2).

In Section 11.1, in the context of normal linear models, we saw that prime candidates for influential observations are those that are distant from the remaining observations in terms of the values of the explanatory variables. The distance of the ith observation from the other observations was measured by the leverage. An analogous definition of leverage for generalized linear models leads to a similar quantity, which was mentioned very briefly in Section 11.3.1. As in the case of normal linear models, the exact formula defining leverage in generalized linear models is not given here. Though the definition is analogous, the details of how it works are slightly different from those for normal linear models. In generalized linear models (essentially because the estimated variance of a response can differ from one point to the next in a way that depends on the parameter estimates), the leverage for a particular point can depend on the responses for all the points, as well as on the values of the explanatory variables. Because of this, it is not usually possible to calculate the leverages in a generalized linear model until after the model has been fitted.

As with normal linear models, observations having unusually large leverages may or may not be influential. As in Section 11.1, the rule-of-thumb guide for 'unusually large' is if the leverage h_i is greater than $2q/n$, where q is the number of regression coefficients estimated in the model. GENSTAT stores the leverages for generalized linear models and makes it easy to produce an index plot of them to help identify the observations that are potentially influential.

An important kind of influence is the effect of the ith observation on the set of estimates of the regression coefficients in the linear predictor. The Cook statistic was used in Section 11.2 to measure this kind of influence in normal linear models. A similar statistic exists for generalized linear models. As for normal linear models, GENSTAT calculates modified Cook statistics. The *modified Cook statistics* for a generalized linear model are defined by

$$C_i = \sqrt{\frac{n-q}{q} r_{Li}^2 \frac{h_i}{1-h_i}}. \qquad (11.9)$$

These statistics are based on approximating the changes between the regression coefficients estimated from all n observations and those estimated with the ith observation deleted. Compare (11.5) with (11.9): they are the same formula apart from the residuals used; in both cases the residuals are based on deleting observation i. (Indeed, for normal linear models, the only difference between the deletion residuals defined by r_i^* and those defined by r_{Li} is that the latter are calculated using an approximate formula.) Relatively large values of C_i indicate influential observations; and, just as in the normal linear case, this can occur because a point has relatively high leverage, because it has a large residual (in absolute value), or some combination of the two. Note that, in the generalized linear model, the Cook statistics are not directly measuring changes in fitted values, as they do in the normal linear model. They are measuring the changes in the linear predictor, which happens to be the same as the fitted values in the normal linear model. In logistic regression, for example, the change in the linear predictor corresponds to changes in the logits.

Usually, the best way to inspect the modified Cook statistics is by means of an index plot. The index plots (and other plots) for the leverages and the Cook statistics can be produced in GENSTAT using the **Further Output** button of the **Generalized Linear Models** dialogue box, followed by using the **Model Checking** button.

Exercise 11.9

(a) We turn again to the data in `wetbirds.gsh`, which you last analysed in Exercise 11.7. If necessary, load the data into GENSTAT again. Carry out a logistic regression of **woodduck** on **grav2**. Does GENSTAT flag any points as having high leverage? Obtain the Model Checking dialogue box and produce an index plot of the leverages. Produce an index plot of the Cook statistics in a similar way. What do you conclude?

(b) Produce index plots of the leverages and Cook statistics for the exponential regression model you fitted to the data in `leuksurv.gsh` in Exercise 11.8. What do you conclude?

(c) In part (b), you should have found that one point was very influential. Looking at the data in the spreadsheet, can you explain why this point has high leverage? Exclude the point from the dataset and refit an exponential regression model to the restricted dataset. What difference does this make to the regression coefficients?

Exercise 11.10

This exercise concerns the bone marrow transplant data from the file `gvhd.gsh`, which we analysed in Chapter 1 and again in Section 9.3. Load the dataset into GENSTAT. In the analysis in Section 9.3, we took logs of the explanatory variable **indx** to produce **lindx**. Do so again.

(a) The final model that we fitted in Section 9.3 contained just two explanatory variables, **lindx** and **donage**. Fit a logistic regression model to the response variable (**gvhd**) using these two explanatory variables. Produce index plots of the leverages and Cook statistics. You should see a pattern in the plots. Look at the original data (ignoring all the columns except **indx**, **donage** and **gvhd**, since the other variables have not been used in the analysis), and see whether you can provide an explanation for the pattern.

(b) Another possibility for these data would have been not to transform **indx**. It was transformed essentially because it has a very skewed distribution. For instance, two points (36 and 37) have much higher values of **indx** than all the others. In an analysis where **indx** is not transformed, would you expect points 36 and 37 to have high leverage? Fit a logistic regression model with **indx** and **donage** as explanatory variables. Do points 36 and 37 indeed have high leverage?

(c) To investigate further how leverage works in logistic regression, fit a logistic regression model to the response variable **gvhd** using the single explanatory variable **indx**. Store the leverages in a variate **lev**, using the same method as in Exercise 11.2, and produce a plot of **lev** against the explanatory variable **indx**. In the light of what you have found in previous parts of this exercise, can you explain why the plot looks as it does?

Partly because of the rather strange behaviour of leverages in binary regression that you saw in the previous exercise, and partly because the response variable in binary regression can take only two values, the leverages and Cook statistics do not always give a very satisfactory measure of the potential or actual influence of datapoints on the regression coefficients of binary regression models. In most other kinds of generalized linear models, the leverages and Cook statistics are more helpful.

11.4   Recommended use of model diagnostics

Residuals, leverages and modified Cook statistics should routinely be displayed, preferably in index plots, after fitting a normal linear or generalized linear model. Plots of residuals against fitted values, and normal plots of residuals, should be produced for normal linear models. A half-normal plot of residuals may instead be useful for generalized linear models, particularly for drawing attention to outliers. Normal plots remain useful in some models where the distribution of residuals is approximately normal. Standardized deviance residuals are recommended for plotting (and they are the GENSTAT default).

Observations identified as outliers or as influential should not necessarily be discarded. If possible, they should be checked for errors or discussed with those who have more knowledge about the origin of the data and the plausibility of the unusual observations. In the absence of any other information, separate analyses should be performed, with and without the suspect observations; and if their results are substantially different, both analyses should be reported.

Note that it is not always sensible to ignore points that have high leverage but low influence (as measured by their Cook statistic). Such points generally follow the pattern of the others, for otherwise they would have a large residual and a large Cook statistic. Thus they do not significantly affect regression coefficient estimates. However, they can, particularly if they have very high leverage, have a marked effect in decreasing the standard errors of the estimated regression coefficients. Thus, if they are actually erroneous in some way, they may make the model appear to fit better or the predictions appear to be more precise than is really the case.

If some outlying or influential observations are found to be erroneous or unrepresentative and are omitted, then it is advisable to go back to the model-building stage to check that the original conclusions regarding the selection and transformation of explanatory variables, the inclusion of quadratic and interaction terms, etc., are unaffected by omitting the observations. If it has been found necessary to change the linear predictor, then the model diagnostics need to be repeated for the new model. There may be several such cycles of model building and diagnostics before a satisfactory model is found.

Finally, you should be aware that there is a much wider range of diagnostic quantities and methods available than there is space to discuss in this book. The methods that have been covered are the most used and arguably the most important, but you should be aware that there are other methods available, for instance, for assessing the goodness of fit of a particular link function or for checking whether there are strong linear relationships between the explanatory variables (i.e. multicollinearity, which can have a major effect on the quality of parameter estimates and on the accuracy of prediction from a model). You saw in Exercise 9.10 that GENSTAT flags multicollinearity in certain circumstances.

The following exercise is to give you further practice at many of the model-checking techniques that we have discussed in this chapter.

Exercise 11.11

The exercise concerns the ship1907 dataset, which you met in Exercise 10.3. Load the dataset into GENSTAT, transform the explanatory variable **tonnage** to ltonnage by taking logs, and fit a Poisson regression model with ltonnage as the (only) explanatory variable. (This model was found to fit adequately in Exercise 10.3.) Produce appropriate plots of diagnostic quantities and report your conclusions.

Summary of methodology

1 In normal linear models, points that have values for the explanatory variables that are very different from those for other points have the potential to dominate a regression analysis. These points are said to have high leverage. Leverage is defined in an analogous but slightly more complicated way for generalized linear models.

2 High-leverage points only have the potential to dominate the analysis. Observations that actually do dominate the analysis are called influential points.

3 Cook's distance is a widely used method of measuring how influential an observation is, in normal linear or generalized linear modelling. GENSTAT calculates a modified form of this, called the modified Cook statistic.

4 The leverages and Cook statistics can be plotted using index plots to identify which points have high leverage or influence.

5 There are several types of residuals for generalized linear models. The ones that are used most often in practice are standardized deviance residuals and deletion residuals. Deletion residuals are also known as likelihood residuals.

6 Plotting residuals against fitted values is often unhelpful for GLMs, because most such models, other than normal linear models, will yield plots exhibiting strong patterns, even if the model is a good one. Index plots and half-normal plots are used instead. For some GLMs, normal plots are useful.

7 Advice on what to do if you discover influential observations or outliers is given in Sections 11.2 and 11.4.

12

Loglinear models for contingency tables

In Chapters 9 and 10, you learned about generalized linear models and saw how to use GENSTAT to fit them to many different types of data. This chapter continues that thread, but involves a different type of data from those you have been working with so far. Here we shall concentrate on data in the form of *contingency tables*: that is, tables of *counts* showing how often within a given sample each combination of the different values of various discrete random variables occurs. In Chapters 1 and 9 of this book, you have met contingency tables which record the values of two variables at a time (so-called *two-way* contingency tables), and you will be aware that such tables are often laid out in such a way that the rows correspond to the values of one of the variables and the columns to the other. There are several methods for analysing such data, including *Fisher's exact test* (for 2 × 2 tables) and the *chi-squared test*. The methods covered in this chapter will deal with tables of this kind, and in a sense they provide alternative versions of the chi-squared test for contingency tables. If that was their only use, they would have little importance in data analysis; however, one of their main strengths is they can be extended easily to more complicated contingency tables involving more than two discrete variables.

The analysis of two-way contingency tables by means of a chi-squared test generally involves looking for evidence of association between the two discrete variables involved. Often the two variables are treated on an equal footing, rather than labelling one as an explanatory variable and one as a response variable. You might be wondering how such an analysis can fit into the framework of generalized linear models, where one of the variables is clearly labelled as a response variable. The idea is to treat the *counts* in the cells of the contingency table as values of the response variable, rather than using either of the variables defining the rows and columns. A *loglinear model* is then set up, which relates the distribution of the counts to the values of the explanatory (row and column) variables; thus testing for an association between these variables amounts to testing for an interaction between them in the loglinear model. Section 12.1 shows you how to do this, as well as how to perform a chi-squared test and Fisher's exact test, in GENSTAT. The ideas of Section 12.1 are considerably extended and generalized in Sections 12.2–12.4.

12.1 Two-way contingency tables

We shall begin by looking at an example of the simplest kind of contingency table data: a 2×2 table.

Example 12.1

A study was carried out on 65 patients who had received sodium aurothiomalate (SA) as a treatment for rheumatoid arthritis. Of these patients, 37 had shown evidence of a toxic effect of the drug. The patients were also classified according to whether or not they had impaired sulphoxidation capacity; the researchers thought this might be linked to toxicity in some way. The data are given in Table 12.1. Notice that the row and column ('marginal') totals are also given, along with the overall total count (65).

Table 12.1

Impaired sulphoxidation	Toxicity		Row totals
	Yes	No	
Yes	30	9	39
No	7	19	26
Column totals	37	28	65

Source: Ayesh, R., Mitchell. S. C., Waring, R. H., Withrington, R. H., Seifert, M. H. and Smith, R. L. (1987) 'Sodium aurothiomalate toxicity and sulphoxidation capacity in rheumatoid arthritic patients'. *British Journal of Rheumatology*, **26**, 197–201.
Dataset name: `toxic`.

It looks pretty obvious from the table that toxicity is not independent of impaired sulphoxidation. In the next exercise, you will investigate this using several different statistical tests.

Exercise 12.1

(a) Load the dataset stored in `toxic.gsh` into the GENSTAT spreadsheet. Compare the spreadsheet with Table 12.1, to ensure you know how the data in the spreadsheet correspond to the table. (GENSTAT expects contingency table data to be represented as in the spreadsheet in order to fit loglinear models.)

(b) In part (d) you will test whether the row and column factors are associated, by fitting a (loglinear) generalized linear model. But you should already know how to test this hypothesis, using Fisher's exact test and/or the chi-squared test for contingency tables. Let us investigate how to carry out those procedures in GENSTAT.

An awkwardness here is that, for these tests, GENSTAT requires the data to be in a different format than for the loglinear modelling approach. In fact, it requires the data to be laid out in a structure that it calls a *table*. To do this, choose the **Stats|Summary Statistics|Summaries of Groups (Tabulation)** menu item. In the resulting dialogue box, choose **count** as the **Variate** and **impsulph** and **toxicity** as the **Groups**. Click on the **Save** button and, in the resulting dialogue box, check **Totals** and enter **toxtab** in the corresponding **In** field. This will save the totals of the values of **count**, for each combination of values of impsulph and toxicity, in a table called **toxtab**. In this case, there is actually only one value of count for each combination of values of impsulph and toxicity, and it is this value that is saved. Click on **OK** to close the Summary by Groups Save Options dialogue box. Back in the **Summary by Groups** dialogue box, ensure things are set up so that the same table is printed that we are going to save in **toxtab**; do this by ensuring that Totals is the only item checked under Type of Summary and that **Set Margin** is *not* checked. Click on **OK**; click on **Cancel**. Look at the Output window; you should see a table of data there, laid out in a similar fashion to Table 12.1 (without the marginal totals).

Now choose the Stats|Statistical Tests|Contingency Tables menu item. In the resulting dialogue box, the **Test** field should read **Chi-square test**. Enter **toxtab** as the **Table**, and make sure **Pearson** is chosen as the **Method**. This will calculate the test statistic by the original method developed by Karl Pearson (which is probably the one you have seen before). Click on **OK** (do not click on Cancel yet) and look at the Output window. What do you conclude?

Go back to the **Contingency Tables** dialogue box and choose **Maximum Likelihood** instead of **Pearson** as the **Method**. Click on **OK**; again, do not click on Cancel yet. How do the results differ from those with the Pearson test statistic?

(c) Return to the **Contingency Tables** dialogue box. Choose **Fisher's exact test** as the **Test**, and check that **toxtab** is still in the **Table** field. Click on **OK**; click on **Cancel**. Look at the **Output** window. What do you conclude?

(d) Choose the Stats|Regression Analysis|Generalized Linear menu item. In the resulting dialogue box, change the **Analysis** field to **Log-linear modelling**. Choose count as the **Response Variate**, and enter **impsulph+toxicity** as the **Model to be Fitted**. Leave the other field blank. Click on **OK**. This fits a loglinear model in which the two factors involved (impsulph and toxicity) are assumed to have zero interaction. Examine the output.

Now return to the **Generalized Linear Models** dialogue box and change the model to **impsulph*toxicity**. Do this by clicking on **Change Model**, entering impsulph.toxicity in the **Terms** field, and clicking on **Add**. This fits a model which includes the interaction between the row and column factors. Look at the output from this model. In the analysis of deviance table, what is unusual about the residual deviance? Using the difference between the regression deviance values for the two models you have fitted (as described in Chapters 9 and 10 and given in the GENSTAT output), test whether the interaction term is necessary. What do you conclude?

Look back at the value of the **Maximum Likelihood** test statistic you found from a chi-squared test in part (b). Can you find this number in the output from the loglinear modelling?

(e) Refit the loglinear model (with an interaction term), if it was not the last model you fitted. (Note that carrying out a **Chi-square test** with the **Maximum Likelihood** method actually fits a generalized linear model, so if you did that after fitting the loglinear models in part (d) then you will have to fit the loglinear model again here.) In the **Generalized Linear Models** dialogue box, click on the **Save** button. In the resulting dialogue box, check **Fitted Values** and enter fits in the **In** field. Click on **OK**. This will store the fitted values from the loglinear model in a variate called fits. Use the Stats|Summary Statistics|Summaries of Groups (Tabulation) menu item, as you did with the raw data in part (b), to output the fitted values as a table in the Output window. (There is no need to save the resulting table.) Look at the fitted values. Where have you seen them before?

Refit the loglinear model with no interaction term, and save and output the fitted values in the same way as above. Check that the resulting fitted values are the same as the expected values you would get if you performed a chi-squared test on the data in Table 12.1.

Hint   The chi-squared expected values are the expected values (E_i) under no association, given by multiplying together the appropriate marginal (row and column) totals in Table 12.1, and dividing by the overall total.

■

Thus, if we treat the counts in the cells of a contingency table as values of a response variate, and if we treat the categorical variables defining the rows and columns as explanatory factors, we can fit a particular kind of generalized linear model, and use a test for the presence of interaction between the explanatory variables as a test for association between them. You will see in the next section exactly what form the generalized linear model takes, and why this works. (If you want a clue, you could look again at the output for the GLMs you fitted in the last exercise. It tells you that you have used a Poisson model with the log link function, and you should remember from Chapter 10 that the log link function is the canonical link function for Poisson models. What this has to do with contingency tables remains to be seen.)

But before that, let us clear up a point about the chi-squared test and then look at another example. GENSTAT allows you a choice of two different test statistics for the chi-squared test. Denoting the observed values in the cells by O_i and the expected values (calculated in the usual way as in Solution 12.1(e)) by E_i, the usual chi-squared test statistic is $\sum (O_i - E_i)^2 / E_i$, with the sum being over all the cells in the contingency table; and this is what GENSTAT calculates if you ask for the **Pearson** option for the **Method** in the **Contingency Tables** dialogue box. You should know that, under the null hypothesis of no association, this quantity has an approximate χ^2 distribution with $(r-1)(c-1)$ degrees of freedom, for a contingency table with r rows and c columns. The test statistic you get if you ask GENSTAT for the **Maximum Likelihood** option instead is $2 \sum O_i \log (O_i / E_i)$. This quantity, as was hinted at in Exercise 12.1(d), is actually the residual deviance from fitting a loglinear model with no interaction, or equivalently the deviance difference between a model with the interaction included and a model with the interaction excluded. (Actually, with the appropriate fitted values E_i, this formula gives the residual deviance for any loglinear model.) GENSTAT calls it the 'likelihood chi-square value' because, as you saw in Chapter 10, the deviance is actually a difference between log-likelihoods. Since it is a deviance, it also has an approximate χ^2 distribution, and it turns out that it again has $(r-1)(c-1)$ degrees of freedom. (With no interaction in the model, there are $r-1$ d.f. for the row treatment, $c-1$ d.f. for the column treatment, and hence $(r-1)(c-1)$ d.f. for the residual.) Thus, from the point of view of testing for association in a two-way contingency table, you can think of GENSTAT's 'likelihood' test statistic merely as an alternative to the usual test statistic that works in much the same way. Indeed, for most tables, particularly when the overall sample size n is reasonably large, the values of the two statistics are likely to be close to each other.

Now try the following exercise, which looks at a rather larger (but still two-way) contingency table.

Exercise 12.2

A study was carried out in which 671 tiger beetles were classified in two ways, according to their colour pattern (bright red or not bright red) and the season in which they were found (early spring, late spring, early summer or late summer). The data are given in Table 12.2.

Table 12.2

Season	Colour pattern		Row totals
	Bright red	Not bright red	
Early spring	29	11	40
Late spring	273	191	464
Early summer	8	31	39
Late summer	64	64	128
Column totals	374	297	671

Source: Sokal, R. R. and Rohlf, F. J. (1981) *Biometry*, 2nd edition, New York, W. H. Freeman.
Dataset name: `beetles`.

(a) Analyse these data using a chi-squared test with both Pearson and maximum likelihood methods, as you did in Exercise 12.1(b). Report your conclusions.

(b) Test for independence of colour pattern and season by fitting loglinear models, as you did in Exercise 12.1(d). Report your conclusion.

(c) Suppose, however, that for some reason, GENSTAT was unable to fit the colour*season model. Given what was the case in Solution 12.1(d), how can you still make the test of independence, given only the result of fitting colour+season?

12.2 Sampling models

In this section,[1] you will learn about the particular generalized linear model that is usually used to fit contingency table data.

The situation we shall consider is that where individuals (i.e. people, or beetles, or whatever the units of observation are) are chosen at random from some population, with the size of the sample (n) being fixed in advance. Then each unit in the sample is investigated to see which cell of the contingency table it falls into. (This sampling situation is reasonably common, but it is not the only possibility, as will be seen later.) This involves introducing a probability distribution, the *multinomial distribution*, that you may not have met before.

[1] This section does not involve the use of GENSTAT; you do not need your computer to work through it.

To begin with, let us forget temporarily that the cells in the contingency table are laid out in a particular way. Let us just say that there are k such cells. If k were only 2, we would be in a familiar situation, where individuals are chosen at random and fall into one of two categories. The commonest way to model such data would be as Bernoulli trials; in other words, we would make the assumptions that:

- individuals are independent of each other;

- the probability of falling into a given cell is the same for all individuals.

As you know, it turns out that the number of individuals with values in a given cell has a binomial distribution $B(n, p)$, where p is the probability of falling in that cell and n is the total number of individuals.

How can this be extended to more than two cells? Label the k cells from 1 up to k. Let us assume as before that individuals are independent of one another, and that the probabilities of falling into given cells are the same for everyone. Denote these probabilities by $p_1, p_2, \ldots, p_k$, so that the probability that an individual falls into cell j is given by p_j. Then $p_1 + p_2 + \cdots + p_k = 1$. The number of individuals that finish up in any given cell is a random variable; let us denote these random variables by $N_1, N_2, \ldots, N_k$. We know that $N_1 + N_2 + \cdots + N_k = n$, because there are n individuals in all. What we need is the joint probability distribution of the random variables $N_1, N_2, \ldots, N_k$.

To make things concrete, suppose that $k = 3$ and $n = 4$. We have three random variables, N_1, N_2 and N_3, with $N_1 + N_2 + N_3 = 4$. What we want is the value of $P(N_1 = n_1, N_2 = n_2, N_3 = n_3)$. This has to be zero unless $n_1 + n_2 + n_3 = 4$. Let us look at some specific cases.

First, what about $P(N_1 = 4, N_2 = 0, N_3 = 0)$? This event can happen only if all four individuals end up in the first cell. The probability that each of them ends up in this cell is p_1, and we're assuming they are independent, so the probability they all end up there is p_1^4.

Now what about $P(N_1 = 3, N_2 = 1, N_3 = 0)$? This happens if three of our four individuals end up in the first cell, with the other one in the second cell. The probability that the *first* three end up in the first cell and the *last* ends up in the second cell is $p_1^3 p_2$. But the probability that the *last* three end up in the first cell and the *first* ends up in the second cell is also $p_1^3 p_2$. All together there are four ways in which one of the four individuals can end up in cell 2 with the other three in cell 1, so $P(N_1 = 3, N_2 = 1, N_3 = 0) = 4p_1^3 p_2$.

In general, the probabilities we need all involve the number of possible ways of arranging a certain number of individuals so that given numbers fall into given cells, multiplied by some powers of probabilities. In fact, for this specific case:

$$P(N_1 = n_1, N_2 = n_2, N_3 = n_3)$$
$$= \begin{cases} \dfrac{4!}{n_1! n_2! n_3!} p_1^{n_1} p_2^{n_2} p_3^{n_3} & \text{if } n_1 + n_2 + n_3 = 4 \\ 0 & \text{otherwise} \end{cases}$$

Here the term $4!/(n_1! n_2! n_3!)$ can be shown to give the number of ways of arranging four items in three cells with n_1 in the first cell, n_2 in the second and n_3 in the third.

In general, if random variables $N_1, N_2, \ldots, N_k$ have a joint distribution defined in this way, they are said to have a multinomial distribution. The formal definition is as follows.

Box 12.1

The integer-valued discrete random variables $N_1, N_2, \ldots, N_k$ are said to have a **multinomial distribution** with parameters n and $p_1, p_2, \ldots, p_k$ (where $p_1 + p_2 + \cdots + p_k = 1$) if their joint probability mass function is:

$$P(N_1 = n_1, N_2 = n_2, \ldots, N_k = n_k)$$

$$= \begin{cases} \dfrac{n!}{n_1! n_2! \cdots n_k!} p_1^{n_1} p_2^{n_2} \cdots p_k^{n_k} & \text{if } n_1 + n_2 + \cdots + n_k = n \\ 0 & \text{otherwise} \end{cases}$$

The multinomial distribution provides a model for the number of individual items falling into each of k cells, where there are n items in all and where items fall into cells independently of each other, with the probability of an item falling into cell j taking the value p_j for all items.

Exercise 12.3

We developed the multinomial distribution by extending the ideas behind the binomial distribution to more than two categories. Check that the binomial distribution indeed fits into the generalized form, as follows. Let $k = 2$ in Box 12.1. Write down the joint probability mass function of N_1 and N_2, and hence show that $N_1 \sim B(n, p_1)$ and $N_2 \sim B(n, p_2)$.
Hint   To obtain the probability mass function of N_1 from that of N_1 and N_2 jointly, you need to sum the latter over values of N_2.

So we now have a probability model for individuals falling into cells, such as the cells in a contingency table. But it isn't a very nice model, on the face of it. It gives the joint distribution of a whole set of random variables, which are certainly not independent since they have to add up to a given total (n). How are we going to turn this into something we can use in a generalized linear model? It's done by means of something you might call either a simple but powerful bit of probability theory or (if you prefer) a crafty trick.

What has made things awkward here is the constraint that the cell counts have to add up to n, the fixed sample size. Suppose n were not fixed in advance, but instead the researchers just collected data on individuals as they passed at random, recording which cell each of them fell into, until it was time for the researchers to go home for their tea. Then the sample size would be random, not fixed. It would also be plausible to treat the cell counts as independent of one another, and it might also be reasonable to model each of them as having a Poisson distribution (with, possibly, a different mean for each cell). We would then be in the realm of Poisson regression, which can be treated as a GLM, as you saw in Chapter 10. Can we not just make life easier for ourselves by pretending that we really did collect the sample like this, rather than fixing the sample size in advance?

Well, we can. This sounds like cheating, but it is not, because it turns out that if one looks at it in the right way, the Poisson model leads to exactly the same probability distributions as the multinomial model. In the Poisson model, the number of individuals in the jth cell ($j = 1, 2, \ldots, k$) is a random variable, N_j, with a Poisson distribution, with parameter λ_j (i.e. $N_j \sim \text{Poisson}(\lambda_j)$). Counts for different cells are independent of one another. The total number of individuals in all the cells (i.e. the total sample size) is also a random variable, N, given by $N = N_1 + N_2 + \cdots + N_k$, and since the sum of independent Poisson variables has a Poisson distribution, $N \sim \text{Poisson}(\lambda_1 + \lambda_2 + \cdots + \lambda_k)$. Now, for our multinomial setup, we are not interested in any values of the sample size other than the value, n, which we fixed in advance. So let's do something similar for the Poisson model, by looking at the joint distribution of $N_1, N_2, \ldots, N_k$ *conditional* on $N = n$, i.e. the distribution of $N_1, N_2, \ldots, N_k$ when we know that $N = n$. It turns out that this distribution is multinomial! This result is both rather remarkable and reasonably easy to prove, so it is presented here as the only theorem and proof in this book. We do not, though, make such rigorous use of the theorem as a mathematician would!

Box 12.2

Theorem 1 Suppose $N_1, N_2, \ldots, N_k$ are independent Poisson random variables with means $\lambda_1, \lambda_2, \ldots, \lambda_k$, respectively. Let $N = N_1 + N_2 + \cdots + N_k$, so that $N \sim \text{Poisson}(\lambda)$, where $\lambda = \lambda_1 + \lambda_2 + \cdots + \lambda_k$. Then the joint distribution of $N_1, N_2, \ldots, N_k$ conditional on $N = n$ is multinomial with parameters n and $p_1, p_2, \ldots, p_k$, where $p_j = \lambda_j / \lambda$.

Proof

Since the N_j are independent, their joint probability mass function is the product of their individual probability mass functions, i.e.

$$P(N_1 = n_1, N_2 = n_2, \ldots, N_k = n_k) = \prod_{j=1}^{k} \frac{e^{-\lambda_j} \lambda_j^{n_j}}{n_j!}$$

$$= e^{-(\lambda_1 + \lambda_2 + \cdots + \lambda_k)} \prod_{j=1}^{k} \frac{\lambda_j^{n_j}}{n_j!}$$

$$= e^{-\lambda} \prod_{j=1}^{k} \frac{\lambda_j^{n_j}}{n_j!}.$$

Because $N \sim \text{Poisson}(\lambda)$, $P(N = n) = e^{-\lambda} \lambda^n / n!$.

We want the joint distribution of $N_1, N_2, \ldots, N_k$ conditional on $N = n$. This can be found by calculating $P(N_1 = n_1, N_2 = n_2, \ldots, N_k = n_k | N = n)$ for all possible values of $n_1, n_2, \ldots, n_k$. This conditional probability is perhaps more complicated than probabilities you have met before, but generally it behaves in the same way. In general, the conditional probability of an event A conditional on another event B is given by $P(A|B) = P(A \text{ and } B)/P(B)$. (There is more on this in Exercise 12.5.)

Therefore

$$P(N_1 = n_1, N_2 = n_2, \ldots, N_k = n_k | N = n)$$
$$= \frac{P(N_1 = n_1, N_2 = n_2, \ldots, N_k = n_k, N = n)}{P(N = n)}.$$

But when $n = n_1 + n_2 + \cdots + n_k$,

$$P(N_1 = n_1, N_2 = n_2, \ldots, N_k = n_k, N = n) = P(N_1 = n_1, N_2 = n_2, \ldots, N_k = n_k),$$

so that

$$P(N_1 = n_1, N_2 = n_2, \ldots, N_k = n_k | N = n) = \frac{P(N_1 = n_1, N_2 = n_2, \ldots, N_k = n_k)}{P(N = n)}$$

$$= \frac{e^{-\lambda} \prod_{j=1}^{k} \frac{\lambda_j^{n_j}}{n_j!}}{e^{-\lambda} \frac{\lambda^n}{n!}}$$

$$= \frac{\prod_{j=1}^{k} \frac{\lambda_j^{n_j}}{n_j!}}{\frac{\lambda^n}{n!}}.$$

But when $n = n_1 + n_2 + \cdots + n_k$, $\lambda^n = \prod_{j=1}^{k} \lambda^{n_j}$, so that

$$P(N_1 = n_1, N_2 = n_2, \ldots, N_k = n_k | N = n) = \frac{n!}{n_1! n_2! \cdots n_k!} \prod_{j=1}^{k} \left(\frac{\lambda_j}{\lambda} \right)^{n_j}.$$

This is the joint probability mass function for the multinomial distribution with parameters n and $p_1, p_2, \ldots, p_k$, as required. ∎

This result means that in some respects Poisson models for categorical data behave like multinomial models. In fact, it turns out that, if one fits a Poisson generalized linear model to data from a contingency table (with the canonical, log, link function), the maximum likelihood estimation process ensures that the fitted values for the counts in the cells do indeed add up to the actual total count that was observed in the first place. Therefore it makes *no difference* whether or not we fix the total sample size, n, in advance. If the sample size really is fixed, then a correct model for the data is the multinomial model; but the mles and fitted values turn out to be exactly the same as if we had used a Poisson model instead. In the Poisson model, the cell counts are independent of one another, and everything fits nicely into the standard GLM framework. (A third, different, sampling scheme for contingency tables, which also results in precisely the same answers – and which can therefore be treated as a Poisson GLM – will be discussed in Example 12.3.) Therefore, having introduced the multinomial distribution as a sensible model for contingency table data, we can largely forget about it in practice!

Box 12.3

To model contingency table data where the total sample size is fixed in advance, fit a Poisson generalized linear model with a log link function (the canonical link function for Poisson models) to the cell counts.

Why a log link function? There are some mathematical reasons why it is sensible; but, perhaps more importantly, it gives results that are relatively easy to interpret. To illustrate this, consider fitting a Poisson generalized linear model, with a log link function, to the data from a 2×2 contingency table like the one in Example 12.1. There is a factor, at two levels, representing the classification defined by the rows (i.e. the sulphoxidation status), and another two-level factor for the columns (i.e. the toxicity). Suppose we fit a model with no interaction between these factors. There is an indicator variable, x_r say, taking the value 0 for counts in the first row and the value 1 for counts in the second row, and another, x_c say, taking the value 0 for counts in the first column and the value 1 for counts in the second column. The linear predictor will take the form $\mu + \beta_r x_r + \beta_c x_c$ (r and c for row and column). Thus, the model for a 2×2 table is that the count in cell j, N_j, is Poisson with mean λ_j, where

$$\lambda_j = \exp(\mu + \beta_r x_r + \beta_c x_c)$$

because we are using a log link function. Equivalently, $\log \lambda_j = \mu + \beta_r x_r + \beta_c x_c$. Therefore, the log of the mean count in the cell in the first row and the second column (for example) is $\mu + \beta_c$, and the mean count in that cell is $e^{\mu + \beta_c}$.

Exercise 12.4

(a) What are the mean counts in the other three cells of the table? What is the mean of the total count for the first row of the table? Calculate the other row and column means, and show that the mean total count for the whole table is $e^{\mu}(1 + e^{\beta_r})(1 + e^{\beta_c})$.

(b) Now let us use Theorem 1 and start treating the data as multinomial. The cell probabilities for the multinomial model are given in the theorem as $p_j = \lambda_j / \lambda$, where λ is the sum of the λ_j. For the cell in the first row and the second column, $\lambda_j = e^{\mu + \beta_c}$; and, from part (a), $\lambda = e^{\mu}(1 + e^{\beta_r})(1 + e^{\beta_c})$. Hence the probability of falling into this cell is

$$p_j = \frac{\lambda_j}{\lambda} = \frac{e^{\beta_c}}{(1 + e^{\beta_r})(1 + e^{\beta_c})}.$$

(The e^{μ} terms cancel out.) Find the corresponding probabilities for the other cells of the table.

(c) You now have the probabilities for all four cells. By adding up the probabilities of the two cells in the first row, show that the probability of falling in the first row is $1/(1 + e^{\beta_r})$. Show in a similar way that the probability of falling in the second column is $e^{\beta_c}/(1 + e^{\beta_c})$. Check that the probability of falling in the cell in the first row and the second column is the product of the probability of falling in the first row and the probability of falling in the second column.

Translating Exercise 12.4 back into the terms of Example 12.1, the probability of falling in the first row is the probability that a patient has impaired sulphoxidation capacity, and the probability of falling in the second column is the probability that a patient does not show toxicity. Thus, according to this model, the probability that a patient has impaired sulphoxidation capacity *and* does not show toxicity is the product of these two marginal probabilities – which is what one would expect if the row and column variables were independent.

In fact, a result like this applies in every cell: the probability of falling in any given cell is the product of the probabilities of falling in the appropriate row and the appropriate column. (This corresponds to what you saw in Exercise 12.1(e), where the fitted values for the no-interaction model could be calculated by the chi-squared test method of multiplying the row and column totals together, and dividing by the overall total.) In other words, under this model, the row and column variables are independent. This is because no interaction term is included in the model. Thus testing for independence is simply a matter of fitting a Poisson generalized linear model with a log link function, with two explanatory factors defining the rows and column of the table, and testing whether the interaction term is necessary. Such a model is called a *loglinear model* because the logs of the cell counts are predicted by a linear predictor involving the row and column factors. In fact, although this won't be explicitly shown, the logs of the cell probabilities are predicted by a linear predictor also. (The linear predictors for log counts and log probabilities differ only in their constant term.) Oversimplifying (but not all that much), the log link function is useful because the hypotheses of interest for contingency tables are to do with independence, which is a multiplicative property, and taking logs turns the multiplication into addition, which can be represented by linear predictors.

Having got enough of the relevant theory under our belts, let us turn back to data analysis!

12.3   Loglinear models in practice

Example 12.2

In his book *Analyzing Tabular Data* (1993), Nigel Gilbert quotes some results from a survey carried out in the Oxfordshire town of Banbury in 1967, originally reported by Stacey *et al.* In the survey, a sample of 6% of the population was interviewed and asked a number of different questions. The data given in Table 12.3 exclude those who said they voted Liberal or who did not vote.

How are the three factors in this $2 \times 2 \times 2$ contingency table (occupational class, vote and gender) related to one another? One way to investigate this would be to pick on one of the variables (vote seems the most sensible) as a response variable, and to investigate how well it is explained by the other two, using logistic regression. (In Section 12.4 we look at the relationship between logistic regression and loglinear modelling.) Instead, however, let us treat them all on an equal footing and fit a loglinear model.

The process of choosing an appropriate loglinear model is in some ways rather different from the modelling you have done before (though the general principles are the same). This is because it is always possible to fit a model, within the multinomial family being used, that fits the data *exactly*. This model, called the *saturated model*, is that where all the interactions (of whatever order) between the various factors involved are included. (You met the saturated model before in Section 10.3.) You saw in Exercise 12.2 that we can test whether some other model is significantly worse than the saturated model by fitting the other model and comparing its residual deviance with a chi-squared distribution.

(With the other types of generalized linear model we have met, it is often not possible to fit the data exactly within the family of models being considered. In these cases, the best model that can be found might still not fit the data all that well (for example, if there is overdispersion). Such problems cannot arise with loglinear models.)

Exercise 12.5

(a) Load the banvote data into the GENSTAT spreadsheet. Check how it compares with the data given in Table 12.3.

Table 12.3 Numbers voting Labour or Conservative, by gender and occupational class, in Banbury in 1967.

Occupational class	Vote			
	Conservative		Labour	
	Male	Female	Male	Female
Non-manual	140	152	50	50
Manual	109	136	215	159

Source: Gilbert, N. (1993) *Analyzing Tabular Data*, London, UCL Press; derived from Table XV of Stacey, M., Batstone, E., Bell, C. and Murcott, A. (1975) *Power, Persistence and Change*, London, Routledge & Kegan Paul.
Dataset name: banvote.

Choose the Stats|Summary Statistics|Summaries of Groups (Tabulation) menu item. Enter count as the Variate and, *in this order*, vote, class and gender in the Groups field. Make sure the Totals item is the only one checked under Type of Summary. Click on OK. (You do not need to save any tables this time.) Comment on how the table produced compares with Table 12.3.

(b) Choose the Stats|Regression Analysis|Generalized Linear menu item. In the resulting dialogue box, choose Log-linear modelling in the Analysis field. Choose count as the Response Variate, and enter class+vote+gender as the Model to be Fitted. Click on OK. This fits a loglinear model in which the three factors involved (class, vote and gender) are assumed to be independent. By looking at the residual deviance in the analysis of deviance table, and comparing it with a chi-squared distribution with degrees of freedom given by the residual degrees of freedom, decide whether this no-interaction model fits the data adequately, compared with the saturated model.

(c) Fit a loglinear model which includes the main effects of each of the three factors, as well as all the two-factor interactions, by entering

class+vote+gender+class.vote+class.gender+vote.gender

as the Model to be Fitted in the Generalized Linear Models dialogue box. This model no longer assumes that the three factors are independent, but it is still a little simpler than the saturated model (which would include the three-factor interaction class.gender.vote as well). Again by using the residual deviance, check whether this model fits the data adequately.

You should have found that a model with no interactions does not fit adequately, but that a model with all the two-factor interactions does fit adequately. This raises the question of whether we can find a model that also fits the data adequately but which has fewer terms in it. There are six terms in the model of part (c) (three main effects and three interactions); can we leave any of them out?

In fact, we shall not consider leaving out any of the three main effects. Many statisticians recommend that one should not fit models which include the interaction of two factors without including the main effects of the corresponding factors (and, similarly, that one should not include a three-factor interaction without including all the main effects and two-factor interactions of the factors concerned, and so on). Models which always include all the lower-order interactions and main effects corresponding to any interaction they include are known as *hierarchical* models. We shall be using only hierarchical models, because they are often easier to interpret than are non-hierarchical models and because some GENSTAT commands (e.g. STEP) do not allow non-hierarchical models to be fitted in many circumstances.

(d) Fit, in turn, the three models that include all the main effects and two of the two-factor interactions. You can do this most easily using the **Change Model** button in the **Generalized Linear Models** window. Test whether the model with all of the two-factor interactions is significantly better than any or all of these models.

(e) Refit the best-fitting model that you found in part (d), namely the model with all three main effects together with the **class.vote** and **vote.gender** interactions. Save the fitted values from this model in a variate called **fits**. Clearly these fitted values are not the same as the original counts, because the residual deviance for this model is not zero. Check this by printing them out in a table, as in part (a). (You will need this table in part (f).)

In Section 12.1, you saw that, in a two-way table for a model that left out the two-way interaction between the factors, the fitted values were calculated (as for the chi-squared test) by assuming that the row and column totals should be the same as in the original data. What are the corresponding data totals here for the two two-factor interactions (**class.vote** and **vote.gender**) left in the model? And what are the corresponding sets of fitted values? To answer these questions in the case, for example, of the **class.vote** interaction, use the **Stats|Summary Statistics| Summaries of Groups (Tabulation)** menu item to produce a tabulation of the original data, ignoring the **gender** factor, by asking for the **Totals** of **count** with **vote** and **class** as the **Groups**. Click on **OK** but not on **Cancel**. This produces a two-way table, classified by **vote** and **class**, in which each cell contains the total of the counts in the two corresponding cells (one for males and one for females) in the original data table. Check with Table 12.3 that this is indeed what has happened. Now produce a similar table of the fitted values, by going back to the **Summary by Groups** dialogue box and changing the **Variate** field to **fits**. How does the resulting table compare with that for the original data?

If you like, you can check that the same thing happens for the other two-way table (i.e. **vote** by **gender**) corresponding to the other two-factor interaction that appears in the model.

Finally, you should check that the two **class** by **gender** tables, corresponding to the omitted **class.gender** interaction, are *not* the same.

What can one do with a loglinear model after fitting it? Here is one example of what one can do.

Suppose you chose a person at random from the adult population of Labour and Conservative voters in Banbury in 1967 (i.e. the population from which the sample in Table 12.3 was drawn). Suppose that you find out that this person is a woman and a Conservative voter. What, according to the model, is the probability that the person's occupational class is non-manual? That is, what is

$$P(\text{class} = \text{nonman} \mid \text{vote} = \text{con}, \text{gender} = \text{female})?$$

In what follows, you will see how to answer this question. We shall answer it in a slightly roundabout fashion, in order to demonstrate a link between loglinear models and logistic regression.

From the table of fitted values (classified by all three factors) that you found in part (e), we know that the fitted value for the count in the cell for non-manual female Conservative voters is 156.6. Since the total number of people in the sample is $n = 1011$ then, according to the model, 156.6 out of the sample of 1011 are non-manual, female and Conservative. Thus is makes sense to take the fitted value for $P(\text{class} = \text{nonman}, \text{vote} = \text{con}, \text{gender} = \text{female})$ as $156.6/1011 = 0.1549$. (In fact it can be shown, essentially from Theorem 1, that this is the maximum likelihood estimate of the required probability.)

(f) In a similar way, find

$$P(\text{class} = \text{manual}, \text{vote} = \text{con}, \text{gender} = \text{female}).$$

From these two fitted probabilities, one can find the conditional probability we want in several ways. One method is as follows. If we write p for the probability we want, then

$$p = P(\text{class} = \text{nonman} \mid \text{vote} = \text{con}, \text{gender} = \text{female})$$

and

$$1 - p = P(\text{class} = \text{manual} \mid \text{vote} = \text{con}, \text{gender} = \text{female}).$$

Now, we know that the voter in question is female and voted Conservative. Thus, according to the model, she belongs to a subset of the voters that forms a proportion $P(\text{vote} = \text{con}, \text{gender} = \text{female})$ of the total population. Some of this subset are in the non-manual occupational class. They make up a proportion $P(\text{class} = \text{nonman}, \text{vote} = \text{con}, \text{gender} = \text{female})$ of the total population, so they make up a proportion

$$\frac{P(\text{class} = \text{nonman}, \text{vote} = \text{con}, \text{gender} = \text{female})}{P(\text{vote} = \text{con}, \text{gender} = \text{female})}$$

of the subset of female Conservative voters.

Thus

$$P(\text{class} = \text{nonman} \mid \text{vote} = \text{con}, \text{gender} = \text{female})$$
$$= \frac{P(\text{class} = \text{nonman}, \text{vote} = \text{con}, \text{gender} = \text{female})}{P(\text{vote} = \text{con}, \text{gender} = \text{female})},$$

and similarly

$$P(\text{class} = \text{manual} \mid \text{vote} = \text{con}, \text{gender} = \text{female})$$
$$= \frac{P(\text{class} = \text{manual}, \text{vote} = \text{con}, \text{gender} = \text{female})}{P(\text{vote} = \text{con}, \text{gender} = \text{female})}.$$

(These are special cases of the formula $P(A|B) = P(A \text{ and } B)/P(B)$ mentioned in the proof of Theorem 1.) Hence

$$\frac{p}{1 - p} = \frac{P(\text{class} = \text{nonman}, \text{vote} = \text{con}, \text{gender} = \text{female})}{P(\text{class} = \text{manual}, \text{vote} = \text{con}, \text{gender} = \text{female})}.$$

(The $P(\text{vote} = \text{con}, \text{gender} = \text{female})$ terms cancel out.)

(g) Using this result, what is the fitted value of $p/(1 - p)$ from the model? Hence find the value of p, the required conditional probability.

We have found the probability we wanted from a ratio of the sort used in logistic regression. If this conditional probability were the only one that mattered to us, we could simply have found it by logistic regression in the first place (though the answer would have been slightly different – see Section 12.4). But the loglinear model can be used to find *any* conditional probability involving the three factors in the original table, whichever one we choose to treat as a response variable.

■

In general, the fitted values in a loglinear model are such that the corresponding tables of total counts are the same as in the original data. (All this is true of hierarchical loglinear models, at any rate.) For the Banbury example, if we had fitted, say, **class + vote + gender + vote.gender**, then the totals of the fitted values for each class, for each voting group and for each gender would be the same as for the original data, as would the totals in each vote–gender combination, but (because the **class.vote** interaction does not appear in the model) the totals of fitted values and original data would *not* match for each class–vote combination. If we had fitted **class + vote + gender + class.vote + class.gender + vote.gender + class.vote.gender**, then the fitted values for every combination of class, vote and gender would match those in the original data. In other words, the fitted values would be the same as the data – which makes sense, because this is the saturated model. In fitting a loglinear model, GENSTAT essentially constrains the fitted values so that their totals match the data in the way they should. In some cases, there is an explicit formula to say what the fitted values should be; you will know what the formula looks like for two-way tables, because one uses it to calculate the expected values for a chi-squared test. However in most cases one needs to use an iterative method, and in fact GENSTAT always does this. You are spared the details of how the iteration works.

You probably noticed that you were not asked to produce residual plots for the Banbury data, or indeed for any other data to which you have fitted a loglinear model. This is because in very many cases, with loglinear models, such plots do not show you anything very useful. Questions of the model not fitting the data do not arise in the same way that they do with some other GLMs, because there is always a model in the family being considered (i.e. the saturated model) that fits the data exactly, so the question of having to transform something or use a different distribution does not usually arise. GENSTAT still flags points (i.e. table cells) with large residuals, or which have high leverage, as it does with all GLMs. You can use this flagging (or look at the Cook statistics) to investigate whether certain cells are causing a simple model to fail to fit. If this does occur, then one might be able to go back and check to see if these cell counts have been recorded correctly.

Problems with the modelling assumptions do arise with multinomial models, but it is often impossible to investigate them once the data have been recorded as a contingency table. For instance, there may be dependence between successive observations; but the order of the observations is not recorded when they are summarized in a table. To test for lack of independence of this sort, one needs to go back and look at the original data sequence, and methods for doing so are beyond the scope of this book.

In the previous section, you met a model for the situation where the data arose as a single sample of fixed size from a population. But you have probably met contingency tables where this sampling model was not valid, such as the following.

Example 12.3

In an experiment to explore the issue of whether people are generally more helpful to females than to males, eight students approached people and asked if they could change a 5 p coin into 1 p and 2 p coins. Altogether 100 people were approached by male students and 105 by female students. These numbers (100 and 105) were essentially fixed in advance. The results of the experiment are shown in Table 12.4.

Table 12.4

Gender of student	Help offered	Help not offered	Total
Male	71	29	100
Female	89	16	105

Source: Sissons, M. (1981) 'Race, sex and helping behaviour', *British Journal of Social Psychology*, **20**, 285–292.
Dataset name: `helping`.

You probably know from your previous studies that one can analyse these data using the chi-squared test, or Fisher's exact test, just as if the overall total rather than the two row totals had been fixed in advance. In this case, rather than the null hypothesis being one of no association between the row and column factors, it would state that the probability of being offered help was the same for males as for females; but the calculations would be the same. One could also fit a binomial (logistic) regression model, treating gender as the explanatory variable.

Can we still fit a loglinear model? The cell counts no longer have a multinomial distribution, because the row totals are fixed, and for a multinomial distribution only the overall total is fixed. But (assuming the results for females are independent of those for males) we can think of the data for female students as resulting from one multinomial distribution (actually binomial in this case because there are only two classes) and those for male students as resulting from another, independent, multinomial distribution. Since they are independent, the overall likelihood would be the product of two multinomial likelihoods: such a model is an example of a *product multinomial* model. We cannot apply Theorem 1 directly; but a similar theorem applies and, subject to a proviso that we shall come to, we can fit a product multinomial distribution, just as we fitted the multinomial distribution, by pretending that the counts are Poisson (and using a log link function).

The proviso is that we cannot fit a model that would lead to total counts for male students and for female students that would not be equal to 100 and 105 respectively, because those numbers are fixed. But we can ensure this by always including in our model the main effect of the factor **gender**, which defines the rows in Table 12.4. Then the fitted values will always add up to 100 for males and 105 for females, the same as for the data. If the main effect term for **gender** were omitted, we would be saying that the distribution for the cell counts was not affected at all by gender, so the fitted totals for males and females would both be equal to 102.5 (= (100 + 105)/2). (Actually, even when only the overall total is fixed in advance, it is unusual to find the main effects omitted from a loglinear model even though hypotheses about main effects are not usually of much interest. In this sort of analysis, hypotheses of interest are usually about *relationships* between variables, which correspond to interactions rather than main effects. However, if only the overall total is fixed, it is statistically legitimate to fit a model with main effects omitted; while, if other totals are fixed as well, it makes no statistical sense to omit the corresponding effects. For similar reasons, it would make no statistical sense to omit the constant term in a loglinear model for multinomial data.) Thus we can proceed exactly as before, as long as we take care never to omit the term for the fixed marginal totals from our model.

To be very precise, the total deviances in the loglinear models are incorrect, because they are defined in terms of differences in log-likelihood between the model that is fitted and a null model, and here the correctly defined null model would involve fitting a different mean count for the male students than for the female students (because of the fixed totals). The total deviance as calculated by GENSTAT is worked out using a null model that fits the same mean for males and females, because there is no way of telling GENSTAT that the row totals are fixed. However, this means that the total deviances calculated by GENSTAT differ from the correct total deviances by a constant amount, and hence (since the residual deviances are calculated correctly) the regression deviances calculated by GENSTAT are incorrect by the same fixed amount. All the tests one might make involve either comparing the (correct) residual deviance with a chi-squared distribution or comparing the difference between two regression deviances with a chi-squared distribution. When taking the difference, the constant amounts by which each of them is wrong cancel out, so the test is still correct.

Exercise 12.6
Load the data from `helping.gsh` into the GENSTAT spreadsheet and check that they look like Table 12.4. Carry out a test of the null hypothesis that the probability of being offered help is the same for males as for females, using a chi-squared test (with Pearson test statistic), using Fisher's exact test, and by fitting loglinear models. Report your results.

The next exercise shows a more complicated example of the same sort of thing, where chi-squared tests would not be straightforward.

Exercise 12.7
The data in Table 12.5 come from an experiment on the effect, upon their mortality the following autumn, of planting longleaf and slash pine seedlings half an inch too high or half an inch too deep in winter. The experimental design fixed the number of pine seedlings of each type planted at each depth at 100.

Here we really have four different binomial distributions, one for each combination of type of seedling (longleaf or slash) and depth of planting, and it looks pretty obvious that they differ in terms of the probability of survival. But how is that probability related to the planting depth and the type of seedling? We could analyse the data by fitting a logistic regression model; but instead, let us fit a loglinear model.

(a) Load the data from `pineseed.gsh` into the GENSTAT spreadsheet and check how they are laid out. You will see that there are three factors, **depth**, **seedtype** and **mortalty**, each at two levels.

Table 12.5

Depth of planting	Longleaf seedlings			Slash seedlings		
	Dead	Alive	Total	Dead	Alive	Total
Too high	41	59	100	12	88	100
Too low	11	89	100	5	95	100

Source: Wakeley, P. C. (1954) 'Planting the southern pines', *U.S. Department of Agriculture Forest Service Agricultural Monographs*, **18**, 1–233.
Dataset name: pineseed.

The design of the study fixed the counts for each of the four combinations of **depth** and **seedtype** at 100; so we can only fit loglinear models if they use fitted values that preserve these fixed totals. In order to fix the totals for *combinations* of the two factors, we therefore must always include the terms **depth+seedtype+depth.seedtype** (or **depth*seedtype** for short) in the model. Also, as usual, there will be no interest in looking at models that do not include the main effect of the other factor, **mortality**. This leaves five models (including the saturated model, **depth*seedtype*mortalty**) that we might consider. List the five.

(b) Use GENSTAT to fit the four non-saturated models that you listed in part (a). Do any of them fit adequately, compared with the saturated model? Which model seems to be the best? What do you conclude about the mortality of pine seedlings?

Examples like the pine seedling data indicate, in passing, one reason why it can be awkward to use the STEP command in fitting loglinear models. One must set it up in such a way that it does not omit from the model terms that are there because certain totals are fixed by the design of the study. The command *can* be set up to do this, but it is rather messy. Also, in complicated problems where there can be interactions of many orders, automatic stepwise fitting does not always work well. Some authors recommend a process whereby one decides first what order of interactions to work with and then uses a stepwise procedure within a particular order of interaction. One might find, for instance, that a model including all the two-factor interactions, but no higher-order interactions, does not fit, but that a model with all the three-factor interactions does fit. One might then decide to leave all the two-factor interactions in, but to use a stepwise method to decide on a limited number of three-factor interactions to include. Let us see how this might work out in practice.

Exercise 12.8

A survey was carried out in Denmark, in which a sample of employed men aged between 18 and 67 were asked whether, in the preceding year, they had carried out work on their home which they would previously have employed a craftsman to do. (Source: Edwards, D. E. and Kreiner, S. (1983) 'The analysis of contingency tables by graphical models', *Biometrika*, **70**, 553–565. Dataset name: danish.) All together 1591 men responded. They were classified according to five categorical variables:

- age (factor **age**): under 30, 31–45, over 45;

- type of work (factor **work**): skilled, unskilled, office;

- tenure (factor **tenure**): rent, own;

- type of accommodation (factor **acctype**): apartment, house;

- response to the question on do-it-yourself work (factor **resp**): yes, no.

These categorizations lead to a five-way contingency table containing $3 \times 3 \times 2 \times 2 \times 2 = 72$ cells. The counts in these cells are given in the datafile danish.gsh. We could fit a logistic regression model to these data (and, indeed, we shall in the next section), because there is a binary response variable, **resp**. But, for now, let us find an appropriate loglinear model, treating all the factors on an equal basis.

(a) Load the danish data into the GENSTAT spreadsheet and check how they are laid out.

Here no marginal totals other than the overall sample size are fixed in advance, so there are no resulting constraints on the models we are allowed to fit. Begin by fitting a model with only main effects for each factor, by entering

age+work+tenure+acctype+resp

as the Model to be Fitted in the Generalized Linear Models dialogue box. Does this model fit adequately?

(b) The main-effect-only model fits very badly. Fit a model including all the main effects and the two-factor interactions only: a relatively easy way to write this is as

(age+work+tenure+acctype+resp).(age+work+tenure+acctype+resp)

which works because GENSTAT uses the rules of algebra to expand the expression, and interprets terms like **age.age** as **age**, and treats things like **age.work+work.age** simply as **age.work**. Does this model fit well? If not, go on to fit a model including all the three-factor interactions as well, and so on. Stop when you have a model that fits adequately.

The remainder of this question involves the most computationally burdensome part of the whole book. If your computer is not especially fast, you will be at no great disadvantage if you simply read through the rest of the exercise and look at the associated solution.

(c) You should have found that a model going as far as two-factor interactions does not fit the data, but a model with three-factor interactions does fit. Use stepwise fitting to see which, if any, of the three-factor interactions can be left out. To do this, you need to start from the model with all main effects, two factor interactions and three-factor interactions included; so fit that model again, if it was not the last model you fitted. Then you will need to use the STEP command, in the Input Log in interactive mode as usual, telling GENSTAT just to use the three-factor interactions in its steps. Since there are ten such interactions, the command is going to be messy. There are several ways to write the model specification, but the most obvious is just to list all the three-factor interactions, as follows.

```
STEP [MAX=20;IN=4;OUT=4] age.work.tenure+age.work.acctype
     +age.work.resp+age.tenure.acctype+age.tenure.resp
     +age.acctype.resp+work.tenure.acctype+work.tenure.resp
     +work.acctype.resp+tenure.acctype.resp
```

(This command is so long that you might find it difficult to read if you type it into the **Input Log** all on one line! You can go on to a new line before the end of the command, if you wish, by typing a backslash \ followed by <Return> and then continuing with the rest of the command on the next line. The backslash tells GENSTAT that the command is incomplete. Do not put a backslash at the end of the last line of the command, or GENSTAT will sit there waiting for yet more!) Try this STEP command. (It may take a long time!) What model does GENSTAT end up with? Check its residual deviance to see whether it fits well enough.

(d) Now you should try a stepwise fitting process that is not restricted to looking at three-factor interactions. However, as in most loglinear modelling situations, it would make little sense here to omit the main effects. So what we shall do is the same as in part (c), except that we shall allow two-factor interactions to be dropped from the model. Go back to the **Generalized Linear Models** dialogue box and refit the model including all main effects, two-factor interactions and three-factor interactions, as you did in part (b). Then the stepwise fitting command is as follows.

```
STEP [MAX=30;IN=4;OUT=4;FACTORIAL=3]
     (age*work*tenure*acctype*resp)
     -(age+work+tenure+acctype+resp)
```

The age*work*tenure*acctype*resp part of the command tells GENSTAT to fit (in a stepwise fashion) the main effects of all of the factors, together with their interactions right up to fifth-order. We do not want to go above third-order interactions, and the FACTORIAL=3 part of the command ensures that GENSTAT does not bother with fourth- and fifth-order interactions. (Strictly speaking, this specification is unnecessary here – the way GENSTAT is set up to perform the STEP command in this context, it will not fit interactions beyond third order anyway, unless you ask for them specifically. However, putting it in makes it explicit what we want to do, and there are other circumstances where STEP *would* go beyond third order interactions if FACTORIAL=3 were not included.) Finally, including -(age+work+tenure+acctype+resp) in the command tells GENSTAT not to use the main effects in the stepwise fitting (so that the main effects will remain in the model, since they were in the last model you fitted). Here STEP is working with 10 three-way and 10 two-way interactions, so the previous rule of thumb for choosing MAX gives MAX $= 2 \times 20 = 40$. However, another rule of thumb that is just as sensible is to guess some reasonable number (e.g. 30) for MAX; then, if your guess is too small and GENSTAT is still stepping when the MAX number of steps is reached, you can simply use the STEP command again with a higher value for MAX.

Carry out this stepwise fitting. Do you end up with the same model as in part (c)?

12.4   Logistic and loglinear models

In the previous section, you fitted loglinear models in several examples where the data could be thought of as including a binary response variable. In these cases, you could have fitted a logistic regression model instead. Remember that the logistic regression model is a GLM involving a binomial distribution and logistic link function. What is the relationship between logistic regression and loglinear modelling, for datasets to which both methods can be applied? This section aims to answer this question.

Exercise 12.9

The datafile danishlr.gsh contains the same data as in danish.gsh, except that it is laid out for logistic regression instead of loglinear modelling. (It is a little tedious to convert from one format to the other in GENSTAT, so it has been done for you.) Open both danishlr.gsh and danish.gsh at the same time, and check how they match up. Basically, each row in danishlr.gsh contains the counts from two of the rows in danish.gsh, namely the two rows that have the same values of the explanatory variables age, work, tenure and acctype but different values of resp, and in danishlr.gsh these two counts are contained in variates called yes and no, corresponding to the values of resp in danish.gsh. The explanatory factors in danishlr.gsh are distinguished from those in danish.gsh by being called age1 rather than age, and so on. This is simply so that GENSTAT can tell the two versions apart, if you happen to read both of them into the GENSTAT server at the same time.

Update the server so that it knows about the variables in danishlr.gsh. Calculate a new variate, n, as the sum of yes and no. This will be the binomial n for these data. Fit a logistic regression model, entering n as the **Number(s) of Subjects**, yes as the **Numbers of Successes**, and age1*work1*tenure1*acctype1 as the **Model to be Fitted**. Because of the way GENSTAT is set up to perform logistic regression, this fits all the main effects and two- and three-factor interactions, but *not* any higher-order interactions, so in this situation *not* the four-factor interaction. See if the model can be simplified by using (in the Input Log in interactive mode) the following command.

    STEP [MAX=30;IN=4;OUT=4] age1*work1*tenure1*acctype1

What model do you finish up with? (You will need to refer to the output from this fit in the next exercise.)

How does this match up, if at all, with the model you fitted in Exercise 12.8? To see, let us first investigate how things might work in a simpler model. For the data from the experiment on helping that you met in Example 12.3, the loglinear model analysis involved looking for an interaction between the factor help (representing the response) and the factor gender (representing the explanatory variable). Since there was evidence of this interaction, we concluded that gender did have an effect on propensity to help. If you analysed these data using logistic regression, the presence of a main effect of gender would be evidence of the same effect. So it seems that two-factor interactions involving the response factor, in a loglinear model, correspond to main effects of the explanatory variables in a logistic regression.

In fact, this is so in general. Suppose one analyses the same dataset by logistic regression and by loglinear modelling, and (a very important proviso) suppose that all the fitted loglinear models include the main effect of the response variable and a saturated model for all the explanatory variables – that is, they all include all the main effects for the explanatory variables and the response variable, and all the interactions, of whatever order, involving just the explanatory variables. Then the results from the two modelling approaches will *match exactly*, if one identifies the main effects of explanatory variables in the logistic regression with interactions between the response variable and the explanatory variables in the loglinear model, two-factor interactions between explanatory variables in the logistic regression with three-factor interactions between the response variable and the two corresponding explanatory variables in the loglinear model, and so on. (In fact this exact matching may not occur if there are too many cells with zero counts in the wrong places in the contingency table.) To say that the results match exactly here means that the residual deviances will be the same for corresponding models, and that corresponding parameter estimates will be the same as well. The reason all this works is that the two models involved can be shown to amount to the same thing, subject to the proviso about a saturated model on the explanatory variables. This is because they both involve logs of probabilities; you are spared the details. The reason for the proviso is to fix the fitted values of the total counts for all combinations of levels of the explanatory factors in the loglinear models to be the same as in the data. The logistic regression essentially makes the assumption that these counts are fixed, because it treats the values of the explanatory variables as fixed rather than setting up a probability model for them.

To see how this works, let us work again with the data on Danish do-it-yourself.

Exercise 12.10

(a) The model you fitted in Exercise 12.8(d) does not satisfy the proviso, because it does not involve all the three-factor interactions between explanatory variables, let alone their four-factor interaction. None the less, note the terms included in this model that involve the response factor, and check how they match up with the terms in the logistic regression model you ended up with in the stepwise fitting in Exercise 12.9.

(b) Now fit a loglinear model that *does* satisfy the proviso. To do this, first load the data from danish.gsh into GENSTAT again, if necessary. The simplest model that fulfils the proviso is one that includes the main effect of **resp** and all the main effects and interactions involving **age**, **work**, **tenure** and **acctype**. Remembering to choose **Log-linear modelling** in the **Analysis** field in the Generalized Linear Models dialogue box, fit this model by entering **count** in the **Response Variate** field and **resp+age*work*tenure*acctype** as the Model to be Fitted. (For log-linear models, unlike logistic regression, GENSTAT does by default fit interactions of higher order than three, so this model specification works.) Check the output; you should find this model does not fit well.

Now try a stepwise model fitting procedure, stepping only through models that fulfil the proviso, by using the command

```
STEP [MAX=30;IN=4;OUT=4;FACTORIAL=5] resp.(age*work*tenure*acctype)
```

in the **Input Log** in interactive mode. The model specification here allows GENSTAT to step with terms that are an interaction between **resp** and one or more of the explanatory factors, but it does not specify either the main effect of **resp** or any terms involving only explanatory factors (since these terms are all in the model started from, they remain in it as required for the proviso). The FACTORIAL=5 part of the command tells GENSTAT to consider fitting all interactions of order up to five between the five factors in the model, rather than stopping at order three (which would be the default).

What model do you finish up with? Check that it corresponds to the final model in Exercise 12.9. Look at the parameter estimates from both models and check how they match up. Interpret the final model.

Exercise 12.11

Imagine that, instead of a loglinear model, you were going to analyse the data on pine seedlings that you met in Exercise 12.7 using logistic regression, with mortalty as the response variable.

(a) Explain why all the models you fitted in Exercise 12.7 satisfied the proviso described above.

(b) The loglinear model that we ended up by fitting in Exercise 12.7(b) was depth*seedtype+mortalty+mortalty.depth+mortalty.seedtype. What logistic regression model does this correspond to? Explain why, if you had used logistic regression to model these data, you would have ended up with exactly this model.

It seems, then, that when we have data with a binary response variable and explanatory variables that are all categorical, we have a choice of ways to analyse them. We can use either logistic regression or loglinear modelling. We can even ensure that they come to the same answer by including certain terms in the loglinear models. What are the pros and cons of the two approaches?

A key advantage of logistic regression is that it generally takes fewer computing resources, in terms of storage space and time, than loglinear modelling. This is essentially because it does not model the relationships between the explanatory variables, but just takes them as given in the data. (Particularly if your computer is not very fast, you may have found that the loglinear analyses of the Danish data took a long time.) Another advantage is that logistic regression can deal with continuous explanatory variables as well as categorical ones. Indeed, in some cases where there are explanatory variables with many ordered categories, the dataset might be too large to analyse using a loglinear model, but it might still be possible to analyse them using logistic regression by treating these explanatory variables as continuous.

The advantages of loglinear modelling are essentially the mirror image of those of logistic regression. Loglinear modelling *does* involve modelling the relationships between the explanatory variables, and this may be of some interest or importance in certain situations. Some (but far from all) statisticians would argue that one should always fit a saturated model for the explanatory factors; however, if you do not, you may end up with a simpler model for the data overall. (There can be difficulties in fitting a saturated model for the explanatory factors if certain combinations of the explanatory factors do not occur in the data, either by chance or by design. For these combinations, the loglinear model says nothing about the value of p for the response variable. There are ways of dealing with this difficulty, but they are beyond the scope of the book.) The loglinear approach works just as well for categorical response variables with more than two categories, for which the usual logistic regression approach that you have studied in the book so far does not work. Finally, it is also applicable to the case where there is no response variable and all variables are treated on an equal footing.

Summary of methodology

1 Data that arise when a fixed number of individuals fall into cells at random, independently of one another, can be modelled by the multinomial distribution.

2 A theorem on the relationship between the multinomial and Poisson distributions ensures that multinomial data can be fitted by using a generalized linear model with a Poisson response distribution and a log link function. The resulting models for counts in the cells of contingency tables are known as loglinear models.

3 One can interpret the presence or absence of interactions in a loglinear model in terms of independence of variables. The fitted values in a loglinear model are such that they match the totals in the original data corresponding to the terms included in the model.

4 In loglinear modelling, the saturated model (which includes all possible main effects and interactions) fits the data exactly, with zero residual deviance. One can assess whether an unsaturated model fits the data adequately by comparing its residual deviance with a chi-squared distribution with the appropriate number of degrees of freedom. The residual deviance is given by

$$2 \sum O_i \log \left(\frac{O_i}{E_i} \right)$$

where the O_is are the observed values in the cells of the table, the E_is are the fitted values under the model, and the sum is over all cells in the table.

5 Datasets with a binary response variable and a set of explanatory variables that are all categorical can be modelled either by logistic regression or by loglinear modelling. Provided (a) all possible combinations of levels of the explanatory variables are represented in the data and (b) the loglinear models include the main effect of the response variable and a saturated model for the explanatory variables, the results of these two analyses will match exactly (except when there are zeros in unhelpful places in the contingency table).

13

Further data analyses

In this final full chapter, you will get to analyse four rather different but interesting datasets using techniques from all parts of the book.

13.1 Agglomeration of alumina

This section concerns data on a process in chemical engineering. The response variable (agglom) is a measure of the agglomeration of aluminium trihydroxide (AL(OH)$_3$) crystals (or alumina crystals, for short), i.e. the difference between the percentage of crystals exceeding a certain size on leaving and on entering the agglomerator tank. The process involved is called the Bayer precipitation process, and there are nine other measured variables, the explanatory variables, namely:

(a) the flow rate of fine seed (kilolitres/hour) [flowfine];

(b) the solids concentration (grams/litre) [sol];

(c) the flow rate of pregnant liquor (kilolitres/hour) [flowpreg];

(d) the caustic concentration in the pregnant liquor (grams/litre) [causpreg];

(e) the ratio of alumina concentration to caustic concentration in the pregnant liquor [ratpreg];

(f) the temperature (°C) [temp];

(g) the caustic concentration in the overflow (grams/litre) [causover];

(h) the ratio of alumina concentration to caustic concentration in the overflow [ratover];

(i) the solids concentration in the overflow (grams/litre) [solover].

Interest lies in understanding how the explanatory variables affect the agglomeration of alumina crystals. The data are given in Table 13.1.

Exercise 13.1
Looking at Table 13.1, which variables are (to be treated as) continuous and which are categorical? Which kind of technique would you expect to be able to use to analyse these data?

Table 13.1

Case	agglom	flowfine	sol	flowpreg	causpreg	ratpreg	temp	causover	ratover	solover
1	57.8	28.50	414	642	251	0.61	93.6	254	0.55	34
2	58.3	28.50	437	636	251	0.60	92.4	255	0.55	44
3	59.3	27.00	342	643	249	0.59	90.0	249	0.55	27
4	50.5	27.75	386	614	248	0.60	91.2	250	0.53	37
5	58.3	27.75	430	647	251	0.60	92.4	250	0.54	35
6	59.2	29.25	405	683	249	0.60	93.6	253	0.54	33
7	55.5	35.25	390	653	246	0.60	92.4	255	0.53	37
8	56.9	30.00	350	662	250	0.60	92.4	254	0.54	30
9	56.8	26.25	492	718	244	0.60	90.0	255	0.54	31
10	60.1	27.75	428	607	244	0.60	91.2	251	0.54	38
11	60.4	22.50	478	557	240	0.61	92.4	250	0.53	42
12	73.1	26.25	532	619	248	0.66	92.4	254	0.53	35
13	66.1	30.00	504	626	251	0.60	94.8	259	0.54	35
14	61.1	30.00	483	603	244	0.61	91.2	256	0.54	39
15	65.4	32.25	391	633	251	0.61	96.0	256	0.54	30
16	67.5	29.25	345	621	255	0.58	94.8	263	0.54	33
17	58.8	25.50	322	549	251	0.60	93.6	261	0.53	31
18	63.3	0.00	359	632	251	.59	92.4	255	0.52	35
19	60.5	26.25	385	685	245	0.61	94.8	254	0.55	27
20	63.0	25.50	345	631	239	0.60	93.6	246	0.55	30
21	48.2	26.25	406	575	240	0.60	92.4	241	0.54	32
22	61.7	30.00	417	562	240	0.60	92.4	245	0.54	36
23	59.0	32.25	393	620	238	0.61	92.4	241	0.54	36
24	48.7	24.75	471	645	240	0.60	94.8	239	0.55	37
25	55.0	30.75	444	620	244	0.61	94.8	250	0.54	32
26	51.2	34.50	389	589	239	0.61	93.6	240	0.54	41
27	40.2	33.75	451	646	241	0.59	92.4	244	0.52	39
28	46.5	39.00	406	744	241	0.60	92.4	248	0.54	33
29	38.9	42.75	386	748	243	0.60	94.8	245	0.54	33
30	50.5	12.00	374	728	244	0.60	94.8	250	0.53	32
31	42.4	36.00	456	696	239	0.60	93.6	246	0.54	35
32	26.5	48.00	420	679	240	0.61	93.6	251	0.44	41
33	33.8	42.00	396	654	240	0.59	90.0	246	0.53	49
34	31.3	29.25	433	636	233	0.60	92.4	244	0.50	84
35	34.5	27.75	431	644	234	0.60	92.4	244	0.53	37
36	36.4	26.25	406	632	231	0.60	93.6	234	0.53	29
37	27.0	33.75	356	647	235	0.59	93.6	241	0.55	32
38	31.0	34.50	393	658	236	0.59	94.8	244	0.53	32
39	43.2	45.75	282	630	239	0.60	94.8	238	0.54	35
40	39.3	48.75	362	647	241	0.60	94.8	248	0.53	38
41	51.3	59.25	292	553	241	0.61	93.6	246	0.53	44
42	27.1	21.75	328	576	253	0.49	87.6	250	0.47	31
43	32.3	38.25	439	594	233	0.58	91.2	218	0.46	33
44	57.8	29.25	384	648	240	0.60	88.8	243	0.55	30
45	54.5	35.25	318	603	243	0.60	90.0	250	0.54	34
46	33.7	37.50	351	585	245	0.59	90.0	248	0.53	32
47	68.5	39.00	325	599	244	0.59	91.2	248	0.52	37
48	60.2	39.00	341	648	246	0.59	91.2	245	0.53	30
49	51.6	38.25	322	652	248	0.60	92.4	250	0.54	33
50	63.8	36.75	294	639	241	0.60	92.4	250	0.55	32
51	58.9	35.25	313	548	246	0.60	91.2	249	0.54	36

Source: Sommer, S. and Staudte, R. G. (1995) 'Robust variable selection in regression in the presence of outliers and leverage points', *Australian Journal of Statistics*, **37**, 323–36.
Dataset name: chemeng.

Load the chemeng data into GENSTAT and produce a scatterplot matrix. The appropriate (interactive) command is

```
DSCATTER flowfine,sol,flowpreg,causpreg,ratpreg,temp,
         causover,ratover,solover,agglom
```

Comment on what you see (with particular regard to those scatterplots with agglom on the vertical axis).

It seems that we might have to consider our diagnostics somewhat carefully in this example. But as a starting point, forget about the fact that there are unusual points and do Exercise 13.2.

Exercise 13.2

With nine explanatory variables in the dataset, it seems that variable selection is in order. Looking up Example 6.2 of Chapter 6 for the relevant GENSTAT commands if necessary, investigate which variables should appear in the model. You have already done one part of this in Exercise 13.1 where you looked at the individual relationships between agglom and the explanatory variables in the scatterplot matrix. Further tasks include: obtaining and looking at the correlations between explanatory variables; performing the multiple regression of agglom on all nine explanatory variables and seeing which seem to be most important; and running GENSTAT's stepwise regression facility, starting both from the full model and from the null (no-explanatory-variable) model. What set of explanatory variables do you come up with?

It has already been noted that there are various extreme points that might be having a big effect on our modelling of this dataset. To investigate this in more depth, let us concentrate on the regression of agglom on the four explanatory variables flowpreg, causpreg, ratpreg and ratover preferred in Exercise 13.2.

Exercise 13.3

(a) First, if necessary, reperform this four-variable multiple regression. Next, use the facilities provided by the Model Checking button, in the dialogue box produced by the Further Output button in the Linear Regression dialogue box, to produce a plot of residuals against fitted values, and leverage and Cook statistic index plots. Discuss what you find.

(b) With the assistance of the data (in the GENSTAT spreadsheet or in Table 13.1) and the scatterplots (Solution 13.1), find out what it is about units 32 and 43 that is causing their behaviour. Which other units have particularly unusual values of ratpreg and/or ratover?

(c) Use the Spread|Restrictions|Add Selected Rows menu item to remove rows 32 and 43 from the analysis temporarily. Repeat the stepwise regression on the reduced dataset, allowing the inclusion of any of the nine original explanatory variables. Which variables are selected to be in the model now? Comment on the effect of removing rows 32 and 43. By looking at appropriate diagnostic plots for the set of variables now chosen, comment on any remaining important single datapoints.

(d) Given that we have already identified units 12 and 42 because of their unusual values of **ratpreg**, it seems worthwhile to investigate whether they are primarily responsible for the inclusion of **ratpreg** in the chosen model. Redo part (c) with rows 12 and 42 restricted also, and discuss what happens.

Unit 16 – which is flagged at the end of Solution 13.3 – has the highest values of **causpreg** and **causover**. But it turns out that removing row 16 as well does not make a great deal of difference to the analysis. (You need not bother to check this.) Along the way, you may also have noticed point 34 and its extraordinarily high value of **solover**, but this had no particular effect on the analysis either.

The bottom row of the scatterplot matrix when points 12, 32, 42 and 43 are removed is given as Figure 13.1.[1] This might be useful, in conjunction with Table 13.1 and the figure in Solution 13.1, in answering Exercise 13.4(a).

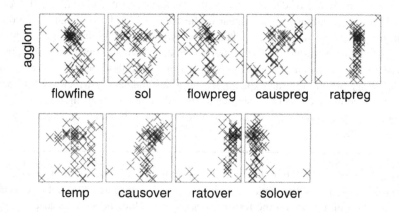

Fig. 13.1

Exercise 13.4

(a) What is the final model fitted by the analysis of Exercise 13.3? What could you say to the chemical engineers about the range of applicability of this model in the light of the datapoints that you removed in this analysis? Refer to the scatterplots as appropriate.

(b) Obtain a 95% confidence interval for the mean response (**agglom**) when **causpreg** and **ratpreg** take the values 244 and 0.60, respectively. What source of variability has not been taken into account in this confidence interval?

Hint Look at Example 6.3 for appropriate GENSTAT commands.

---

[1] The row of plots will not be split like this on your computer.

13.2 Prostatic cancer

The treatment regime to be adopted for patients who have been diagnosed as having cancer of the prostate depends strongly on whether or not the cancer has spread to the surrounding lymph nodes. A surgical procedure can be performed to ascertain the extent of this 'nodal involvement', yielding a binary variable **nodal** which takes values 1 if the cancer has spread and 0 otherwise. It would be very useful, however, if such surgery could be avoided and a combination of other indicators, measurable without the need for surgery, could be found to predict well whether or not the cancer has spread. To this end, five other variables were collected as potential predictors: the age of the patient (**age**, in years); the level of serum acid phosphatase (**acid**, in King–Armstrong units); the result of an X-ray examination (**xray** = 0 if negative, 1 if positive); the size of the tumour as determined by a rectal examination (**tsize** = 0 if small, 1 if large); and a summary of the 'pathological grade' of the tumour determined by biopsy (**tgrade** = 0 if less serious, 1 if more serious). Data collected on 53 patients with prostatic cancer on whom the nodal involvement operation had been carried out is given in Table 13.2.

Exercise 13.5
Identify the response variable and the explanatory variables in this situation. What kind of modelling approach should you take to analysing data of this sort?

Exercise 13.6
Perform a regression of the type identified in Solution 13.5 on these data (using all the explanatory variables), and explore the leverage and Cook statistic diagnostics. Focusing on the latter, what do you find? Explain this finding with the help of a scatterplot of **nodal** against **acid**.

Perform the rest of this analysis without point 24 in the dataset.   (Use the Spread|Restrictions menu item.) Although there will be no further mention of this patient in the remainder of the section, do not forget that this patient's unusual values should be investigated further and given a prominent place in any reports for, and discussions with, the medics involved.

Let us next think about model selection. There are five basic explanatory variables. There are no high correlations between these variables, so no reason to leave any of them out on these grounds. But what about other potentially important derived variables? There are two ways in which we might expand our set of explanatory variables. One is that we might consider functions of the individual explanatory variables, such as square terms, implying quadratic dependence (on the logit scale). You need not investigate this here; rest assured that there is little need for such terms in this example. The other important possibility, especially given the presence of three binary explanatory variables, is interactions between explanatory variables. To this end, we might consider binary–binary interactions such as **tgrade.tsize**, which allows for different effects of **tsize** at different levels of **tgrade** and vice versa. But we can also consider interactions such as **acid.tgrade** between a continuous

Table 13.2

Age of patient	Acid level	X-ray result	Tumour size	Tumour grade	Nodal involvement
66	0.48	0	0	0	0
68	0.56	0	0	0	0
66	0.50	0	0	0	0
56	0.52	0	0	0	0
58	0.50	0	0	0	0
60	0.49	0	0	0	0
65	0.46	1	0	0	0
60	0.62	1	0	0	0
50	0.56	0	0	1	1
49	0.55	1	0	0	0
61	0.62	0	0	0	0
58	0.71	0	0	0	0
51	0.65	0	0	0	0
67	0.67	1	0	1	1
67	0.47	0	0	1	0
51	0.49	0	0	0	0
56	0.50	0	0	1	0
60	0.78	0	0	0	0
52	0.83	0	0	0	0
56	0.98	0	0	0	0
67	0.52	0	0	0	0
63	0.75	0	0	0	0
59	0.99	0	0	1	1
64	1.87	0	0	0	0
61	1.36	1	0	0	1
56	0.82	0	0	0	1
64	0.40	0	1	1	0
61	0.50	0	1	0	0
64	0.50	0	1	1	0
63	0.40	0	1	0	0
52	0.55	0	1	1	0
66	0.59	0	1	1	0
58	0.48	1	1	0	1
57	0.51	1	1	1	1
65	0.49	0	1	0	1
65	0.48	0	1	1	0
59	0.63	1	1	1	0
61	1.02	0	1	0	0
53	0.76	0	1	0	0
67	0.95	0	1	0	0
53	0.66	0	1	1	0
65	0.84	1	1	1	1
50	0.81	1	1	1	1
60	0.76	1	1	1	1
45	0.70	0	1	1	1
56	0.78	1	1	1	1
46	0.70	0	1	0	1
67	0.67	0	1	0	1
63	0.82	0	1	0	1
57	0.67	0	1	1	1
51	0.72	1	1	0	1
64	0.89	1	1	0	1
68	1.26	1	1	1	1

Source: data of G. Ray, in Brown. B. W. (1980) 'Prediction analyses for binary data', in Miller. R. J., Efron. B., Brown, B. W. and Moses, L. E. (eds) *Biostatistics Casebook*, New York, John Wiley.
Dataset name: `prostat`.

variable, **acid**, and a binary variable, **tgrade**; this allows for different effects of **acid** at the different levels of **tgrade**. In fact, each interaction term is a product of the explanatory variables concerned.

What you are asked to do next is to perform stepwise regression allowing interactions to enter the model if required.

Exercise 13.7

(a) First, perform the following 'full model' binary regression: choose Stats| Regression Analysis|Generalized Linear; let Analysis be Modelling of binomial proportions. (e.g. by logits), Number(s) of Subjects be 1, Numbers of Successes be nodal, and Model to be Fitted be age*acid*xray*tgrade*tsize; click on OK. The result will be a mess and a fault message will come up! Do not worry about this: to tidy up, make the Input Log active and, in interactive mode, locate the FIT command which is the final command generated by GENSTAT above, and insert the term FACTORIAL=2 (with an appropriate semicolon) somewhere inside the square brackets. (FACTORIAL was used in the STEP command in Exercises 12.8 and 12.10.) Press <Return> to reperform the regression. What property have you changed (from what to what) to enable the regression to be performed successfully the second time?

(b) Now perform a stepwise regression, starting from the full model just obtained, and with the FACTORIAL=2 restriction: i.e. use the following command.

```
STEP [MAX=30;IN=4;OUT=4;FACTORIAL=2] age*acid*xray*tgrade*tsize
```

Note which variables are in the chosen model, along with the regression deviance.

(c) Repeat part (b) starting from the null model.

(d) The variables chosen in (c) are a subset of those chosen in (b). Test whether the larger model is necessary by comparing the deviance difference with the appropriate χ^2 distribution.

(e) Obtain again the seven-variable model chosen in part (b) by using acid*tgrade*tsize-acid.tgrade.tsize+xray in the Model to be Fitted field. Notice how this works: the first term fits all main effects and two-way interactions associated with **acid**, **tgrade** and **tsize**, which are six of the seven variables we want; it also fits the three-way interaction between these variables which is unwanted and hence subtracted by the second term; and then **xray** is added to complete the model. Look at the leverage and Cook statistic information for this model. Are you happy with its fit?

Although we have made a good start at looking for an appropriate model for these data, it is clear that there remains scope for further investigation and refining of the model. If you were working directly with the investigators on this project – and you therefore also knew more about the background to the problem and hence could use this to inform the modelling process – such further investigation should indeed be carried out.

However, for the purposes of this section, let us proceed as if the model selected in Solution 13.7 is indeed a particularly good one for this situation, and consider its interpretation. (The model is *a* fairly good one. But there may be other good ones too. And one might also not wish to confine further attention to just a single chosen model; for example, one might like to look at average predictions over several good ones.)

Exercise 13.8

(a) Write down the model fitted in Solution 13.7.

(b) Suppose a new patient with prostatic cancer was aged 60, with acid level 0.55, a negative X-ray result, and a small and less serious tumour. Using your calculator, what is the probability (according to the model) that the patient's cancer has spread to the lymph nodes? (Remember that if $\text{logit}(p) = q$, say, then $p = e^q/(1 + e^q)$.) Check your answer using the PREDICT command in GENSTAT.

(c) Suppose now that the patient actually had a large tumour rather than a small one. By what factor would this increase the odds (i.e. $p/(1-p)$) of having nodal involvement? What value results for $P(\text{nodal} = 1)$ in this case? Again, check your answer using GENSTAT.

(d) Repeat part (c) for the case where the patient has a more serious grade of tumour as well as a large one. Is the result at all surprising?

13.3 Ground cover and apple trees

In Chapter 8 you learned about the use of blocking in designed experiments to reduce the impact of variability between experimental units. The basic idea is that responses from experimental units within the same block are likely to be closer than those from units in different blocks. For example, in an agricultural field experiment, the basic fertility of the ground is very likely to affect the response, so the experimental plots of ground are very often grouped into blocks of similar fertility. However, there are other ways of allowing for variability between experimental units. This is an example.

An experiment was carried out many years ago, at East Malling Research Station in Kent, on the effect of different ways of forming ground cover under apple trees. At the time, the usual method in England was to keep the ground in the plantation clear during most of the growing season, but to let weeds grow up towards the end, ready to dig into the soil when cultivation was restarted the next spring. The object of the experiment was to compare the yield of apples using this method with the yield when various permanent crops such as grass or clover were grown beneath the trees. There were in all six different ground-cover treatments (including the 'control', namely the standard ground-cover treatment described above). Each experimental unit was a group of several trees (all of the variety Barnack Beauty), and the response variable was the total weight of crop (in pounds) gathered over a four-year period. The experimenters took account of variations in soil characteristics, such as fertility, over the experimental site by grouping the experimental units into four randomized blocks, of six units each, on the basis of their location in the orchard.

What the blocking did not explicitly take account of was the fact that the apple trees used for the experiment were very old, and therefore differed considerably in size, shape and general cropping pattern. Thus, for instance, one would expect certain trees to continue giving relatively high yields, year after year, regardless of any ground-cover treatment. Some of the variability might well be taken account of by the blocks that were used, but it is very likely that much of it is not. In fact, the experimenters had some information about the yield of the trees before the experiment: records had been kept of the total yield of crop (in bushels, a unit of volume) in the four years immediately before the experimental ground covers were begun. One possibility would have been to use a different blocking structure, perhaps with two blocking factors, one to take account of the cropping pattern of the trees (set up on the basis of the past cropping records) and another to take account of soil characteristics (which would affect how the ground-cover plants grew). However, an alternative approach was used. The experimenters explicitly took account of the past cropping records in their analysis by adding an extra explanatory variable. An additional continuous explanatory variable of this sort is often referred to as a *covariate*, and the technique involved is sometimes called *analysis of covariance* (sometimes abbreviated to ANOCOVA).

The data, including values of the covariate, are given in Table 13.3. Treatment O is the 'control' (standard ground cover), and A to E are the experimental ground-cover treatments.

Table 13.3 Yields of apples before (x, in bushels) and during (y, in pounds) an experiment on ground-cover treatments.

Treatment	Blocks							
	I		II		III		IV	
	x	y	x	y	x	y	x	y
O	7.6	222	10.1	301	9.0	238	10.5	357
A	8.2	287	9.4	290	7.7	254	8.5	307
B	8.2	271	6.0	209	9.1	243	10.1	348
C	6.8	234	7.0	210	9.7	286	9.9	371
D	5.7	189	5.5	205	10.2	312	10.3	375
E	6.1	210	7.0	276	8.7	279	8.1	344

Source: Pearce, S. C. (1983) *The Agricultural Field Experiment*, Chichester, John Wiley. The data were first reported by Professor Pearce in 1953.
Dataset name: apple.

Although the experiment was not run in quite the same way as those you have met before in this book, it is possible to analyse the data using methods you should be familiar with.

Exercise 13.9

(a) For the data in Table 13.3, identify the response variable and the explanatory variables. In each case, say whether the variable is continuous or categorical.

(b) What model might it be appropriate to start with for your analysis of these data?

The approach here, then, will be to fit a normal linear (multiple regression) model, using one continuous and two categorical explanatory variables. This is different from the models you have usually fitted to data from designed experiments before, where *all* the explanatory variables (for blocks, plots, rows, columns, treatment factors and so on) were categorical, and where we usually used the GENSTAT ANOVA commands to analyse the data. It is also a little different from those multiple regressions including both categorical and continuous explanatory variables that we have performed. The reason is that here we are not particularly interested in estimating the effects either of the block factor or of the covariate (previous yield); they are put into the model essentially to produce more precise estimates of the explanatory variable we really are interested in, the ground-cover treatment.

(This approach is essentially what we have always done with blocking factors, but we have not applied it to continuous explanatory variables before.)

However, the fact that we do not really care what the coefficients of the blocks or of the covariate are makes no difference to how the model is fitted. This experimental design is balanced, and if it were not for the covariate you would know exactly how to analyse it using the GENSTAT ANOVA commands. In fact it is easy to add the covariate into such analyses, as you will soon see; but before then, let us go ahead and analyse the data using the GENSTAT regression commands.

Exercise 13.10

(a) Load the data from the file apple.gsh into the GENSTAT spreadsheet and update the server. Check how the data in the spreadsheet correspond to those in Table 13.3. Use the Stats|Summary Statistics|Summaries of Groups (Tabulation) menu item to produce a table of the means and the variances of the responses for each treatment group. Produce boxplots of the responses for each treatment. On the basis of this brief exploration of the data, do you think that different ground covers lead to different yields of apples?

(b) Now investigate how the covariate, **before**, affects things. Produce a table of the means and variances of the covariate values for each treatment group. Given that the treatments were allocated to the experimental units at random (within blocks), do you have any comments on the covariate means and variances?

Investigate the relationship between the covariate and the response variable by producing a scatterplot of these two variates, with the response variable on the vertical axis and with the different treatments distinguished by different colours. Do you think it would be reasonable to assume a linear relationship between the covariate and the response variable?

(c) Carry out a test of the null hypothesis that there is no difference in the mean response for different ground-cover treatments, as follows. Choose the Stats|Regression Analysis|Linear menu item. In the resulting dialogue box, change the Regression field to General Linear Regression, and enter weight as the Response Variate. The Model to be Fitted requires a little thought. We want to test whether there is an additional effect of the **treat** factor after the effects of **block** and **before** have been taken account of. Because **treat** is a factor with six levels, the table of coefficients and t values in the regression output will have several (in fact five) rows for this factor, and we want an overall test of whether they can all be taken to be zero. This can be done by producing an accumulated analysis of variance table at the end of the output, which essentially shows an F test for the extra effect of adding each variable to the model, given the other variables previously in the model. Thus we have to enter **treat** last in the model. (For this purpose, the order of **block** and **before** does not matter, as long as they both come before **treat**.) Enter block+before+treat as the Model to be Fitted. Click on Options; in the resulting dialogue box, check the Accumulated option and click on OK. Then click on OK in the Linear Regression dialogue box to perform the regression. Report on the results of this analysis. Comment on the values for the treatments in the table of estimates of regression coefficients.

(d) Produce a plot of the fitted model by clicking on **Further Output** in the Linear Regression dialogue box, then on **Fitted Model**, and, in the resulting dialogue box, choosing **before** as the **Explanatory Variable** and **treat** as the **Grouping Factor**. Does the model appear to fit reasonably? Why are the lines parallel? What term would you have to add to the model to allow it to fit non-parallel lines?

(e) Produce appropriate diagnostic plots to investigate whether the model you have fitted is adequate. What do you conclude?

(f) Calculate a 95% confidence interval for the mean increase in yield that would result if the ground cover for an experimental unit were changed from the control (Treatment O) to Treatment D.

(g) Suppose you were asked to find the expected yield for an experimental unit of apple trees 'like those in the experiment', with ground-cover treatment A, say. This question is not terribly well defined. You have seen that the yield differs considerably, depending on the previous yield (**before**) from the trees in question, and the model that has been fitted includes terms for this covariate (and for the blocks). But one possible interpretation is to find the mean yield at the average value of the covariate observed in the experiment, and to average over all four blocks.

Expected responses can be calculated in GENSTAT using the PREDICT command interactively. When you have used this before, you have explicitly given values for all the explanatory variables in the model. But, if you do not tell GENSTAT the values of all explanatory variables in the PREDICT command, it will average over the values of those you miss out.

Thus

```
PREDICT treat;2
```

will produce a mean yield for Treatment A (which is the second level of **treat**) and

```
PREDICT treat
```

will produce a table of expected values for all levels of **treat**.

Type these commands into the **Input Log** in interactive mode and note the results (for comparison with another analysis you will do later). How do the means compare with the actual treatment means you found in part (a)? Can you explain the differences in terms of the covariate means you found in part (b)?

(h) Finally, to investigate further why it was a good idea to use a covariate in this experiment, check what would have happened if the covariate had not been used by repeating the regression analysis in part (c) but without using **before** in the model at all. What happens in the test of the null hypothesis of no treatment differences? How does the residual mean square in this model compare with that in the model of part (c)?

In this experiment, the use of the covariate has had a major impact. It has reduced the residual mean square considerably, allowing differences between treatments to be estimated much more precisely. In general, this is the main aim of using covariates;

like blocking, they can allow for heterogeneity in experimental units. In addition, the presence of the covariate in the model has allowed us to estimate the effect of changing the ground cover on an experimental unit. This could not be done so accurately from the treatment means, which in this case were markedly affected by the fact that the randomization happened to give high-yielding trees to the control treatment. The somewhat unsatisfactory residual distribution casts some doubt on the accuracy of the estimates we have produced, but overall the general picture of treatment differences seems reasonably clear.

For experiments like this, which have the property of first-order balance, analysis of covariance can also be performed using the GENSTAT analysis of variance commands. Let us see how to do this.

Exercise 13.11

Choose the Stats|Analysis of Variance menu item. In the resulting dialogue box, choose **One-way ANOVA (in Randomized Blocks)** as the **Design**. Enter **weight** as the **Y-Variate**, **treat** as the **Treatments**, and **block** as the **Blocks**. (So far, this is the same as you would do for an analysis ignoring the covariate.) Click on the check box next to **Covariates** to choose it. This makes the **Covariates** field available; enter **before** in it. Click on **OK**. Check the resulting output and see how it compares with the output from the regression analysis of Exercise 13.10. Produce the standard residual plots, and comment.

The ANOVA analysis is slightly more complicated than the regression analysis, though, as far as the estimates of treatment differences are concerned, the model is the same and the results are the same. An advantage of using the analysis of variance commands rather than the regression commands (which does not apply in this case) is that they can deal more easily with experiments that have a complicated blocking structure, with effects being estimated in different strata. But, as usual, the analysis of variance approach will not work with unbalanced data.

Before leaving this experiment, it is worth reading what Professor Pearce had to say about how the experiment came to be done.

> The trees were very old. They occupied a field between the East Malling Research Station and the railway, and had become virtually derelict, a fact that incited passing travellers to ungenerous comment. In fact, the Station purchased them almost to save its good name. Four years were spent rehabilitating them, during which time volume records were taken on the fruit. Then someone had the idea of using them for an experiment on ground cover, that being a fashionable topic of study when applied to aged trees.

The process of planning experiments is not always as straightforward as one might think!

13.4   Epileptic seizures

The data of interest in this section come from a clinical trial of a new drug for reducing the number of seizures suffered by patients with epilepsy. The experiment was carried out on 15 patients attending the Westmead Hospital in Sydney, Australia. The patients were monitored for two periods (coded **period** = 0 for the first and 1 for the second). During one of these periods they took their usual drug (coded **treat** = 0) and during the other the new drug (**treat** = 1). Seven patients were given the new drug during the first period and the old drug in the second, and the other eight were given the new drug in the second period and the old drug in the first. There was a sufficient gap between time periods for there to be no residual effect of the treatment given in the first period carrying over to the second period. Patients were randomly allocated to treatment orderings. Such a trial is called a *crossover* trial. The data are given in Table 13.4.

Table 13.4

Patient	Seizures	Exposure	Treat	Period
3	152	56	1	0
8	10	56	1	0
10	23	56	1	0
12	3	56	1	0
13	3	56	1	0
14	19	56	1	0
15	64	56	1	0
1	160	56	0	0
2	110	56	0	0
4	12	54	0	0
5	12	56	0	0
6	10	42	0	0
7	81	56	0	0
9	21	56	0	0
11	1161	56	0	0
3	149	56	0	1
8	0	56	0	1
10	37	56	0	1
12	1	56	0	1
13	22	56	0	1
14	34	56	0	1
15	127	56	0	1
1	117	56	1	1
2	6	56	1	1
4	5	56	1	1
5	7	55	1	1
6	8	49	1	1
7	52	56	1	1
9	4	56	1	1
11	854	56	1	1

Source: unpublished data provided by A. D. Lunn.
Dataset name: `seizure`.

The **seizures** column refers to the number of epileptic seizures experienced by the patient in the specified period when receiving the specified treatment. There is also an **exposure** time variable (in days) which is the length of the period of observation. For most patients this is 56 days, but a few were observed over slightly shorter time spans. So, for instance, patient 3 in period 0 was observed to suffer 152 seizures in 56 days, while patient 6 in period 1 was observed to suffer 8 seizures in 49 days.

Exercise 13.12

For the moment ignore the **exposure** variate. Identify the response variable and the explanatory variables. Load the data into GENSTAT. Make a suitable plot of **seizures** against **patient** and comment on the apparent importance or otherwise of the **patient** factor. (You might have to transform the data.) Include the different levels of **treat** in an appropriate way on your plot, and hence comment on whether the new drug seems better than the old. What kind of model comes to mind as a good first stab at modelling these data?

In trying to fit the model mentioned in Solution 13.12, we immediately come up against the difficulty of trying to take proper account of the information (in **exposure**) on differing exposure times. Rather than simply fitting a Poisson response distribution where, for the ith response, $Y_i \sim \text{Poisson}(\lambda(x_i))$, it will take **exposure** properly into account if we think of $\lambda(x_i)$ as a rate parameter which has to be multiplied by the exposure time t_i to obtain a mean count

$$Y_i \sim \text{Poisson}(\lambda(x_i)t_i).$$

In the usual Poisson regression context, we would take $\lambda(x_i)$ to be a function of a form such that

$$\log \lambda(x_i) = \alpha + \beta_1 \, \textbf{treat} + \beta_2 \, \textbf{period} + \beta_{3.\,j(i)} \, \textbf{patient}_{j(i)}. \qquad (13.1)$$

Here, the i continues to refer to the ith response and the $j(i)$ refers to the patient number of the ith response. (There are two responses for each patient.) But now, using (13.1) and the fact that $\log[\lambda(x_i)t_i] = \log \lambda(x_i) + \log(t_i)$, we need to use GENSTAT to fit

$$\log E(Y_i) = \log t_i + \alpha + \beta_1 \, \textbf{treat} + \beta_2 \, \textbf{period} + \beta_{3.\,j(i)} \, \textbf{patient}_{j(i)}. \qquad (13.2)$$

In order to incorporate the $\log t_i$ terms with a coefficient constrained to be 1 – a so-called *offset* – you will need to click on the **Options** button in the **Generalized Linear Models** dialogue box and enter these terms in the **Offset** field in the **Generalized Linear Model Options** dialogue box.

Exercise 13.13

Fit the Poisson regression model with offset, as given by (13.2). Are you disconcerted by any GENSTAT warnings?

What we have here is a serious case of overdispersion, a phenomenon referred to briefly at the end of Section 10.5.2. Remember that for a Poisson distribution the variance is the same as the mean (both being equal to the parameter λ). What GENSTAT is trying to tell us is that, even when the effects of the explanatory variables on the mean are taken into account, the response distribution appears not to be a Poisson distribution, but to have a variance much greater than the mean.

The main pointer to this overdispersion explanation is the value of the residual mean deviance in the Summary of analysis table in the GENSTAT output. If there is no problem with overdispersion, this value is near one, and certainly less than two. Here it is 10.38. (You might like to check that all the residual mean deviances in earlier Poisson or binomial regressions were indeed less than two.)

In the absence of further example-specific information that might lead us to modify our model (such as some covariate), let us take the first of two approaches to dealing with overdispersion mentioned briefly at the end of Section 10.5.2. This is to introduce a further parameter controlling the degree of dispersion, and to let GENSTAT estimate this parameter also. In fact, we introduce this parameter ϕ by letting the variance become of the form $\phi E(Y_i)$. To do this, click on **Options**, then make **Estimate** checked under **Dispersion Parameter** in the resulting dialogue box. Click on **OK**, then on the next **OK**. This causes the Poisson-based analysis to be performed again but with an estimated dispersion parameter taking account of dispersion. (Note that, as part of the output, GENSTAT tells you that the dispersion parameter ϕ is estimated to be 10.4.)

Exercise 13.14

(a) The effects of just two of the explanatory variables are of substantive interest. Which are they?

(b) You are now going to consider whether one (or both) of the two explanatory variables identified in part (a) can be dropped from the model. To do this, you will have to cope with the overdispersion as you did in Exercise 10.9(c). In particular, to test whether either of these two explanatory variables can be omitted, you need to use the **Change Model** button appropriately, and seek out the `approx F pr.` values in the Change lines in the output. These refer to approximate tests of the differences in deviances taking overdispersion into account. Can either of these two explanatory variables be removed from the model?

(c) Obtain the Poisson regression model preferred in the solution to part (b), if it was not the last model you fitted. Using the **Change Model** button, test whether the interesting explanatory variable here is necessary to the model. Which explanatory variables are in the final model you have chosen?

The fact that the period factor does not appear in the model chosen in Solution 13.14 is reassuring. We do not have information about precisely when the time periods were, but we would not in general expect any difference between numbers of seizures due to time period in addition to any treatment effects. In some crossover trials, one might expect a period effect because of, say, a learning effect. Suppose that the experiment were a psychological one in which it was desired to compare two ways of performing a task. Regardless of which method was used in which order, there may well be a learning effect simply due to getting used to performing the task at all (or even due to getting used to being in a psychological experiment). This might result in an improved performance in the second period regardless of 'treatment'.

By fitting just the **treat, period** and **patient** main effects in this example (and ending up with just two of them in our final model), we have ignored any possible interactions between these factors. It is important for the validity of crossover trials that there is no period by treatment interaction: if the different treatments had different effects in the different periods, it would tell us little about overall treatment efficacy. You cannot test for such an interaction directly in the current setup. There is some information available in the block (**patient**) totals about this interaction, in much the same way as certain effects could only be estimated in the blocks stratum in Chapter 8, but you need not worry about the details of this. It turns out that there is indeed no need to include a **period.treat** interaction in the model.

Exercise 13.15
There are no **patient.period** or **patient.treat** interactions in our model either. Why?

Exercise 13.16
(a) In the output connected with Exercise 13.14, even after accounting for overdispersion, there remained a GENSTAT warning that two units, numbers 9 and 24, had large standardized residuals. Identify these two units in terms of patients and describe what has been brought to our attention.

(b) Remove units 9 and 24 from the spreadsheet (by making the spreadsheet active and using the Spread|Restrictions menu item) and repeat the analysis of Exercise 13.14(b) and (c) omitting this outlier. How is the analysis affected?

(c) Describe the overall conclusions of your study of the `seizure` data.

Overdispersion makes a difference to standard errors and hence to confidence intervals, etc. Exercise 13.17 draws your attention to the big difference the overdispersion makes in this case.

Exercise 13.17
Consider the treatment effect, i.e. the estimated regression parameter $\widehat{\beta}$ associated with **treat**, in the model containing just **patient** and **treat** when patient 2 (rows 9 and 24) is removed from the analysis. When the dispersion parameter ϕ is estimated, what is the value of $\widehat{\beta}$ and its estimated standard error? Rerun the model with offset, **patient** and **treat**, but with the dispersion parameter returned to 1 (by clicking on Fix under Dispersion Parameter in the Generalized Linear Model Options dialogue box, and filling the Value field with 1). What is the value of $\widehat{\beta}$ and its estimated standard error in the case where the dispersion parameter is fixed at 1? Can you account for the difference between the two cases in terms of the value $\widehat{\phi} = 4.28$ found in Exercise 13.16?

Summary of methodology

1 It is often possible to improve the accuracy of estimates of treatment effects in the analysis of experimental data by including further variables known as covariates in the analysis, to account for some of the variability between experimental units. The analysis involves fitting the usual sort of linear model, and is called analysis of covariance.

2 In some circumstances it is appropriate to include a term with a fixed, known, regression coefficient in a linear model. Such a term is known as an offset.

3 Overdispersion in Poisson (and binomial) regression can be taken account of in some circumstances by estimating a dispersion parameter.

Postscript

Congratulations! You have reached the end of the book, or at least the end of the material which is gone into in detail. During your studies, you will have become aware of the wide range of problems that you are now equipped to address thanks to the power of the normal linear model and its extension to generalized linear modelling. In this final brief essay, a few ways are indicated in which this methodology can be extended to cope with yet further complications that arise in many real practical problems. There are no exercises nor any need to use GENSTAT here either.

The authors have chosen three topics to describe briefly. The methodology mentioned in each case can in fact be applied using GENSTAT – though usually not from the menu system – but you will be spared any details.

Ordinal responses

Suppose that the response variable Y in a regression study is coded 0 for no disease, 1 for mild disease, and 2 for severe disease. Had Y taken only two values, perhaps no disease and disease, you could apply the binary regression methods of Chapter 9. But here the response variable has three categories, and ordered categories at that.

One approach to the analysis of such data is an attractive extension of logistic regression modelling. Let $p_j = P(Y = j)$, $j = 0, 1, 2$. The cumulative probabilities are defined as $\pi_j = P(Y \leq j)$, $j = 0, 1, 2$. This means that $\pi_0 = p_0$, $\pi_1 = p_0 + p_1$ and $\pi_2 = p_0 + p_1 + p_2 = 1$. Now, we can reduce the problem to two (by no means independent) binary regression problems: in the first one, Y_1, say, takes the value 1 with probability π_0 and the value 0 otherwise; and in the other, Y_2, say, takes the value 1 with probability π_1 and the value 0 otherwise. That is, Y_1 is an indicator of no disease against some disease (whether mild or severe), and Y_2 is an indicator of mild or no disease against severe disease.

A model for this situation can then be written as a pair of logistic regression equations of the form

$$\log\left(\frac{\pi_j}{1 - \pi_j}\right) = \alpha_j + \beta x, \quad j = 0, 1, \tag{P.1}$$

where x represents an explanatory variable. (As usual, if there were more than one explanatory variable, you would use $\sum_k \beta_k x_k$ instead of βx.) These two logistic regression equations could have had different slopes as well (i.e. β_j instead of β); but the above form, indicating parallel regression lines on the logistic scale, is often found to be adequate. If so, the model is called the *proportional odds model*, because the ratio of the odds of the event $Y \leq j$ at two values of x, say x_1 and x_2, is

$$\frac{\pi_j(x_1)/(1 - \pi_j(x_1))}{\pi_j(x_2)/(1 - \pi_j(x_2))} = \exp(\beta(x_1 - x_2)) \tag{P.2}$$

which has no dependence on j, the particular category under consideration.

In their famous book, referred to in Section 10.4, Nelder and McCullagh apply this model to some data where the disease in question is pneumoconiosis and the single explanatory variable is the log of the time spent at the coal-face by each of a large number of coal-miners in the 1950s. The estimated parameter values were $\beta = -2.60$, $\alpha_0 = 9.68$ and $\alpha_1 = 10.58$. What does this fitted model imply? Well, for miners who had spent five years at the coal-face (for whom the value of the explanatory variable is $\log 5$), the odds of having pneumoconiosis (at all, the case $j = 0$ in (P.1)) is estimated to be one in $\exp(9.68 - 2.60 \log 5)$, i.e. 1 in 244. Each doubling of the exposure time leads to the odds of having the disease being multiplied, using (P.2), by $2^{2.60} = 6.1$. So for 10 years at the coal-face, the risk is about 1 in 40, and for 20 years it is about 1 in 7. For severe pneumoconiosis, the equivalent rates are approximately 1 in 600, 1 in 100 and 1 in 16, with these rates changing by the same proportional amounts since the exposure time increases in the same way.

Smoothing: generalized additive models

Now let Y be any of the response types considered earlier in the book, satisfying a GLM with mean relationship of the form $g(E(Y)) = \alpha + \beta x$ for a single explanatory variable x and link function g, as in Chapter 10. For all the usefulness of the linear predictors, as exhibited throughout this book, it is sometimes better to allow much more general functions on the right-hand side, i.e. to allow $g(E(Y)) = f(x)$ for some function f. (You would be right to complain that there is no real need for a general g and a general f here – they could be rolled into one function – but this formulation, with perhaps a canonical choice for the link function g, remains more standard and convenient.)

Non-linear regression, which is a big area of statistics, allows you to think of a variety of non-linear functions that could be used for f. An example might be $f(x) = \alpha + \beta/(1+\gamma x)$; this is the form taken, for instance, by the Michaelis–Menten law of chemical kinetics. Non-parametric regression, another big area of statistics, allows you to set f equal to a smooth function obtained even more directly from the data without the imposition, by the investigator, of any particular parametric form for f.

The basic idea of non-parametric regression – or *smoothing* for short – is to perform some kind of local averaging of the data; for instance, one might use a moving average estimate, in which case the regression function at point x would be estimated by the average of the responses for individuals whose explanatory variable values are reasonably close to x. The curve estimated in Figure P.1 is the result of a considerable sophistication of this idea, details of which need not concern you now. Figure P.1 is a non-parametric estimate of the regression of loss on strength, based on the rubber dataset first analysed in Chapter 3 by (normal) linear regression. Here, the linear parametric form has not been imposed, and the data seem to suggest a rather more complicated relationship: perhaps loss first increases

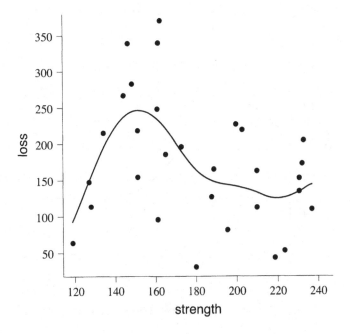

Fig. P.1

with increasing **strength**, then reaches a peak and decreases again, to a somewhat higher level than it started at, and finally flattens out in the right-hand half of the plot (if we don't take too much notice of any remaining wiggles).

In the rubber dataset, there are in fact two explanatory variables: **strength** and **hardness**. The smoothing idea could be extended directly to yield a general non-parametric estimator $\widehat{m}(x_1, x_2)$, say, where $E(Y) = m(x_1, x_2)$. For p explanatory variables, one could contemplate the very general model $E(Y) = m(x_1, x_2, \ldots, x_p)$. However, it turns out that, for p more than about three or four, this general formulation is very hard to estimate well (without restrictive parametric assumptions). So, simpler alternatives have been suggested. One is the *generalized additive model* (GAM) given by

$$E(Y) = \alpha + m(x_1) + m(x_2) + \cdots + m(x_p). \tag{P.3}$$

This formulation is still very general but not as general as the previous formulation: non-additive functions of two or more variables are not allowed. If each m function in (P.3) were actually of the form $m(x_i) = \beta_i x_i$ then the GAM would reduce to the multiple linear regression model. If each m is a general smooth function, then the GAM is at its most general. And mixtures of parametric and non-parametric ms are possible too. If some ms are linear and others are general smooths, then we have a partial linear model. The result of applying such a procedure to the rubber data with a linear dependence on **hardness** (which is what the data supports) and a smooth dependence on **strength** yields Figure P.2. Note that the difference between Figures P.1 and P.2 is that the former gives the mean **loss** when regressed on **strength** alone, while the latter is the dependence of **loss** on **strength** when a linear dependence of **loss** on **hardness** is present also. The main difference the presence of **hardness** makes is to remove much of the estimated increase in **loss** for small **strength**.

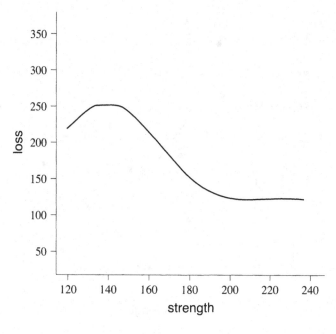

Fig. P.2

Censoring in survival data

In Section 10.5.3, you analysed a dataset (leuksurv) in which the exponential distribution was used as response distribution in a GLM relating the time to death (in weeks) of patients suffering from a form of leukaemia to two explanatory variables, the white blood count and something called 'the AG factor'. This was an example – the only one in the book – of *survival analysis*, so called because it concerns the survival of patients with a certain disease under certain conditions. Survival analysis is a huge subarea of the huge area of medical statistics. It is also very closely related to a very large field in engineering, that of reliability analysis (or reliability for short). As mentioned in Section 10.5.3, components, such as computer chips, and their relationship to explanatory variables, are a typical concern of reliability analysis.

But if, as in Exercise 10.8, you are able to analyse survival data using the ordinary techniques of generalized linear modelling, why should survival analysis, or reliability, be a special field? There are two main reasons for this.

The more important is a feature missing from the leuksurv data, that of *censoring*: for some individuals, the experimental or observational period is over before the event of interest has happened. In this case, you would not observe the exact length of survival, but would just know that the individual survived at least as long as the observation period. For example, for many diseases and/or treatments, patients may well survive for many years before death or recurrence of the disease. (It was only because none of the unfortunate patients in the leuksurv dataset survived very long that this complication did not arise there, so that you were able to analyse these data by standard means.) Likewise, the lifetime of non-faulty computer chips may well be several years, but answers are required relating to their reliability long before they all fail.

It is not right to ignore censored data because you would then be being too pessimistic by ignoring the individuals who do best. But, provided the censoring mechanism (the reason for the cut-off) is independent of the responses, it is not difficult to incorporate censoring into GLM fitting using maximum likelihood, and programs like GENSTAT will do this for you.

The second characteristic feature of survival data that make them unsuited to the ordinary techniques of generalized linear modelling is that responses, which tend to be continuous and skewed, also have to be non-negative. This means that, in general, the normal distribution will not be an appropriate response distribution. The exponential distribution and its generalization the gamma distribution are appropriate candidates; but, in many cases, for continuous non-negative data, better response modelling is achieved by using other distributions which are not mentioned in this book. For such distributions, it is generally not the case that they are members of the exponential family of distributions, mentioned in Chapter 10 as having particularly helpful properties. None the less, researchers have used their ingenuity over the years to accommodate these distributions in a GLM context too.

Here ends the indication of what else can be done in the framework of generalized linear modelling. If only the book had been longer, you could have had the opportunity to delve deeply into topics like these as well!

Solutions to the Exercises

Solution 1.1

Functions (a), (b) and (d) are linear in the parameters; only function (c) is non-linear, and this is because of the way β enters the model.

Solution 1.2

(a) As for recipient age, a lower donor age may be advantageous with respect to GvHD. However, there remains a considerable degree of overlap, and indeed two quite high donor ages are associated with zero GvHD (in the recipient).

This conclusion should not be too surprising given the relationship between recipient and donor ages observed in Figure 1.1.

(b) In the case of index, smaller values appear to be associated with zero GvHD, although again there is overlap between the two boxplots.

Solution 1.3

Contingency tables for the dependence of GvHD on **donmfp** are given in count form to the left and proportion form to the right.

		donmfp		
		0	1	2
GvHD	1	7	2	8
	0	9	9	2

		donmfp		
		0	1	2
GvHD	1	0.44	0.18	0.80
	0	0.56	0.82	0.20

Of the patients with male donors, a little less than half developed GvHD. Few of the patients with female donors who had not been pregnant contracted GvHD, while many of those with female donors who had been pregnant did so. Again, without reading too much into the results, it seems that **donmfp** could be an important explanatory variable affecting GvHD. (The fact that the pattern of dependence on **donmfp** is so similar to that on **type** is an intriguing coincidence!)

Solution 2.1
The histograms in (a) and, perhaps to a lesser extent, (b) appear to have, approximately, a bell shape. The histogram in (c) does not: it is much too asymmetric or skewed.

Solution 2.2
The superimposed normal curves point up any discrepancies there may be, but in these cases any discrepancies seem slight at most. Given the vagaries of random variation, most statisticians would be happy to use the normal distribution as a good model for both these datasets.

Since the fitted curve in (a) is the density of the $N(2.299, 0.1689)$ distribution, $\bar{x}$ must be 2.299 and $s = \sqrt{0.1689} = 0.411$.

Solution 2.3
The normal probability plots in (a), (b) and (c) do nothing to change the view that (a) and (b) are acceptably normal – the points on the plots approximately follow straight lines – and that (c) is not. The skewness in the magnetic susceptibility data shows up in the probability plot as a distinct curvature in the plot of the points.

Solution 2.4
The histogram of the logged magnetic susceptibility data, with fitted normal curve, given in Figure 2.6(a), could well be read as indicating a certain amount of negative skewness, which would mean that perhaps the transformation – which started with positively skewed data – has 'overdone it' a bit.

The normal probability plot of the logged magsus data, Figure 2.6(b), is perhaps less indicative of skewness. Any curvature in this plot is only slight.

It is worth seeing if the square root transformation gives different answers.

The histogram of the data transformed in this way, with fitted normal curve, given in Figure 2.7(a), could be interpreted as being more symmetric, and hence 'more normal', than the log-transformed data. The corresponding normal probability plot, Figure 2.7(b), is pretty straight.

A feature of both of the probability plots based on the magsus data is the single low and two high magnetic susceptibility values, which may be somewhat anomalous relative to the rest of the values; in Chapter 4, the term 'outliers' is (re)introduced for such values.

Whatever particular transformation of magnetic susceptibility data best achieves normality – and, for example, a cube root looks good too – it is interesting to note that the geologists who work with magnetic susceptibility take logs as their standard transformation.

Solution 2.5

(a) For the breaking strength data, $\bar{x} = 2.299$, $s^2 = 0.1689$, and the required percentage point is $t_{0.975}(49) = 2.01$ (since $n = 50$ and $\alpha = 0.05$). The resulting 95% confidence interval is $2.299 \pm 2.01\sqrt{0.1689/50}$, which yields

$$(2.18, 2.42).$$

(b) For the forearm data, $\bar{x} = 18.80$, $s^2 = 1.255$, and the required percentage point is $t_{0.95}(139) = 1.656$. The resulting 90% confidence interval is $18.80 \pm 1.656\sqrt{1.255/140}$, i.e.

$$(18.64, 18.96).$$

Solution 2.6

(a) The t statistic has value $(2.299 - 2)/\sqrt{0.1689/50} = 5.145$ which, when compared with the t distribution on $n - 1 = 49$ degrees of freedom, is a very large value. Since we are asked to assume that the alternative hypothesis is $H_1 : \mu \neq 2$, we need to perform a two-sided test. What exactly you can say about the SP for this test depends on your tables. Those the authors used show that $SP < 0.01$, i.e. the t statistic is larger than $t_{1-\alpha/2}(49)$ for $\alpha = 0.1, 0.05, 0.02$ and 0.01. If you are using a software package, you should be able to obtain the exact SP which is 4.70×10^{-6}, an extremely small value. There is very strong evidence in the data that the mean breaking strength is not 2.

(b) A one-sided test is required, since the alternative hypothesis is now $H_1 : \mu > 2$. The required SP for the t test is one-half of the two-sided SP. So we can say either that $SP < 0.005$ (from the authors' tables) or, exactly, $SP = 2.35 \times 10^{-6}$, and the evidence is even stronger against the null hypothesis, and in favour of the alternative hypothesis that $\mu > 2$.

Solution 2.7

(a) The 90% confidence interval for the mean male forearm length does not contain the hypothesized value 18.5. The SP must therefore be less than 0.1. There is certainly a small amount of evidence that the mean length is not 18.5, but without knowing more about the SP we cannot be precise about how much evidence there is.

(b) The t statistic is $(18.80 - 18.5)/\sqrt{1.255/140} = 3.169$ on 139 d.f. Again, from tables, we can say that the $SP < 0.01$. (In fact, the exact SP is 0.0018.) There is therefore considerable evidence against the null hypothesis. (The value of the SP reflects how far outside the confidence interval the hypothesized value is.)

Solution 2.8

(a) At least as a first approximation, normality does not seem entirely unreasonable for these differences, given that there is not an enormous number of them, although various aspects of the plot are not entirely supportive of the distributional assumption.

(b) For the differences, $\bar{x} = -0.156$ and $s^2 = 1.028$, so that the t statistic based on the differences is $(-0.156 - 0)/\sqrt{1.028/25} = -0.7693$. This is to be compared with the t distribution with 24 d.f. However, something unexpected is going on! The mean of these differences – as reflected in the t statistic – is negative, whereas our alternative hypothesis is for a positive mean. It is probably best to report this occurrence along with the fact that the two-sided SP is greater than 0.1. (It is, more exactly, 0.45.) It is very clear that there is no evidence whatsoever that primitive societies are the same as Western societies in terms of times of weaning and toilet training.

The one-sided test is appropriate because of the prior knowledge about Western societies.
(Note that details would have been slightly different had we calculated the differences as weaning age minus toilet-training age rather than vice versa, but the same conclusion would have been reached.)

Solution 2.9

(a) The sample means are:

   carpeted: 11.20;   uncarpeted: 9.79.

   The airborne bacteria level in carpeted rooms seems to be higher than in uncarpeted.

(b) The sample variances are:

   carpeted: 7.17;   uncarpeted: 10.30.

   These variances are quite similar: the larger is about 1.4 times the smaller. A two-sample t test seems a defensible way to proceed.

(c)
$$S_p^2 = \frac{7(7.17) + 7(10.30)}{14} = \frac{1}{2}(7.17 + 10.30) = 8.735.$$

The test statistic takes the value $(11.20 - 9.79)/\left(\sqrt{8.735}\sqrt{\frac{1}{8} + \frac{1}{8}}\right) = 0.9542$, and is compared with the t distribution with $8 + 8 - 2 = 14$ d.f. The (two-sided) SP is greater than 0.1 (in fact, 0.36) and there is no evidence that there is any difference in bacteria levels between the carpeted and uncarpeted rooms.

Solution 2.10

(a) Given that sample sizes are only 12 in each group, it seems not unreasonable to assume normality. Certainly, the differences in the first group follow normality well; those in the second are more dubiously normal. The sample variances are 0.0053 and 0.0167 respectively: the variance in Group 2 is about three times that in Group 1! The equal variance assumption is, therefore, in some doubt, although it is not entirely unreasonable to proceed with the t test.

(b) The mean differences are 0.0775 and 0.0708 for Groups 1 and 2 respectively.

$$S_p^2 = \frac{11(0.0053) + 11(0.0167)}{22} = \frac{1}{2}(0.0053 + 0.0167) = 0.0110.$$

The test statistic is $(0.0775 - 0.0708)/\left(\sqrt{0.0110}\sqrt{\frac{1}{12} + \frac{1}{12}}\right) = 0.1565$ with 22 d.f., yielding a (two-sided) $SP > 0.1$ (actually 0.88). Provided the t test is reliable, there is no evidence of any difference between the differences in Groups 1 and 2.

Solution 2.11
The values of c_L and c_U used are $c_L = c_{0.025}(49) = 31.55$ and $c_U = c_{0.975}(49) = 70.22$. The confidence limits are thus $49 \times 0.1689/70.22 = 0.1179$ and $49 \times 0.1689/31.55 = 0.2623$. Therefore the 95% confidence interval for σ^2 is

$$(0.1178, 0.2623).$$

Notice that this interval, unlike t intervals, is not symmetric about the point estimate 0.1689.

Solution 2.12
From Solution 2.10, $s_1^2 = 0.0167$ and $s_2^2 = 0.0053$ (using the convention that gives a ratio > 1). Under the null hypothesis that $\sigma_1^2 = \sigma_2^2$, we want the SP in the obtained direction which is given by $P(F > 3.151)$ where $F \sim F(11, 11)$ and $3.151 = s_1^2/s_2^2$. Tables tell us that $0.05 > SP > 0.025$, and hence the SP for the two-sided test satisfies $0.10 > SP > 0.05$. (In fact, SP (obtained direction) $= 0.0344$, and the SP for the two-sided test is 0.0688.) There is some indication that the variances may not be equal, but the evidence is marginal at most. This is a similar conclusion to that arrived at in Solution 2.10(a), suggesting that the rule of thumb used there gave an appropriate message in this case.

Solution 2.13
There are 107 bees in total, of which 66 fed off lined flowers and 41 off plain ones (these are the counts of ones and zeros, respectively, in Table 2.5). So the estimate of p is $66/107 = 0.617$.

Solution 2.14
$$E(\widehat{p}) = E\left(\frac{X}{n}\right) = \frac{1}{n}E(X) = \frac{1}{n}np = p \text{ (so } \widehat{p} \text{ is an unbiased estimator of } p).$$
$$V(\widehat{p}) = V\left(\frac{X}{n}\right) = \frac{1}{n^2}V(X) = \frac{1}{n^2}np(1-p) = \frac{p(1-p)}{n}.$$

Solution 2.15
(a) The 95% confidence interval for p turns out to be

$$(0.518, 0.709).$$

(b) There is some evidence that $p \neq 0.5$; certainly $SP < 0.05$. There are indications that more than 50% of bees choose to feed off lined flowers.

Solution 2.16

(a) $\bar{x} = \dfrac{0 \times 220 + 1 \times 113 + 2 \times 23 + 3 \times 8 + 4 \times 1}{365} = 0.5123.$

(b) The bar chart in Figure 2.14 and the plot in Figure 2.15 look very alike, supporting the notion that the Poisson distribution provides a good model for these data.

(c)

$$s^2 = \frac{(0 - 0.5123)^2 \times 220 + (1 - 0.5123)^2 \times 113 + \cdots + (4 - 0.5123)^2 \times 1}{365}$$

$$= 0.5403.$$

This is close to $\bar{x} = 0.5123$, and so provides further evidence that a Poisson model is appropriate.

Solution 2.17

The answer is $(0.452, 0.578)$.

Solution 2.18

Use the relationship

$$\hat{\sigma}^2 = \frac{n-1}{n} s^2,$$

which results from the fact that $\hat{\sigma}^2$ divides the sum of the squares about the mean by n, while the divisor in s^2 is $n - 1$.

For hald,

$$s^2 = 0.1689, \text{ so } \hat{\sigma}^2 = (49/50) \times 0.1689 = 0.1655;$$

for forearm,

$$s^2 = 1.255, \text{ so } \hat{\sigma}^2 = (139/140) \times 1.255 = 1.246.$$

There is clearly little difference between the values of s^2 and $\hat{\sigma}^2$, with s^2 being slightly the larger.

Clearly, s^2 and $\hat{\sigma}^2$ get closer together as n gets larger. For most practical purposes, unless n is very small, the difference between s^2 and $\hat{\sigma}^2$ is unimportant.

Solution 2.19

S^2 is an unbiased estimator of σ^2, and $\hat{\sigma}^2 = \dfrac{n-1}{n} S^2$; so $\hat{\sigma}^2$ is not unbiased. In fact,

$$E(\hat{\sigma}^2) = E\left(\frac{n-1}{n} S^2\right) = \frac{n-1}{n} E(S^2) = \frac{n-1}{n} \sigma^2,$$

so the bias is

$$E(\hat{\sigma}^2) - \sigma^2 = \left[\left(\frac{n-1}{n}\right) - 1\right]\sigma^2 = -\frac{\sigma^2}{n}.$$

For large n, this bias will be small because its denominator is large. And, as expected from the general theory of mles, the bias tends to 0 as $n \to \infty$, implying asymptotic unbiasedness.

Solution 2.20

(a) Since the mean and variance of the Poisson distribution are both λ, by the central limit theorem, $\widehat{\lambda} \approx N(\lambda, \lambda/n)$. So $V(\widehat{\lambda})$ can be estimated as $\widehat{\lambda}/n$, and it follows that an approximate $100(1 - \alpha)\%$ confidence interval for λ is

$$\widehat{\lambda} \pm z_{1-\alpha/2}\sqrt{\frac{\widehat{\lambda}}{n}}.$$

(b) When $\alpha = 0.1$, the standard normal percentage point used is $z_{0.95} = 1.645$. The confidence interval corresponds to $0.5123 \pm 1.645\sqrt{0.5123/365}$, i.e.

$$(0.451, \ 0.574).$$

The approximate confidence interval here is very similar to the exact one given in Solution 2.17.

Solution 2.21

An approximate 95% confidence interval for the mean is $1.507 \pm 1.96\sqrt{0.1051/63}$ which gives

$$(1.427, \ 1.587).$$

Solution 2.22

(a) Ordinal. The categorical sizes are ordered from smallest to largest.

(b) Quantitative (in fact, continuous).

(c) Ordinal. (The only difficulty here might be the position of 'don't know', which may not mean the same as 'indifferent'.)

(d) Nominal. There is no ordering of the religions.

(e) Quantitative (in fact, discrete).

(f) Ordinal.

(g) Quantitative (assuming the time is measured fairly accurately and not in large chunks, e.g. in five-year intervals, in which case the data may be considered ordinal categorical).

(h) Nominal.

(i) Nominal.

3

Solution 3.1

(a) A list of some of the things that you might have identified in the **Genstat 5** window follows.

- There is an **Output** window and an **Input Log** window. Both have scroll bars and some standard Windows controls. The **Output** window is active, i.e. has a cursor that blinks and a highlighted title bar.

- There is a row of words near the top of the **Genstat 5** window, which looks like the sort of thing you might have seen in other Windows packages. You might have guessed that they would produce menus, and indeed they do.

- Just below the row of words is a row of pictures. You might have guessed that they would work in a similar way to the menus, but it is not obvious what all of them do.

- At the bottom of the **Genstat 5** window is a strip with the word Output in it at the left-hand end, and various other things to the right. You might have guessed that the word Output is telling you that the **Output** window is active (and you would be right).

(b) When you click with the pointer over the menu bar, a menu should appear. In other words, a list of tasks will appear, superimposed on the **Genstat 5** window. Also one of the words on the menu bar will now be highlighted.

(c) When you click with the pointer positioned over the highlighted word on the menu bar, the menu should disappear and the word which was highlighted should return to normal.

(d) When you moved the pointer along the menu bar, the available menus should have appeared one at a time. The menu that is visible will be the one corresponding to the word on the menu bar which is nearest to the pointer.

This is a fast way to scan through what choices are available via the menu bar. You can use it when you know that the task you want to do can be reached through the menu bar, but cannot remember which of the words on the menu bar to click on.

Solution 3.2

(a) The way your GENSTAT is probably set up, the options that are switched on are:

- Show Toolbar;

- Show Statusbar.

(b) When you click on **Show Toolbar**, the **Options** menu disappears and so does the row of pictures below the menu bar.

(c) The **Options** menu has changed in that there is no longer a ✓ next to **Show Toolbar**.

(d) The menu should change back if you click on **Show Toolbar** again. The row of pictures below the menu bar should reappear, after you click on **Show Toolbar**. (And **Show Toolbar** would be ticked again if you looked at the **Options** menu.)

Solution 3.3

(a) When the mouse button was held down over the **A** button, the statusbar should have contained the words Changes Font. After the mouse button has been released, the statusbar should return to what it said before. If you released while still over the **A** button, a dialogue box will have appeared. If so, click on **Cancel** to remove it.

(b) There are four styles available on many computers: normal; **bold**; *italic*; ***bold italic***.

After the style was changed to **bold**, the text in the **Output** window was much darker.

The other styles have the following effects:

- *italic* makes the letters in the text slant to the right;

- ***bold italic*** also makes the letters slant to the right, but it also makes the text darker.

(c) When you click on the downarrow button, a menu will appear. On some computers the menu consists of: **Fixedsys**; **Terminal**; **Courier**; **Courier New**.

When you click on **Terminal**, the text in the **Output** window changes in appearance. It changes to a different font family. Depending on how your computer is set up, the size of the type might have changed as well.

(d) You can choose what you like!

(e) To obtain the **Font** dialogue box using the menu bar, you: click on **Options** from the menu bar; click on **Font Settings**.

Solution 3.4

(a) The command

```
PRINT 1+2
```

causes GENSTAT to work out $1 + 2$ and display the answer. The answer, 3.000, appears in the **Output** window and is labelled 1+ 2. In this context, PRINT means 'display in **Output** window', not print on printer.

(b) The instruction

```
PRINT ednormal(0.975)
```

calculates and displays the required answer, 1.960.

The instruction

```
PRINT ednormal(0.975,0.5,0.95,0.025)
```

calculates and displays the 0.975, 0.5, 0.95 and 0.025 quantiles of the standard normal distribution. The quantiles are 1.960, 0, 1.645 and -1.960, respectively.

(c) The first command of the four, PRINT edt(0.975;10,20), gives the 0.975 quantiles of the $t(10)$ and $t(20)$ distributions. You can tell this by looking at the Output window; there are two numbers printed there, 2.228 and 2.086, which are labelled EDT((0.975; 10)) and EDT((0.975; 20)), respectively.

The second command, PRINT edt(0.975,0.95;10), gives the 0.975 and 0.95 quantiles of the $t(10)$ distribution.

The third command, PRINT edt(0.975,0.95;10,20), gives the 0.975 quantile of the $t(10)$ distribution and the 0.95 quantile of the $t(20)$ distribution (and not both quantiles of both distributions as you might have expected).

The fourth command, PRINT edt(0.975,0.95,0.975,0.95;10,10,20,20), does give the 0.975 and 0.95 quantiles of the $t(10)$ and $t(20)$ distributions.

The numbers to the right of the semicolon are the degrees of freedom, those to the left indicate which quantile to calculate. The semicolon is what GENSTAT uses to know where the list of the required quantiles finishes and the list of the degrees of freedom starts.

(d) The command PRINT edt(0.975,0.95;10,20,30) has three different values for degrees of freedom, but only *two* values for the quantiles. The output from this command contains three values: the 0.975 quantile of the $t(10)$ distribution; the 0.95 quantile of the $t(20)$ distribution; the 0.975 quantile of the $t(30)$ distribution.

What GENSTAT does is to find the longer list of values, in this case the degrees of freedom, and repeat values in the shorter list until there are enough to match the longer list. So in this case, GENSTAT:

- pairs the 0.975 with the 10;
- pairs the 0.95 with the 20;
- pairs 30 with 0.975, because there is no third value for which quantile to find, and so GENSTAT goes back to the first value.

This idea of going back to the start of a list is built into all of the GENSTAT command language. It is even built into the arithmetic, as can be seen in the result of the command PRINT 1,2+3,4,5. In this case there are two lists of numbers to add, but the first list (1, 2) is shorter than the second (3, 4, 5). The addition sign is what separates the two lists in this case. (This is not obvious: you would have been perfectly sensible in expecting the answers 1, 5, 4, 5!)

Usually, either the two lists will be the same length, or one list contains a single value. (You are learning here about a more complicated case, because it is easy to miss a value out of a list accidentally; if you do not know about GENSTAT restarting lists, it is hard to work out what has happened when you miss a value out of a list.)

Solution 3.5

(a) The **Spreadsheet** window appears after you click on the **OK** button.

The **Spreadsheet** window should look like a table with two columns and ten rows. The columns have the names **C1** and **C2**; the rows are labelled 1 to 10. Each cell of the table should contain a symbol similar to an asterisk (*), where one of the numbers from Table 3.1 will go.

There are also scroll bars on the **Spreadsheet** window, but in this case you can probably see all of the values in the spreadsheet without using the scroll bars. You will need to use the scroll bars for larger datasets later in the course.

(b) You weren't asked a specific answerable question for this part.

(c) When you press < Return > at the end of typing 12.1 into the spreadsheet, the rectangle moves on to the next row, ready for you to type the next number.

(d) It is good practice to save your work fairly frequently; then if you make a serious mistake you do not have to start again from the beginning. Saving your work is something you can do quickly once you are used to doing it.

(e) You can change any cell of the spreadsheet by moving to that cell (by clicking on it) and then simply typing the correct value, followed by <Return>.

Solution 3.6

Several lines of GENSTAT commands (numbered in the left hand margin) appear first in the **Output** window. This is because, by default, GENSTAT 'echoes' its internal commands to this window. These commands are followed by summaries (minimum, mean, maximum, etc.) of the values in the **before** and **after** columns. These summaries are the part of the output that is mainly of interest to us.

Solution 3.7

The procedure for loading mydata.gsh directly is as follows. Click on the **Data** menu; click on **Load**; click on **GSH Spreadsheet**. A dialogue box will appear, which works in exactly the same way as the dialogue box in Exercise 4.6 (this dialogue box is called the **Open Genstat Spreadsheet** file dialogue box); you change **File name** to mydata.gsh and click on the **Open** button.

Solution 3.8

(a) When you change **Directories** from c:\lmgen to c:\lmgen\data, the list(s) of files and directories change. Instead of just the datafile mydata.gsh, lots of datafiles are listed.

(b) There are several ways to do this. One is to use the scroll bars or the cursor keys (if necessary) to move through the list of files until you can see water1.gsh, and then click on its name.

Solution 3.9

(a) The name north will be in italics to signify that the column is a factor.

(b) The main change should be that the north column has been replaced by the **region** column, which has the words **south** and **north** in place of the levels 0 and 1.

Solution 3.10

(a) The histogram should look something like Figure 2.1(c). but not identical because of different choices made by GENSTAT and because GENSTAT histograms have a different appearance. As mentioned in Solution 2.1, the data do not look as though they are from a normal distribution, as the histogram is not symmetrical about its peak.

(b) The boundaries have now changed to 1, 2, ... , 14 and the histogram is the same one as in Figure 2.1(c). (Actually, the two histograms are very slightly different. This is because one datapoint happened to be equal to a group boundary value. The software that produced Figure 2.1(c) allocated that datapoint to the group directly above the boundary; GENSTAT allocates such a point to the group directly below.)

To get boundaries at 1, 1.5, ... , 13.5, you would fill in the Limits field as !v(1,1.5...13.5).

(c) The axis style has changed from simply having the two axes to having a box enclosing the histogram.

The title appears centred just above the plotting area.

If we wanted to, we could add labels to the axes and alter where the tick marks appear. Tick marks are the marks indicating the scale along an axis.

Solution 3.11

The histogram looks like the data are from a normal distribution, as does that in Figure 2.1(a).

To obtain this histogram, you could do the following. Load hald.gsh into GENSTAT. Click on Graphics|Histogram. A dialogue box will appear: fill in the Data field as hald; fill in the Limits field as !v(1.25,1.5...3.5); fill in the Title field as Breaking Strength (for example); click on the Axes button; change Style to Box; click on the OK button. Back in the Histogram dialogue box, click on the OK button.

Solution 3.12

(a) GENSTAT will give everything that has a check mark next to it. Depending on how your computer is set up, this is likely to be: the number of non-missing values; the number of missing values; the arithmetic mean; the median and the two quartiles; the minimum and maximum values.

So, the mean would be given (GENSTAT calls it the arithmetic mean), but the variance would not be, since the variance does not have a check mark.

(b) There are two ways to alter the options. The first is to click on the Clear button, so that none of the Options have check marks. You can then click on Arithmetic Mean and on Variance. The second way is simply to click on each option that is checked, except for the Arithmetic Mean, and then to click on Variance. (This is assuming that your version of GENSTAT is set up so that Variance is not one of the default Options.)

From the Output window, the mean is 18.802 and the variance is 1.255.

Solution 3.13

The GENSTAT histograms look similar to those produced in Figures 2.6 and 2.7.

(a) The histogram for the square-rooted magnetic susceptibility looks like what we would expect for a normal distribution.

(b) The histogram for the logged magnetic susceptibility looks a little less like what we would expect for a normal distribution than in the case of the square-rooted data.

For comparison of the two transformations, see Solution 2.4.

Solution 3.14

(a) The null hypothesis is that the mean breaking strength, μ, of linen thread is 2.[1] The alternative hypothesis is that $\mu \neq 2$. The test statistic is $t = 5.15$. The SP is less than 0.001. Notice that GENSTAT calls the SP the Probability level (under null hypothesis). Notice also that GENSTAT indicates that this is a two-sided test by the heading Test for evidence that mean(hald) is different from 2.000.

From this you should conclude that there is overwhelming evidence that the mean breaking strength is not 2. This corresponds to the conclusion in Solution 2.6(a). Looking at the mean given in the **Output** window, which is 2.299, it seems that the mean is greater than 2.

(b) The only difference between the two sets of output is that the heading now reads Test for evidence that mean(hald) is greater than 2.000. In particular, the test statistic still takes the same value. In general, the SP will also change; in fact, it will be smaller. However, this doesn't show up in this case, because GENSTAT only tells us that $SP < 0.001$.

(c) This output is the same as that in part (a) except that it has an additional line containing the confidence interval.

The 95% confidence interval for the mean breaking strength is (2.182, 2.416).

If you were only interested in the confidence interval, you could simply have entered the command as:

```
TTEST [ciprob=0.95] hald
```

In this case, GENSTAT would do a two-sided test of the hypothesis $\mu = 0$ and attach the same confidence interval to the end of this output. (However, if you left method=greater in the command, you would not get the appropriate confidence interval.)

Solution 3.15

(a) The boxplots give the impression that normality is a reasonable assumption, since the medians are towards the centres of the interquartile ranges.

Both the range and the interquartile range are noticeably bigger for the second group than for the first. This suggests that the variances might not be roughly equal; but we shall proceed with the t test anyway, as you did in Exercise 2.10(b).

(b) The means are 0.0775 and 0.0708, while the variances are 0.0053 and 0.0167.

As in Solution 2.10, the rule of thumb gives a little concern that the variances might be too different to be considered the same. The rule of thumb is informal, and in fact GENSTAT does a test of the null hypothesis that the variances are equal. If this test had suggested that the variances were too different, a warning would have been printed

---

[1] Do not worry about normality here: we have already checked this in Exercise 2.3(a).

in the **Output** window, and it wasn't. (If you want to see the warning message, go through the procedure for a two-sample *t* test as before, but fill in the **Data set 2** field as 2*d2.) So it is not unreasonable to use the *t* test here.

The results of the *t* test indicate that the data are consistent with the population mean differences being the same.

Solution 3.16
The easiest way to do this analysis is just to go ahead and ask GENSTAT to do the two-sample *t* test, using **carp** as **Data set 1** and **uncarp** as **Data set 2**. You can use exactly the same method as was used in Exercise 3.15 for **d1** and **d2**.

Having done the *t* test, you can look in the **Output** window, read off the means, read off the variances, and check that they are similar. If you had decided that the *t* test was not appropriate, then you could have ignored the *t* statistic and the other information about the result of the *t* test. If you had decided the *t* test was appropriate, then the results of the *t* test would be before your eyes.

Your answers should match those in Solution 2.9.

4

Solution 4.1

The scatterplot looks like this.

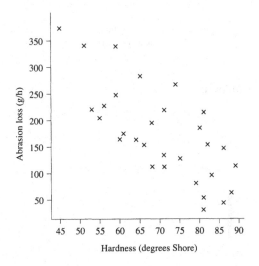

The relationship between the variables appears well described by a straight line, abrasion loss decreasing with increasing hardness. The scatter about the main trend of the data looks reasonably even.

Solution 4.2

As mentioned in Solution 4.1, the relationship between the two variables looks linear, and since the scatter about the line is not greater at some parts of the line than others, the assumption of constant variance looks tenable. It is harder to check on the normality assumption and the independence assumption simply by looking at the scatterplot, but there is no obvious reason to doubt them. Overall, the simple linear regression model would appear to be appropriate.

Solution 4.3

The scatterplot, on scales showing the origin, is shown overleaf. The points do indeed appear to lie close to a straight line through the origin.

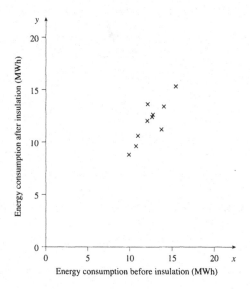

Solution 4.4

The scatterplot is as follows.

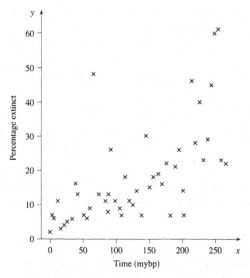

Note that, on this scatterplot, times long ago are far to the right, which conflicts with the usual convention for plots with time on the horizontal axis. If you desire a more conventional plot, you could plot extinction rate against ($-$time) or perhaps against ($265 - $ time).

There is a noticeable upward trend from left to right in the plot; in other words, extinction rates seem to be getting smaller as time moves towards the present day. However, there are several marked outliers (possible 'mass extinctions', including the one at 65 mybp, corresponding to the end of the Cretaceous era, when the dinosaurs also became extinct). Generally, it would seem inappropriate to assume normality for these data. It also appears that the variability of the data is greater for greater extinction rates. Overall, the simple linear regression model is not adequate.

Solution 4.5

(a) The part of the output that you ought to recognize is the bit starting *** Estimates
of parameters ***.

The estimate for the intercept, $\hat{\alpha}$, is in the estimate column and the Constant row;
the value is 550.4. The estimate for the slope, $\hat{\beta}$, is in the same column, but in the
hardness row; its value is −5.337. The rows have these labels because we multiply
the values in hardness by $\hat{\beta}$ and add a constant, $\hat{\alpha}$, to get fitted values.

Thus the fitted line is

$$\text{abrasion loss} = 550.4 - 5.337 \times \text{hardness}$$

where loss is measured in grams per hour and hardness in degrees Shore.

You might well have worked out that s.e. means 'standard error' and that some sort
of t test is referred to in the t(28) and t pr. columns. We shall come back to these
columns, and the rest of the output, later.

(b) When the line is drawn on the scatterplot, it appears to fit reasonably well.

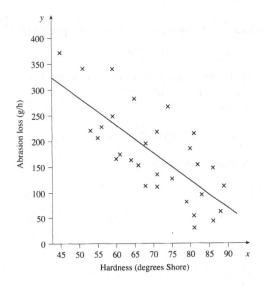

Solution 4.6

(a) The scatterplot (together with the regression line from part (b)) is shown overleaf. The
two variables are related negatively. The relationship seems to be well summarized
by a straight line. The scatter of points about the straight line is fairly regular, though
there is a tendency for points at low aflatoxin levels to be clustered more closely than
points where the aflatoxin level is higher. (This is hardly surprising, given that the
percentage of non-contaminated peanuts cannot rise above 100.) There seems to be
the possibility of producing helpful predictions.

(b) The fitted line is

$$y = 100.0021 - 0.002904x.$$

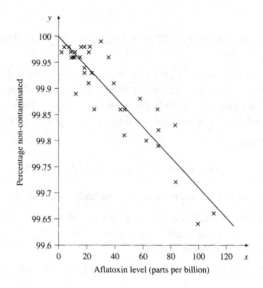

When plotted on the scatterplot, as above, the line appears to fit reasonably well.

Solution 4.7

This time, because there is no constant (intercept), there is only one parameter estimate given in the output, in the `before` row of the `Estimates of parameters` section. Thus the fitted line is

$$y = 0.9574x.$$

On the scatterplot, the fit looks reasonably good.

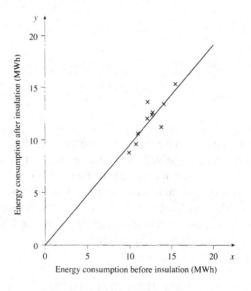

Solution 4.8

(a) There is a lot of scatter in the numbers of taps at each caffeine dose, but (as far as can be seen) the scatter at each point looks reasonably normally distributed and the numbers of taps increase with dose in a reasonably linear manner. So fitting a straight line is not unreasonable.

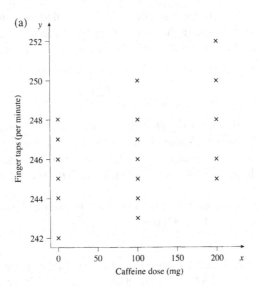

(b) The fitted line is

$$y = 244.75 + 0.0175x.$$

On the scatterplot, it goes where one might expect.

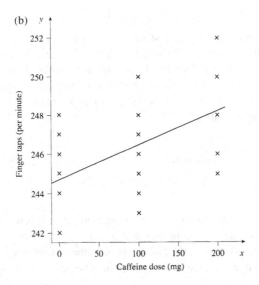

Solution 4.9

Display (4.4) says that

$$\frac{\widehat{\beta} - \beta}{S / \sqrt{s_{xx}}} \sim t(n-2)$$

where $s_{xx} = \sum(x_i - \bar{x})^2$. The denominator, $S / \sqrt{s_{xx}}$, is called the standard error of $\widehat{\beta}$. The sample value of this is the value in the s.e. column of the Estimates of parameters section of the output. The hardness row is the one containing $\widehat{\beta}$ and we can read off the standard error as 0.923, i.e.

$$\frac{s}{\sqrt{s_{xx}}} = 0.923.$$

To calculate the t statistic for testing $H_0 : \beta = 0$, we take $\widehat{\beta}$ and divide it by its standard error. This calculation gives the value printed in the t(28) column, namely -5.78. (The 28 is the number of degrees of freedom, $n - 2$, for this particular dataset.) This t statistic can be compared with the $t(28)$ distribution to give the SP. The SP is given in the t pr. column; it is < 0.001.

The same pattern applies to the Constant row: the standard error of $\widehat{\alpha}$ is 65.8; the t statistic for testing $H_0 : \alpha = 0$ is 8.37; the SP is < 0.001.

Solution 4.10

Both the v.r. value and the t(28) value ($= \widehat{\beta}/\text{s.e.}(\widehat{\beta})$) in the output are test statistics for testing $\beta = 0$, and both yield SPs < 0.001. The connection between the two is that

$$\text{v.r.} = t(28)^2.$$

In this case, $33.43 \simeq (-5.78)^2$. The approximation is caused by rounding error. (One way of noticing this might have been by observing that $\log(\text{v.r.}) = 2 \log(-\text{t}(28))$.) So, really, the same test is being done in two ways, and exactly the same SP must ensue.

In general, it is a fact that if $T \sim t(v)$ then $T^2 \sim F(1, v)$.

Solution 4.11

The value of the t statistic for this test turns out to be 3.57. There are 30 datapoints. Comparing the t statistic with a t distribution with $30 - 2 = 28$ degrees of freedom, the total SP is very small at 0.001. The hypothesis of zero slope is clearly rejected. Alternatively, the v.r. value is $12.77 = 3.57^2$, allowing for rounding error, again on 28 degrees of freedom, and again with an SP of 0.001. These data do provide evidence of an effect of caffeine on finger-tapping rate.

Solution 4.12

The 95% confidence interval for α is $(415.7, 685.2)$. The 95% confidence interval for β is $(-7.2, -3.4)$.

Solution 4.13

(a) The relationship looks reasonably linear (or, at any rate, it is not immediately obvious that it is curved) and it is plausible that a straight line through the origin would fit reasonably well. However, the points are all rather a long way from the origin, so it is difficult to judge from the scatterplot alone whether a line through the origin would be appropriate.

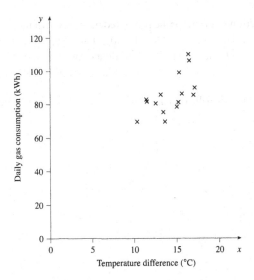

(b) The (unconstrained) fitted line is

$$y = 36.9 + 3.41x$$

where y denotes the average gas consumption in kWh and x denotes the average temperature difference in °C.

The confidence interval for the intercept is found using the method described in Exercise 4.12. The estimate of α is 36.9, and its (estimated) standard error is 17.0. The appropriate t percentage point is given by edt(0.975;13). Therefore the 95% confidence interval for the intercept of the regression line is (0.2523, 73.54). The interval is wide, largely because the data (and hence $\bar{x}$) are a long way from the origin (where $x_0 = 0$), and so the data do not determine the value of the intercept at all accurately. However, the interval does not include the value zero, and so provides some evidence that fitting a line through the origin is inappropriate for these data.

(An alternative type of model, which is often appropriate in situations like this, is to formulate a non-linear model which goes through the origin and behaves like a straight line in the vicinity of the data, but which joins the origin to this straight line using a curve.)

Solution 4.14

(a) The point estimate for the mean percentage of non-contaminated peanuts in batches for which the aflatoxin level is 13.2 parts per million is 99.964 (with an estimated standard error of 0.009). To calculate the confidence interval you could use commands such as:

```
CALC low=yhat-se*edt(0.95;32)
CALC up=yhat+se*edt(0.95;32)
PRINT low,up
```

The 90% confidence interval for the mean response at an aflatoxin level of 13.2 is (99.95, 99.98). This range is an interval for $\alpha + 13.2\beta$.

(b) The 90% prediction interval for the predicted percentage of non-contaminated peanuts at an aflatoxin level of 13.2 is (99.90, 100.0). This range is an interval for $\alpha + 13.2\beta + \epsilon$. It is indeed rather wider than the confidence interval in part (a).

Solution 4.15

(a) The residual plot is as follows (though the colours cannot be shown here).

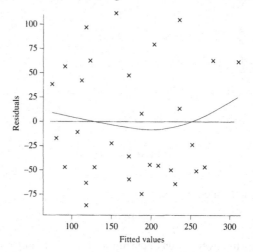

The plot of residuals against fitted values does not indicate any serious grounds for concern about the model. There is no clear sign of a pattern in it. The curved line across the middle gives an indication that the residuals tend to be rather higher at the ends of the range than in the middle, but the tendency is not very great.

(b) The probability plot is as follows.

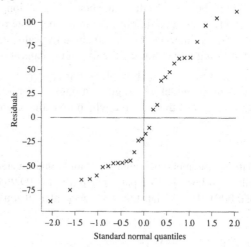

It looks rather curved, indicating that the residuals are rather skewed. Perhaps transforming the data (see Section 4.5) would have led to a more normal distribution for the residuals; however, this improvement in modelling accuracy would have led to a more complicated model.

(c) The composite plot is as follows.

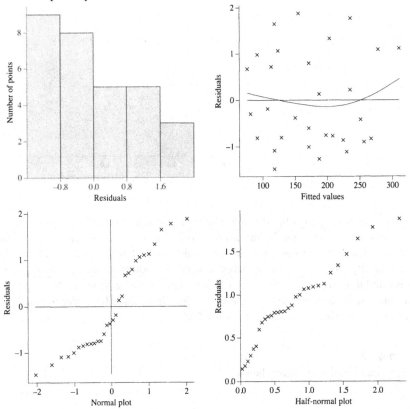

The fitted values plot and the normal probability plot look almost exactly the same as (though smaller than) those in parts (a) and (b). The histogram shows fairly clearly the skewness of the residuals. Thus the composite plot does not (in this case) really tell us anything about the model we had not already noticed.

Solution 4.16
The residual plots look like this.

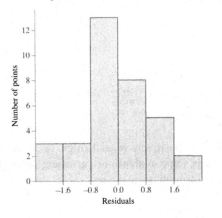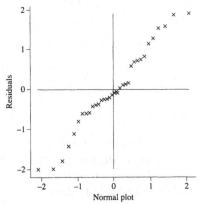

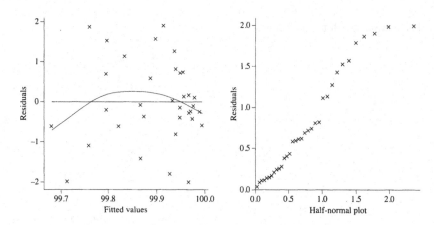

The plot of residuals against fitted values shows no particularly marked sign of curvature, but, like the original scatterplot (Solution 4.6), shows that there is a tendency for the residuals to be smaller at high fitted values, which correspond to low aflatoxin levels. This is reflected in the probability plot, which is reasonably straight but indicates that the extreme residuals (both high and low) are not spread out as much as they would be if they came from a normal distribution. The histogram looks a little skew. Thus the simple linear regression model is not perfect for these data, but the fit is not too bad.

Solution 4.17

(a) The scatterplot is as shown below.

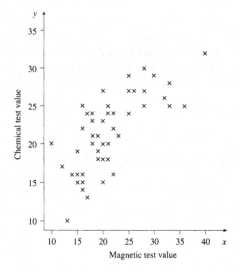

As one might expect, the two variables are positively related (high values of one go with high values of the other). There is some evidence of curvature, in that the chemical test values increase less rapidly for high magnetic test values than they do for low magnetic test values.

The equation of the fitted regression line is

$$y = 8.96 + 0.5866x$$

where y is the chemical test value and x the magnetic test value.

In the composite residual plot, shown below, the plot of residuals against the fitted values emphasizes the way the data appear to deviate from a straight line towards the right-hand end. The probability plot and histogram give no clear signs of lack of normality.

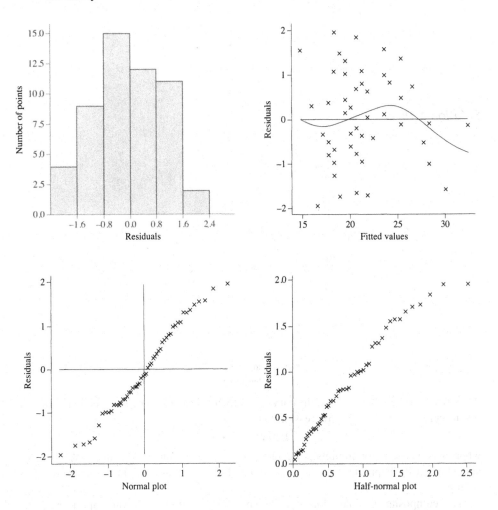

(b) The index plot, shown overleaf, gives an indication that the residuals tend to be negative for those points early in time order and positive for most of the second half of the data. Therefore, there is evidence of some sort of lack of independence between successive residuals.

(c) The simple linear regression model does not fit these data particularly well.

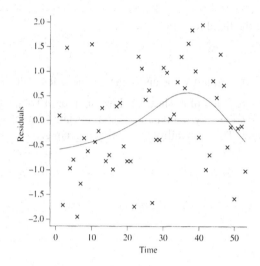

Solution 4.18

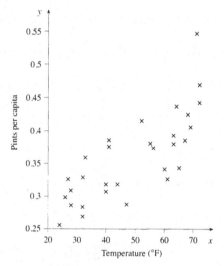

Apart from one very high outlier, the data generally follow the pattern assumed in the simple linear regression model. The regression line is

$$y = 0.2069 + 0.003107x$$

where y is consumption in pints per capita and x is temperature in °F. GENSTAT warns us about the outlier by producing a message saying that its standardized (i.e. Pearson) residual is large.

The composite residual plot (not shown) shows nothing untoward apart from the single high outlier that we noticed in the original scatterplot. However, you may have noticed that the plot of the residuals in time order, shown opposite, is much too smooth for independent, random residuals. For instance, there is a noticeable run of relatively large negative residuals near the start. (The blue curve on the plot does not help to show this interesting structure!) This plot casts doubt on the assumption of independence in the simple linear regression model.

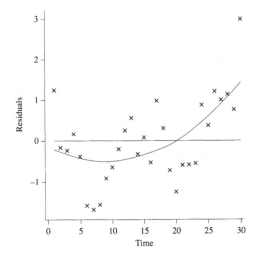

Solution 4.19

The plot of the residuals against tensile strength is as shown.

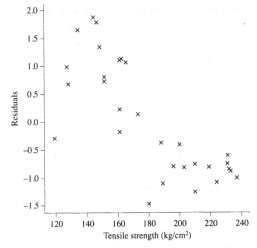

There is a clear negative relationship between the residuals from the fitted line and the tensile strength of the rubber samples. That is, rubber samples with high tensile strength are much more likely to have negative residuals. Putting it another way, suppose you had two rubber samples with the same hardness but with very different strengths. According to the regression model of abrasion loss against hardness, they would be expected to have the same abrasion loss. But this residual plot indicates that the sample with higher tensile strength is much more likely to have a negative residual; in other words, its abrasion loss is likely to be less than that of the sample with lower tensile strength. Overall, therefore, the simple linear regression of abrasion loss against one explanatory variable, hardness, is not adequate to describe the data. We need a more complicated model, involving the other variable, tensile strength, as well.

Solution 4.20

Transforming t to log t straightens the plot very effectively.

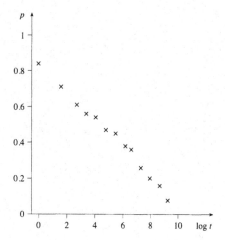

The fitted line is

$$p = 0.8464 - 0.07923 \log t.$$

In fitting this model, GENSTAT warns you that the residuals "do not appear to be random". The residual plots are as follows.

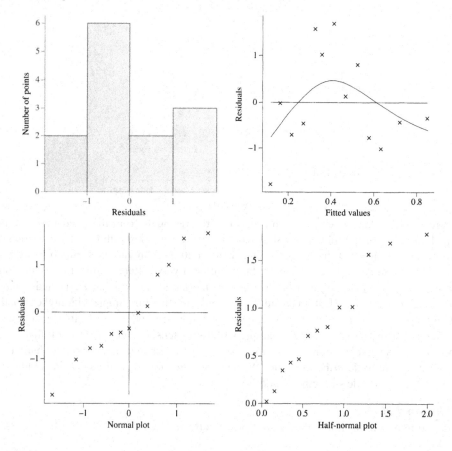

Though the fitted values plot shows some evidence of curvature, generally the model fits reasonably well. (The curvature in the fitted values plot matches the warning message about non-random residuals in the output.)

Solution 4.21

The scatterplot of the data indicates that a simple linear regression model is inappropriate, for two reasons. First, the relationship between the variables is far from linear. Second, it appears that the variability of the tensile strength measurements is greater for high strengths than for low strengths, so the assumption of constant variance is inappropriate.

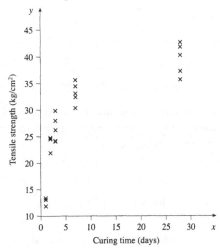

Solution 4.22

A logarithmic transformation of tensile strength appears to deal with the variance, as shown in this scatterplot. (You might instead have felt that a square root or reciprocal transformation is best – there is very little to choose between square root, reciprocal and log in this case.)

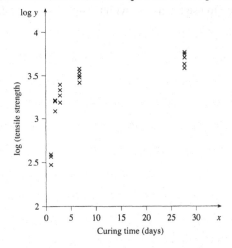

Solution 4.23

(a) Square root and log transformations still produce a noticeably curved scatterplot, but transforming curing time to its reciprocal (i.e. a power transformation with power −1) does the trick, as the following scatterplot shows.

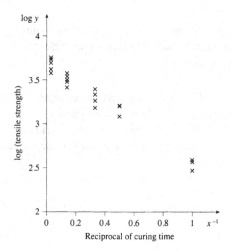

Now the plot looks reasonably linear, and the variance looks reasonably constant. However, the transformation has been arrived at by a roundabout route; it is quite possible that a better transformation exists.

(b) The regression line is

$$\log y = 3.6878 - \frac{1.1455}{x}$$

where tensile strength y is measured in kg/cm^2 and curing time x in days.

The residual plots (not shown) are not totally reassuring about the appropriateness of the simple linear regression model. In particular the histogram and the normal probability plot indicate skewness in the distribution of the residuals. However, these departures from the simple linear regression model do not appear very large, and it may be appropriate to use the model with these transformed data.

Solution 4.24

(a)

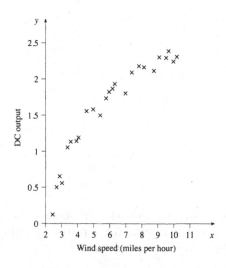

The relationship is clearly non-linear. (However, the scatter of points about the curve in the y direction is fairly constant, so that the variance does not appear to vary with the response.)

(b) In straightening the curve by transforming y, we can only move the points in the y direction (up or down). It is thus necessary to 'pull up' the points at the top right in the scatterplot, or in other words to expand the larger values of y relative to the smaller values. To do this, we must move up the ladder of powers, considering transformations like y^2, y^3 and so on. A plot of y^2 against x, shown below, looks reasonably linear.

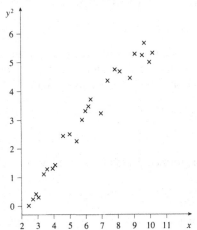

However, another difficulty has arisen regarding the simple linear regression model. The spread of points about the line in the y direction now increases as the response increases, because the transformation has spread out the larger responses. The assumption of constant variance no longer seems appropriate.

(c) To straighten the plot by transforming x, we need to move the points at the top right of the original scatterplot in part (a) towards the left, or in other words to expand the smaller values of x relative to the larger values. To do this, we must move down the ladder of powers, transforming x to something like $x^{1/2}, \log x, x^{-1/2}, x^{-1}, \ldots$. Plotting y against $x^{-1/2}$ produces a reasonably straight line, as shown below.

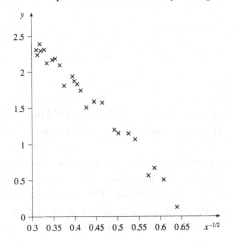

(Plotting y against x^{-1} is more or less as good.) This time, because the points have been moved only in the x direction, the scatter in the y direction has not been affected, and the assumption of constant variance looks appropriate.

(d) The 'best' transformation we have found leaves y unchanged and transforms x to $x^{-1/2}$. For this transformation, the fitted line is

$$y = 4.3897 - \frac{6.416}{\sqrt{x}}.$$

The usual residual plots (not shown) provide no great cause for concern about the appropriateness of the model, though there is some slight suggestion of non-normality in the probability plot and histogram.

Solution 4.25

The scatterplot of the untransformed data is shown below.

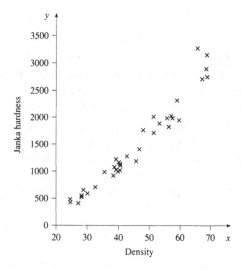

The relationship looks reasonably linear, but there is a clear tendency for the points to be scattered further away from where the regression line would be for timbers with high values of hardness and density. Thus the assumption of constant variance about the regression line is not appropriate.

Transforming hardness in order to bring its higher values closer together, by a log, square root or reciprocal transformation, might reduce the variance of the higher values, but on its own it would produce a curved scatterplot. The curve can be avoided by transforming density as well, in the same direction on the ladder of powers. Plotting log(Janka hardness) against log(density) leads to the following scatterplot.

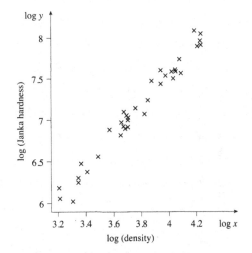

(You are not expected to be able to jump directly to this.)

This is still very linear, but the problem of non-constant variance is no longer evident. For the transformed data, the fitted regression line is

$$\log y = 0.015 + 1.8847 \log x$$

where y is the Janka hardness and x is the density. GENSTAT flags one point as having a large standardized residual, but it does not show up on the plots as being grossly out of line with the others.

Plotting the residuals shows no obvious departure from the model. The normal probability plot is somewhat curved, indicating that perhaps the normality assumption is not exactly satisfied. (The plots are not shown.) But, overall, the simple linear regression model fits the transformed data well.

Solution 4.26

For the northern towns, the regression line is
$$\text{mortality rate} = 1692.3 - 1.931 \times \text{calcium concentration}$$
and the estimated standard error of the slope is 0.848.

For the southern towns, the line is
$$\text{mortality rate} = 1522.8 - 2.093 \times \text{calcium concentration}$$
and the estimated standard error of the slope is 0.566.

Solution 4.27

We have a normally distributed estimator for the difference between the slopes of the two regression lines, whose mean, under the null hypothesis of equal slopes, is 0, and whose estimated standard deviation is 1.0195. The observed value of this estimator is 0.162. Thus the observed value is well under one standard deviation away from its mean under the null hypothesis. It seems that the SP should be high, and there are no grounds for rejecting the null hypothesis of equal slopes.

We cannot calculate an exact SP from the given information merely using the normal distribution, because that calculation would not take into account the fact that the standard deviations involved have been estimated from the samples. (However, this should not actually make too much difference.)

Solution 4.28

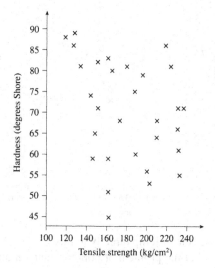

The scatterplot (which you might have plotted with the variables the other way round) is not indicative of any strong relationship between hardness and tensile strength. This is reflected in the correlation coefficient, which is -0.299. Any relationship between hardness and tensile strength is weak at most, the correlation being close to zero (and if there is any weak relationship it seems to be of the negative kind).

5

_____

Solution 5.1

(a) There are two columns in this spreadsheet. The first is a variate called **breadth**, that contains all the skull breadths for both groups of skulls. The second is a factor called **origin**, with labels **Etruscan** and **modern**, identifying which group each of the breadths belongs to.

(b) The resulting output is as follows.

	Nobservd	Mean	Variance
origin			
Etruscan	84	143.8	35.65
modern	70	132.4	33.06

The two groups have different sizes (84 and 70). Their variances are very similar, but their means appear rather different, with the Etruscan skulls being broader on average.

(c) The boxplots are as follows.

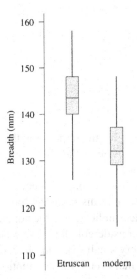

Generally they confirm the impression given in part (b), of similar variances and different means. The distributions in both groups look reasonably symmetric. There is no reason to doubt the assumptions, for t testing, of normal population distributions with equal variances.

(d) The t test statistic is 11.92 on 152 degrees of freedom. The SP is less than 0.001. There is strong evidence that the mean breadths of Etruscan and modern Italian skulls are not the same.

Solution 5.2
The results of this analysis are discussed in the main text, below the exercise.

You should have obtained the results by choosing the **Stats|Regression Analysis|Linear** menu item, ensuring the **Regression** field was set to **Simple Linear Regression**, setting the **Response Variate** field to breadth and the **Explanatory Variate** field to orvar, and then clicking on OK.

Solution 5.3
The boxplots are as follows.

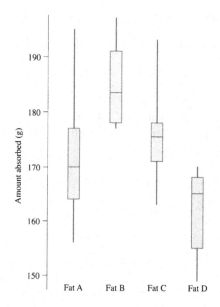

These (and the tabulated means) give the impression that the mean responses for different fats do differ considerably. For instance, the absorption data for Fat B are all higher than all of the absorption data for Fat D.

The assumptions of normal distributions with equal variances look somewhat dubious for these data. Some of the sample distributions look decidedly skewed, and both the boxplots and the table of variances indicate that the population variances may not all be the same. (The Fat A group has a considerably larger variance than the others.) However, the sample size in each group is very small (6), and one would thus expect that the samples would not look particularly normal even if the assumptions are indeed valid.

Solution 5.4
The Summary of analysis table from the regression is as follows.

	d.f.	s.s.	m.s.	v.r.	F pr.
Regression	1	4902.	4902.18	142.20	<.001
Residual	152	5240.	34.47		
Total	153	10142.	66.29		

The Analysis of variance table obtained using the Analysis of Variance command is as follows.

Source of variation	d.f.	s.s.	m.s.	v.r.	F pr.
origin	1	4902.18	4902.18	142.20	<.001
Residual	152	5239.97	34.47		
Total	153	10142.16			

Apart from the row labels, the accuracy to which certain numbers are rounded, and the absence of an m.s.(Total) value in the second table, they are identical.

Solution 5.5
The SP is given as 0.007. Thus there is strong evidence against the null hypothesis that the mean absorption is the same for all four fats. The conclusion is that some of the means differ – but the test does not tell us which.

Solution 5.6
(a) This is a controlled experiment. Each infant would have his/her exercise regime allocated by the experimenter.

(b) This is necessarily an observational study. Water samples cannot be assigned arbitrarily to water masses. (The investigator has control only over the number of samples taken from each mass.)

(c) This is a controlled experiment, in that the investigator assigns diets to families. (Note that this is true even if some families do not follow the diets to which they are assigned – one can think of the explanatory variable as being the diet which a family is *asked* to follow.)

Solution 5.7
In experiment (a), the experimental units are individual infants, the treatments are the different exercise regimes, and the responses are the times when the infants first walk by themselves.

In experiment (c), the experimental units are the families, the treatments are the different diets they are asked to follow, and the responses are the numbers of colds in each family in a year.

Solution 5.8
There are many possibilities. For example, fat lambs might all be assigned to one grass type, thin lambs to the other. Age and sex of the lambs might be important. The available fields of Grass A may be more luxuriant than those of Grass B. And so on.

Solution 5.9
(a) The boxplots are as follows.

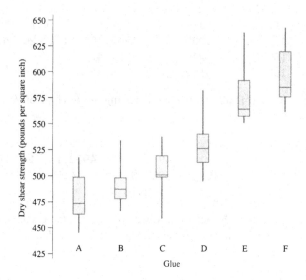

Along with the table of variances, they give some slight indication that the assumption of equal variances may not be appropriate. One glue (E) has a response variance almost double most of the others. Also, some of the boxplots look rather skewed, indicating that normality might not be justified. However, the samples are not large, and so the sample distributions may well not reflect the population distributions very accurately. The indications of skewness and of unequal variances are not enormous. Thus the one-way ANOVA model assumptions are indeed reasonable.

(b) The interpretation of the ANOVA table is discussed in the main text below the exercise.

Solution 5.10

The complete ANOVA table is as follows. (In the ANOVA table produced by GENSTAT, Treatment (voice part) is replaced by the name given in the singers dataset to the explanatory factor, namely voice.)

Source of variation	d.f.	s.s.	m.s.	v.r.	F pr.
Treatment (voice part)	3	81.115	27.038	3.95	0.010
Residual	103	705.670	6.851		
Total	106	786.785			

The numbers in it are worked out as follows.

d.f. column: there are $n = 107$ responses, so d.f.(Total) $= n - 1 = 107 - 1 = 106$; there are $k = 4$ 'treatments', so d.f.(Treatment) $= k - 1 = 4 - 1 = 3$; also d.f.(Residual) $= n - k = 107 - 4 = 103$ (alternatively take d.f.(Total) $-$ d.f.(Treatment) $= 106 - 3 = 103$).

s.s. column: s.s.(Total) $= TSS = 786.785$; s.s.(Residual) $= RSS = 705.670$; so s.s.(Treatment) $= TSS - RSS = 81.115$.

m.s. column: in each case the m.s. is calculated by dividing the corresponding s.s. by the corresponding d.f., so that m.s.(Treatment) $= 81.115/3 = 27.038$ and m.s.(Residual) $= 705.670/103 = 6.851$.

v.r. column: v.r. $=$ m.s.(Treatment)/m.s.(Total) $= 27.038/6.851 = 3.95$.

The SP is $P(F \geq \text{v.r.}) = P(F \geq 3.95)$, where $F \sim F(k-1, n-k) = F(3, 103)$. This can be found using the GENSTAT command PRINT cuf(3.95;3;103), which gives the answer 0.010.

The conclusion is that there is evidence of a difference in mean heights between the different voice parts, but the evidence is not overwhelming.

Solution 5.11

(a) The group variances are, respectively, 15.878, 2.678, 3.511, 2.000 and 3.211. Even allowing for the rather small sample sizes, the huge difference between the variance in the first treatment group (the controls) and the others casts serious doubt on the assumption of equal variances. The same impression is given by the boxplots (which also give some indication of skewness).

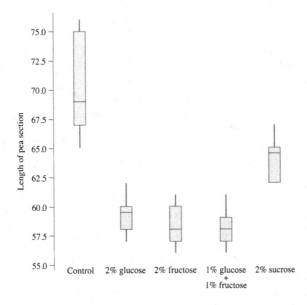

(b) The analysis of variance table is as follows.

Source of variation	d.f.	s.s.	m.s.	v.r.	F pr.
sugar	4	1077.320	269.330	49.37	<.001
Residual	45	245.500	5.456		
Total	49	1322.820			

The SP is very small, providing strong evidence that the different growth mediums result in different mean lengths. (Since the equal variance assumption is in doubt, some doubt about the validity of this conclusion must remain; but the strength of evidence here, and the pattern in the boxplots in part (a), do make it look very improbable that the treatment means are really all the same.)

(c) GENSTAT flags two outliers (2 out of 50, or 4% of the datapoints). They are identified as points 5 and 9, which are both in the first treatment group (the controls). The values of residual/s.e.(residual) for the two points are, respectively, $-5.10/2.22 = -2.30$

and $5.90/2.22 = 2.66$; thus the outliers are not *very* extreme. If this were the only evidence of problems with the model for these data, one might well be inclined not to worry about it. In this case, however, it appears that the problem may be more fundamental than a couple of outliers; we have evidence that the variance of the first treatment group is considerably higher than the others. In the analysis of variance, GENSTAT has assumed equal variances, so it is hardly surprising that it treats the extreme values in the first treatment group as potential outliers.

(d) The residual plots are as follows.

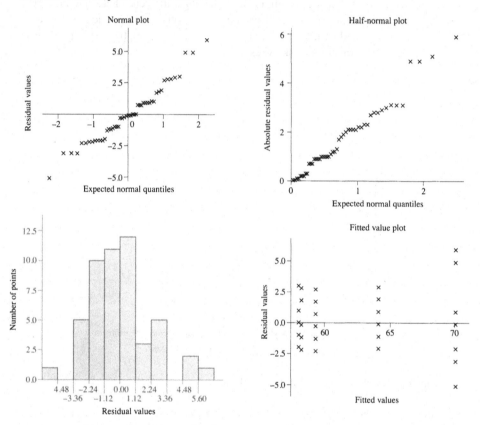

In the half-normal plot, four points (at the upper end) appear detached from the others. Comparison with the normal probability plot indicates that three of these correspond to positive residuals and one to a negative residual. The two most extreme correspond to the two residuals flagged by GENSTAT in the analysis of variance output. The probability plots and the histogram of residuals perhaps throw some doubt on the normality assumption – but one would expect departures from exact normality in samples of this size, even if the population distribution is exactly normal.

The plot of residuals against fitted values is more worrying. It clearly shows that the residuals corresponding to the highest fitted value – those for the control group – are more variable than the others. This is simply reflecting what we know already about the higher variance in this group.

(e) The only major worry about the appropriateness of the one-way ANOVA model assumptions in this case relates to unequal group variances. When the inequality is as great as it appears here, the one-way ANOVA model can produce potentially misleading results. As mentioned in the solution to part (b), the conclusion that the mean lengths are not all the same seems reasonably secure; however, some doubt must remain about the validity of any further analyses we perform (such as those in Section 5.5).

Solution 5.12

(a) The plot is as follows.

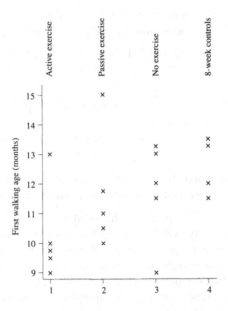

The highest points in the first two treatment groups (active and passive exercise) lie well above the other points in their groups, as does the lowest point in the third group (no exercise). With these points included, the assumption of normal distributions within groups is in considerable doubt.

(b) The ANOVA table is as follows.

Source of variation	d.f.	s.s.	m.s.	v.r.	F pr.
exercise	3	14.778	4.926	2.14	0.129
Residual	19	43.690	2.299		
Total	22	58.467			

It provides an SP of 0.129 for the test of equality of treatment means. This indicates that there is little or no evidence of any differences between the group means.

(c) The two flagged points are the highest responses in each of the first two treatment groups. These correspond to two of the points identified as possible outliers in part (a). However, interestingly, GENSTAT does not flag the dubious point in the third treatment group.

(d) Without points 5 and 12, the ANOVA table is as follows.

```
Source of variation    d.f.      s.s.      m.s.    v.r.   F pr.
exercise                 3       23.057    7.686   7.26   0.002
Residual                17       18.002    1.059
Total                   20       41.060
```

The SP has changed considerably, to 0.002, a value which indicates considerable evidence of difference between the treatment means.

(e) The flagged point is, according to GENSTAT, number 13, which is point 15 in the original dataset. This is indeed the third potential outlier discussed in part (a).

(f) Removing this point as well, the following ANOVA table is obtained.

```
Source of variation    d.f.      s.s.      m.s.     v.r.    F pr.
exercise                 3       27.2500   9.0833   15.80   <.001
Residual                16        9.2000   0.5750
Total                   19       36.4500
```

The SP is now less than 0.001, and the evidence against equality of treatment means has become even stronger. (Also, no further points are flagged.)

(g) Whether or not these three points are included makes a lot of difference to the test outcome. The one-way ANOVA model has not proved robust to these outliers. Looking again at the scatterplot in part (a), what the ANOVA tells us seems reasonable: without the outliers, one might suspect that $\mu_1 < \mu_2 < \mu_3 \simeq \mu_4$; but, with the points included, the position is much less clear. Yet, without further investigation as to why these infants were so different from the others in their treatment groups, we are not justified in throwing them away just because they do not fit the model.

Solution 5.13

(a)
$$\widehat{\theta}_2 = \tfrac{1}{2}(\overline{Y}_1 + \overline{Y}_2) - \tfrac{1}{2}(\overline{Y}_3 + \overline{Y}_4) = \tfrac{1}{2}(\overline{Y}_1 + \overline{Y}_2 - \overline{Y}_3 - \overline{Y}_4)$$
$$= \tfrac{1}{2}(68.90 + 69.90 - 70.72 - 71.38) = -1.65.$$

(b) $V(\widehat{\theta}_2) = \sigma^2 \sum_{j=1}^{4} \dfrac{a_j^2}{n_j} = \sigma^2 \times \dfrac{1}{4}\left(\dfrac{1}{21} + \dfrac{1}{21} + \dfrac{1}{39} + \dfrac{1}{26}\right) = 0.03984\sigma^2.$

(c) The appropriate estimate of σ^2 is m.s.(Residual) $= 6.851$. Thus the estimated $V(\widehat{\theta}_2) = 0.03984 \times 6.851 = 0.2729$, and so the estimated s.e.$(\widehat{\theta}_2) = \sqrt{0.2729} = 0.522$.

(d) $\widehat{\theta}_2$ is $-1.65/0.522 = -3.16$ times its estimated s.e. The number of degrees of freedom of the appropriate t distribution is d.f.(Residual) $= 103$. So the SP is $P(|t(103)| \geq 3.16) = 2P(t(103) \geq 3.16) = 2 \times 0.001 = 0.002$. (Use PRINT cut(3.16;103).) Since this is small, it provides strong evidence that the mean heights of tenors and basses are different. (In fact, basses appear to be taller.)

Solution 5.14

(a)
$$\widehat{\theta}_4 = \overline{Y}_5 - \tfrac{1}{3}(\overline{Y}_2 + \overline{Y}_3 + \overline{Y}_4),$$

which, from the GENSTAT output of the treatment means, yields

$$\widehat{\theta}_4 = 64.1 - \tfrac{1}{3}(59.3 + 58.2 + 58.0) = 5.6.$$

(b)
$$V(\widehat{\theta}_4) = \sigma^2 \sum_{j=1}^{5} \frac{a_j^2}{n_j} = \sigma^2 \times \frac{1}{10}\left[0^2 + \left(\frac{1}{3}\right)^2 + \left(\frac{1}{3}\right)^2 + \left(\frac{1}{3}\right)^2 + 1^2\right]$$

$$= \frac{\sigma^2}{10}\left(\frac{4}{3}\right) = \frac{2\sigma^2}{15}.$$

(c) The appropriate estimate of σ^2 is m.s.(Residual)= 5.456. Thus the estimated $V(\widehat{\theta}_4) = \frac{2}{15} \times 5.456 = 0.7275$, and so the estimated s.e.$(\widehat{\theta}_4) = \sqrt{0.7275} = 0.853$.

(d) $\widehat{\theta}_4$ is $5.6/0.853 = 6.57$ times its estimated s.e. The number of degrees of freedom of the appropriate t distribution is d.f.(Residual) = 45. So the SP is $P(|t(45)| \geq 6.57) = 2P(t(45) \geq 6.57) = 2 \times 0.222 \times 10^{-7} = 0.444 \times 10^{-7}$. Since this is very tiny, it provides very strong evidence that a difference exists between the effect of sucrose and that of glucose or fructose (singly or together).

Solution 5.15

(a) In this dataset, the response variable is the recall score, the treatments are the types of hemispherical dominance (left, right or integrative), and the experimental units are the individual human subjects.

(b) Boxplots of the data by groups are as shown (left).

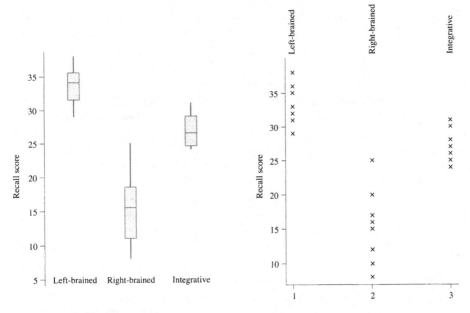

Since the number of datapoints is small, you may have preferred to produce a scatterplot of recall against hemi, as above (right). Both plots give the same impressions.

It appears that recall scores may be rather higher for left-brained individuals than for right-brained individuals. The integrative group appears to be somewhere in between the other two. There seems no reason to doubt the normality assumption of the one-way ANOVA model. The assumption of equal variances is possibly in more doubt: results for the right-brained group seem more spread out than those for the other two groups. However, the difference is not huge, and these are small groups, so we might well not be too concerned about this possible difference in the absence of any other reason to doubt equal variance.

(c) The ANOVA table is as follows.

```
Source of variation    d.f.       s.s.      m.s.     v.r.   F pr.
hemi                     2      1362.33    681.17   44.61   <.001
Residual                21       320.62     15.27
Total                   23      1682.96
```

The F test of the null hypothesis of equality of all three treatment means yields a very small SP, giving strong evidence that the ability to recall tabular information on numbers of doctors depends on hemispherical dominance.

(d) The residual plots are as follows.

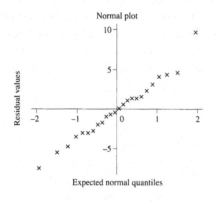

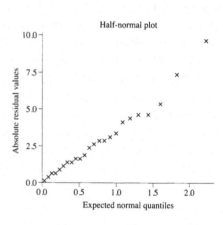

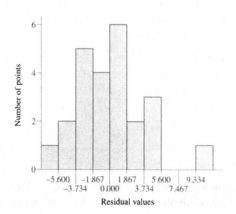

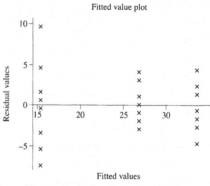

GENSTAT flags two individual residuals, corresponding to the largest and smallest recall scores in the right-brained group. These residuals are, respectively, 2.63 and -2.01 times their standard errors, so neither is really huge. As with the pea data, the flagging may be a symptom of unequal variances rather than an indication of something wrong with these particular datapoints.

The plots show nothing apart from the flagged points and the larger variance in one of the treatment groups, which we already knew about.

Overall, there is some reason to doubt the suitability of the one-way ANOVA model for these data, because of unequal variances – but the departures from the model are not huge, and it seems unlikely that conclusions we might derive from the one-way ANOVA model would be seriously misleading.

(e) A suitable contrast is

$$\theta_5 = \mu_3 - \tfrac{1}{2}(\mu_1 + \mu_2).$$

This is estimated by

$$\widehat{\theta}_5 = \overline{Y}_3 - \tfrac{1}{2}(\overline{Y}_1 + \overline{Y}_2) = 26.87 - \tfrac{1}{2}(33.63 + 15.37) = 2.37.$$

The variance of this estimate is

$$V(\widehat{\theta}_5) = \sigma^2 \sum_{j=1}^{3} \frac{a_j^2}{n_j} = \sigma^2 \times \frac{1}{8}\left[1^2 + \left(\frac{1}{2}\right)^2 + \left(\frac{1}{2}\right)^2 \right] = \frac{3\sigma^2}{16}.$$

The appropriate estimate of σ^2 is $\mathrm{m.s.(Residual)} = 15.27$. Thus, the estimated $V(\widehat{\theta}_5) = \frac{3}{16} \times 15.27 = 2.8631$, and so the estimated $\mathrm{s.e.}(\widehat{\theta}_5) = \sqrt{2.8631} = 1.692$. Thus $\widehat{\theta}_5$ is $2.37/1.692 = 1.40$ times its estimated s.e. The number of degrees of freedom of the appropriate t distribution is $\mathrm{d.f.(Residual)} = 21$. So the SP is $P(|t(21)| \geq 1.40) = 2P(t(21) \geq 1.40) = 2 \times 0.088 = 0.176$. Thus there is hardly any evidence for an overall difference between predominantly single-sided brain users and both-sided brain users.

(f) This analysis has yielded strong evidence for some difference between left-brained, right-brained and integrative people, but none for a difference between integrative people and the mean of the other two types. It remains to investigate further the details of the overall difference that has been identified. From the plot in part (b), one might hypothesize that the most important difference is between left-brained and right-brained people in terms of recall score. In the light of this, one might question the wisdom of combining left-brained and right-brained people together in the formulation of part (e), even though this was a comparison planned before the experiment. (Indeed, perhaps the performance of integrative individuals is like a mixture of the performances of left- and right-brained individuals.) And behind all these conclusions and discussions, there remains some concern that the larger variance in the right-brained group may be rendering the conclusions from the analysis of variance less secure. (One should bear in mind that the larger variance of this group may, in itself, be of interest to the experimenters. However, analysis of variance itself throws no light on this aspect of the data. Remember that, despite its name, analysis of variance, when used in the way you have learned in this chapter, is about differences in *means*, not in variances!)

6

Solution 6.1

The F statistic for the test of whether the regression coefficient of the **strength** variable is zero is 2.74 on 1 and 28 degrees of freedom, giving an SP of 0.109. This is marginal to say the least, and many people would be happy to set $\beta = 0$. The percentage of variance accounted for by this regression (5.7%) is very small.

Solution 6.2

(a) The scatterplots are as below.

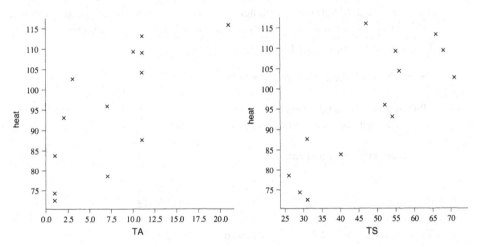

The amount of heat generated appears to go up with each of **TA** and **TS** individually. It is difficult to pronounce too strongly on whether a linear model is entirely appropriate in each case because there are very few datapoints; but, as a first stab, the assumption is not entirely unreasonable.

(b) The GENSTAT output for the regression of **heat** on **TA** is as follows.

```
***** Regression Analysis *****

Response variate: heat
       Fitted terms: Constant, TA
```

```
*** Summary of analysis ***

                   d.f.        s.s.        m.s.       v.r.   F pr.
Regression          1         1450.      1450.1      12.60   0.005
Residual           11         1266.       115.1
Total              12         2716.       226.3
```

Percentage variance accounted for 49.2
Standard error of observations is estimated to be 10.7
* MESSAGE: The following units have high leverage:

```
          Unit      Response    Leverage
           10        115.9        0.52
```

```
*** Estimates of parameters ***

                 estimate        s.e.       t(11)    t pr.
Constant           81.48         4.93       16.54   <.001
TA                 1.869         0.526       3.55   0.005
```

The percentage of variance accounted for is 49.2%. There is considerable evidence ($SP = 0.005$) for a non-zero regression coefficient for the **TA** variable.

For **heat** on **TS**, we get the following.

```
***** Regression Analysis *****

 Response variate: heat
     Fitted terms: Constant, TS
```

```
*** Summary of analysis ***

                   d.f.        s.s.        m.s.       v.r.   F pr.
Regression          1        1809.4      1809.43     21.96   <.001
Residual           11         906.3       82.39
Total              12        2715.8       226.31
```

Percentage variance accounted for 63.6
Standard error of observations is estimated to be 9.08
*MESSAGE: The following units have large standardized residuals:

```
          Unit      Response    Residual
           10        115.90        2.45
```

```
*** Estimates of parameters ***

                 estimate        s.e.       t(11)    t pr.
Constant           57.42         8.49        6.76   <.001
TS                 0.789         0.168       4.69   <.001
```

In this case, the percentage of variance accounted for is greater (63.6%) and the regression SP even smaller.

For the regression of heat on TA and TS together, we get the following.

```
***** Regression Analysis *****

  Response variate: heat
      Fitted terms: Constant, TA, TS

  *** Summary of analysis ***
```

	d.f.	s.s.	m.s.	v.r.	F pr.
Regression	2	2657.86	1328.929	229.50	<.001
Residual	10	57.90	5.790		
Total	12	2715.76	226.314		

```
Percentage variance accounted for 97.4
Standard error of observations is estimated to be 2.41
* MESSAGE: The following units have high leverage:
        Unit     Response   Leverage
         10       115.90      0.55

  *** Estimates of parameters ***
```

	estimate	s.e.	t(10)	t pr.
Constant	52.58	2.29	23.00	<.001
TA	1.468	0.121	12.10	<.001
TS	0.6623	0.0459	14.44	<.001

Even though each individual regression provided fairly strong evidence of linear relationships, the regression on both explanatory variables is considerably stronger: the percentage of variance accounted for has risen to the very respectable 97.4% (and all regression coefficients have $SPs < 0.001$). The two-variable model seems usefully to improve on each single-variable one.

(c) The fitted regression model is

$$\hat{y} = 52.58 + 1.468x_1 + 0.6623x_2$$

where y represents heat, x_1 represents TA and x_2 represents TS. When $x_1 = 15$ and $x_2 = 55$, y is predicted to be

$$\hat{y} = 52.58 + (1.468 \times 15) + (0.6623 \times 55) = 111.03.$$

You were expected to have used your calculator to obtain the value of 111.03 for $\hat{y}$. In GENSTAT, you would have had to enter PREDICT TA,TS;15,55 in the Input Log in interactive mode. You might have been able to work this out from Exercise 4.14, where the PREDICT command made a first appearance (but in a slightly different form).

(d) Even though the percentage of variance accounted for by the two-variable model is very high, something intriguing is shown up by the residual plots.

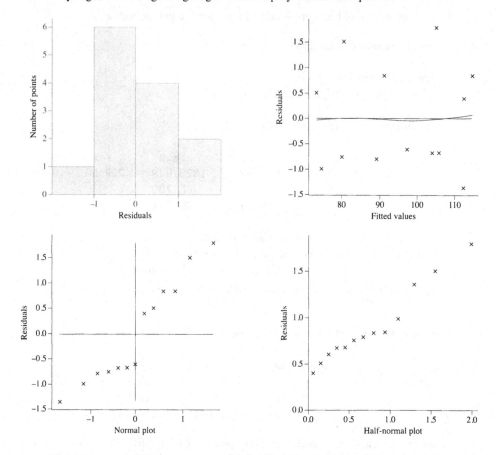

There is a gap! There are no standardized residuals within a distance of about 0.4 from the fitted mean. An alternative model suggested by this is that there may be two separate regression models, quite possibly parallel to one another, that would fit the data even better if it were separated into groups of six (with positive residuals) and seven (with negative residuals), respectively. Further knowledge of the experimental situation would be useful to help understand whether this is sensible, and we shall not consider it further. Note, however, that this is very much a small secondary effect, and the fitted regression model will be adequate for many purposes.

Solution 6.3

(a) The results of the regression are as follows.

```
***** Regression Analysis *****
  Response variate: ventil
      Fitted terms: Constant, oxygen
```

```
*** Summary of analysis ***

                 d.f.          s.s.          m.s.        v.r.   F pr.
Regression        1           75555.       75555.2      528.40  <.001
Residual         51            7292.         143.0
Total            52           82848.        1593.2
```

Percentage variance accounted for 91.0
Standard error of observations is estimated to be 12.0
* MESSAGE: The residuals do not appear to be random;
          for example, fitted values in the range 31.3 to 91.9
          are consistently larger than observed values
          and fitted values in the range -0.6 to 15.7
          are consistently smaller than observed values

```
*** Estimates of parameters ***

                estimate        s.e.      t(51)  t pr.
Constant         -18.45         3.82      -4.84  <.001
oxygen           0.03114      0.00135     22.99  <.001
```

The fit on x_1 alone seems very good on the basis of this output; in particular, the percentage of variance accounted for is 91% (and the SP for β_1 is less than 0.001). When you think about it, as a first approximation to the way ventil increases with oxygen, the straight line is not too terrible: it at least gives the right sort of general trend at the right sort of rate of increase. But the clue to our easily being able to do better is given by GENSTAT's message that 'The residuals do not appear to be random'. A residual plot would confirm what we already in fact know from Figure 6.3, that there is a systematic departure from the straight line.

(b) The Data field needs oxygen, the Save in field needs oxy2 and the Parameters in Equation need to be $c = 0$, $m = 2$. Regressing on oxygen and oxy2 using Multiple Linear Regression in the Linear Regression dialogue box yields the following.

```
***** Regression Analysis *****

  Response variate: ventil
      Fitted terms: Constant, oxygen, oxy2

*** Summary of analysis ***

                 d.f.          s.s.          m.s.        v.r.    F pr.
Regression        2          82339.9       41169.97   4054.90   <.001
Residual         50           507.7         10.15
Total            52          82847.6        1593.22
```

Percentage variance accounted for 99.4

```
Standard error of observations is estimated to be 3.19
*MESSAGE: The following units have large standardized residuals:

        Unit     Response     Residual
         42        89.10        -3.03
         46       111.40        -2.67
         53       144.80         2.59
* MESSAGE: The error variance does not appear to be constant:
          large responses are more variable than small
          responses

  *** Estimates of parameters ***

                  estimate         s.e.       t(50)   t pr.
Constant             24.27         1.94       12.51   <.001
oxygen            -0.01344      0.00176       -7.63   <.001
oxy2          0.000008902  0.000000344       25.85   <.001
```

The percentage of variance accounted for has increased to 99.4% and all the regression coefficients have SPs < 0.001. The model fitted is

$$\widehat{y} = 24.27 - 0.01344x + 0.000008902x^2$$

where x is **oxygen**, and this is the one shown to fit very well in Figure 4.2. Notice that β_2 is very small in absolute terms in the fitted model. None the less, it is clearly an important component of the model (because **oxy2** takes extremely large values).

There is, however, a further warning (aside from the warning about large standardized residuals) that 'The error variance does not appear to be constant'.

(c) The plot of residuals against fitted values in particular (which is shown below) shows the source of GENSTAT's message. There is a sudden increase in variance toward the right-hand end. (This effect is present in Figure 6.3, but went unnoticed: remember that the variance we are talking about is in the vertical direction.)

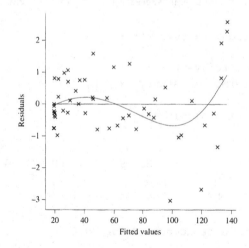

Interestingly, it is not easy to find suitable transformations for remedying this situation. For instance, a log transformation of ventil decreases the variance at the right-hand end at the expense of an increase in variance at the left-hand end! (Many statisticians would take an alternative approach to dealing with non-constant variance in this case, by performing the regression on the raw (untransformed) data using a technique called weighted least squares. This consists of weighting different points differently according to their variability. This technique will not be discussed in this book.)

Solution 6.4

Choose the Stats|Regression Analysis|Linear menu item. In the resulting dialogue box, change the Regression field to Multiple Linear Regression, put sbp as Response Variate, and fill the Explanatory Variates field with all eight explanatory variables separated by commas. The resulting output is as follows.

```
***** Regression Analysis *****
  Response variate: sbp
      Fitted terms: Constant, age, years, weight, height, chin,
                    forearm, calf, pulse

*** Summary of analysis ***
                 d.f.        s.s.        m.s.      v.r.  F pr.
Regression         8        2095.      261.84      2.95  0.015
Residual          29        2575.       88.80
Total             37        4670.      126.21

Percentage variance accounted for 29.6
Standard error of observations is estimated to be 9.42
* MESSAGE: The error variance does not appear to be constant:
           intermediate responses are more variable than small
           or large responses

* MESSAGE: The following units have high leverage:
        Unit     Response     Leverage
          5       140.00        0.53
         39       152.00        0.60

*** Estimates of parameters ***
                estimate       s.e.      t(29)  t pr.
Constant           139.6       53.2       2.62  0.014
age                0.021      0.283       0.07  0.943
years             -0.428      0.211      -2.02  0.052
weight             1.698      0.450       3.77  <.001
height           -0.0595     0.0391      -1.52  0.140
chin              -1.416      0.799      -1.77  0.087
forearm            -0.12       1.34      -0.09  0.929
calf               0.031      0.579       0.05  0.958
pulse             -0.174      0.199      -0.88  0.387
```

(a) Notice that there are 8 d.f. for `Regression`, corresponding to the eight explanatory variables in the model. The SP in the ANOVA table is 0.015, suggesting that there is quite strong evidence against the hypothesis that all the regression coefficients are zero. However, the percentage of variance accounted for is small, suggesting that even after successful modelling of the regression mean, there will remain a high degree of variability about the mean (in particular, the model we produce will have little predictive value – but this does not matter here, since prediction is not the focus of this example).

(b) It is easiest to cast your eye down the list of SPs given here (under `t pr.`) for testing the β_js individually. On this basis, the strongest dependence is on **weight** ($SP < 0.001$) and the constant has SP 0.014. Other leading explanatory variables (SPs less than 0.1) are **years** and then **chin**. It is tempting, therefore to settle on the regression of **sbp** on, say, **weight, years** and **chin** alone. While this may be a sensible model, this is not guaranteed: recall that the values here of regression coefficients are values with all the other variables held fixed; when the other variables are absent from the model, the regression coefficients for **weight, years** and **chin** might be quite different. (You can check this out for yourself if you want.)

(c)

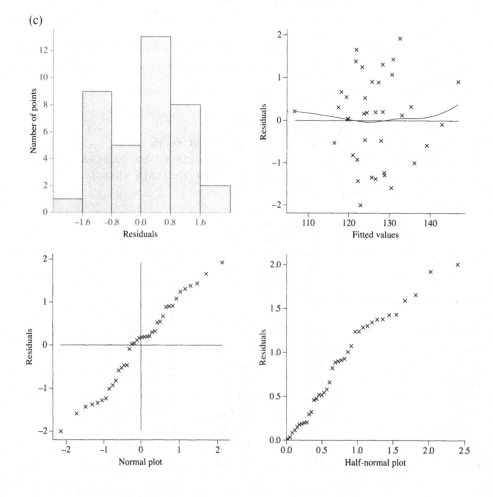

One might feel, from the plot of residuals against fitted values, that the variance increases for the moderate fitted values relative to those at either extreme (and the GENSTAT output flags this). While this is not an unreasonable view to take, it is also possible to argue that, because there are relatively few points with large or small fitted values, assessment of the variability in these regions is very difficult, and an assumption of constant variance is not unwarranted. The normal plot is not completely straight and the histogram looks a little odd, but they do not provide evidence of gross or systematic departures from normality.

Solution 6.5

(a) The regression output is as follows.

```
***** Regression Analysis *****

Response variate: crime
   Fitted terms: Constant, malyth, state, school, pol60, pol59,
                 empyth, mf, popn, race, uneyth, unemid,
                 income, poor

*** Summary of analysis ***
```

	d.f.	s.s.	m.s.	v.r.	F pr.
Regression	13	52931.	4071.6	8.46	<.001
Residual	33	15879.	481.2		
Total	46	68809.	1495.9		

```
Percentage variance accounted for 67.8
Standard error of observations is estimated to be 21.9
*MESSAGE: The following units have large standardized residuals:
         Unit    Response    Residual
          11      167.4        2.68
*MESSAGE: The following units have high leverage:
         Unit    Response    Leverage
          37       83.1        0.68

*** Estimates of parameters ***
```

	estimate	s.e.	t(33)	t pr.
Constant	-692.	156.	-4.44	<.001
malyth	1.040	0.423	2.46	0.019
state	-8.3	14.9	-0.56	0.581
school	1.802	0.650	2.77	0.009
pol60	1.61	1.06	1.52	0.138
pol59	-0.67	1.15	-0.58	0.565
empyth	-0.041	0.153	-0.27	0.791
mf	0.165	0.210	0.78	0.438
popn	-0.041	0.130	-0.32	0.752

race	0.0072	0.0639	0.11	0.911
uneyth	-0.602	0.437	-1.38	0.178
unemid	1.792	0.856	2.09	0.044
income	0.137	0.106	1.30	0.203
poor	0.793	0.235	3.37	0.002

The evidence against the null hypothesis of no linear relationship is strong, according to $SP < 0.001$. The (adjusted) percentage of variance accounted for is reasonable, but not overexciting, at 67.8%. The plot of residuals against fitted values displays somewhat similar characteristics to the one in Solution 6.4, although this time GEN-STAT does not think that there is a non-constant variance. The one possible outlier that GENSTAT flags, with a standardized residual of 2.68, may be worth worrying about, and shows up clearly on the normal plot of residuals, but we shall retain this individual in our analysis. It may also be worth noting that the six smallest fitted values are all associated with positive residuals.

The individual SPs for the estimates of the regression coefficients are smallest for the constant plus (in order) **poor**, **school**, **malyth** and **unemid**. All these have SPs < 0.05. The next smallest is 0.138 (for **pol60**).

(b) The correlation coefficients are as follows.

```
*** Correlation matrix ***
malyth    1.000
 state    0.584    1.000
school   -0.530   -0.703    1.000
 pol60   -0.506   -0.373    0.483    1.000
 pol59   -0.513   -0.376    0.499    0.994    1.000
empyth   -0.161   -0.505    0.561    0.121    0.106    1.000
    mf   -0.029   -0.315    0.437    0.034    0.023    0.514    1.000
  popn   -0.281   -0.050   -0.017    0.526    0.514   -0.124   -0.411
  race    0.593    0.767   -0.665   -0.214   -0.219   -0.341   -0.327
uneyth   -0.224   -0.172    0.018   -0.044   -0.052   -0.229    0.352
unemid   -0.245    0.072   -0.216    0.185    0.169   -0.421   -0.019
income   -0.670   -0.637    0.736    0.787    0.794    0.295    0.180
  poor    0.639    0.737   -0.769   -0.631   -0.648   -0.270   -0.167

         malyth    state   school    pol60    pol59   empyth       mf


  popn    1.000
  race    0.095    1.000
uneyth   -0.038   -0.156    1.000
unemid    0.270    0.081    0.746    1.000
income    0.308   -0.590    0.045    0.092    1.000
  poor   -0.126    0.677   -0.064    0.016   -0.884    1.000

           popn     race   uneyth   unemid   income     poor
```

Most highly correlated of all (0.994) are pol59 and pol60 (police expenditures for 1959 and 1960 respectively). This is not at all surprising! There is no need for both of them in the model and one of them should be dropped. pol59 has the larger SP and, in any case, is the less recent variable and one would expect it to be less important: pol60 is the better candidate for retention.

The variables income and poor are also highly (negatively) correlated (-0.884): both measure a state's wealth, income being higher for richer states while poor is lower, and vice versa. Since income has a larger SP, it appears that it might be unnecessary.

There are several other pairs with (absolute) correlations of 0.7 or so. pol59, pol60 and income actually form a correlated triple; state, school, race and poor are also mutually fairly highly correlated; unemid and uneyth have correlation 0.746 and both measure unemployment. Perhaps only one variable in each of these collections might be necessary in the model.

(c) For the necessary GENSTAT commands, make the obvious changes to those in Example 6.2.

The stepwise regression drops race, empyth, popn, state, pol59, mf, uneyth and income in turn, to leave a model with five explanatory variables, namely malyth, school, pol60, unemid and poor.

This seems very reasonable given what went before. The variables pol60, malyth and school were suggested by the scatterplots; the others suggested by the scatterplots, pol59 and income, have not made the final model because of their high correlations with pol60 and poor, respectively, which have. The five chosen variables have the five smallest SPs in the full multiple regression (though the influence of pol60 did not appear to be great on the basis of its SP, but this was because of its high correlation with pol59). The five variables are sensibly chosen from the groupings suggested by the correlations.

The stepwise regression starting from the null model leads to exactly the same model.

(d) The finally fitted model is

$$\text{crime} = -524.4 + 1.020\,\text{malyth} + 2.031\,\text{school} + 1.233\,\text{pol60}$$
$$+ 0.914\,\text{unemid} + 0.635\,\text{poor}.$$

The fit is fairly good with 69.7% of variance accounted for. The residual plots do not look greatly different from that for the full model, except that the biggest standardized residual is very large at 3.12. (To get the relevant residual plot if the Further Output button in the Linear Regression dialogue box is unavailable, you will first need to rerun the regression with the five chosen variables in the Model to be Fitted field.) This corresponds to state 11, which has a higher crime rate than the values of its other variables would suggest. Further consideration of this outlier and its influence is desirable (see Chapter 11).

We conclude that the main factors influencing crime rate are the number of young males in the population, the level of education, the level of police expenditure, the unemployment level (across all ages because the group used was very highly correlated with the younger age group) and the number of families earning less than half the median income. Notice that the *partial* regression coefficients do not necessarily have the sign one might expect. For instance, the amount of crime appears to go up with the amount of schooling (when all other variables in the model are held constant).

Solution 6.6

First load up crime and regress crime on malyth, school, pol60, unemid and poor. The command

```
PREDICT [pred=yhat;se=se] malyth,school,pol60,unemid,poor;140,110,98,20,180
```

produces a point estimate of 95.19 with a standard error (which is contained in se) of 6.68. You can now input precisely the six command lines given towards the end of Example 6.3. This yields a 95% prediction interval of

$$(50.1, 140.3).$$

(The relevant number of degrees of freedom used in df is 41.)

Solution 6.7

The relevant part of the GENSTAT output is as follows.

```
***** Regression Analysis *****

Response variate: mortalty
      Fitted terms: Constant + calcium + north

*** Summary of analysis ***

                d.f.        s.s.         m.s.      v.r.   F pr.
Regression        2      1248318.     624159.    41.86   <.001
Residual         58       864856.      14911.
Total            60      2113174.      35220.

Change           -1      -342132.     342132.    22.94   <.001

Percentage variance accounted for 57.7
Standard error of observations is estimated to be 122.
* MESSAGE: The following units have large standardized residuals:
        Unit      Response     Residual
          44        1987.        2.57

*** Estimates of parameters ***

                         estimate       s.e.      t(58)   t pr.
Constant                  1518.7        41.3      36.74   <.001
calcium                   -2.034        0.483     -4.21   <.001
north 1                    176.7        36.9       4.79   <.001
```

The SP for the regression is less than 0.001, as is that for each of the regression coefficients. The (adjusted) percentage of variance accounted for is a tiny bit bigger (57.7% versus 56.9%) than for the model with the product variable. This parallel line model seems a good one.

The fitted model for the mean is

$$\text{mortalty} = 1518.7 - 2.034\,\text{calcium} + 176.7\,\text{north}.$$

For the southern towns, the model is

$$\text{mortalty} = 1518.7 - 2.034\,\text{calcium}$$

and, for the northern towns,

$$\text{mortalty} = (1518.7 + 176.7) - 2.034\,\text{calcium} = 1695.4 - 2.034\,\text{calcium}.$$

We have gone further than in Example 4.3 in actually providing fitted lines with the same slope and different intercepts.

Solution 6.8

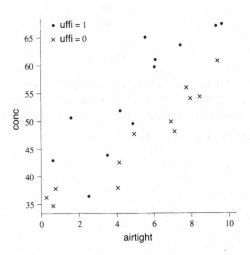

(a) From the scatterplot, straight-line regressions, possibly parallel but with different intercepts, would appear to be a possible model for these data.

(b) This requires the procedure of Example 6.4, based on Simple Linear Regression with Groups. The relevant part of the output is the following.

```
***** Regression Analysis *****

Response variate: conc
     Fitted terms: Constant + airtight + uffi + airtight.uffi
```

```
*** Summary of analysis ***

                 d.f.          s.s.          m.s.       v.r.   F pr.
Regression         3          2051.1        683.72      35.11  <.001
Residual          20           389.5         19.47
Total             23          2440.6        106.11

Change            -1           -20.2         20.21       1.04  0.320

Percentage variance accounted for 81.6
Standard error of observations is estimated to be 4.41
*MESSAGE: The following units have large standardized residuals:
         Unit      Response      Residual
          3         65.06          2.09
         10         36.37         -2.49


*** Estimates of parameters ***

                          estimate        s.e.      t(20)   t pr.
Constant                    33.42         2.47      13.51   <.001
airtight                    2.535         0.408      6.22   <.001
uffi 1                      5.24          3.67       1.43   0.168
airtight.uffi 1             0.632         0.621      1.02   0.320
```

The parameter that determines whether or not the lines are parallel is the coefficient of airtight.uffi. The SP for the test of whether this coefficient is zero is 0.320, suggesting that a model with parallel lines is indeed appropriate.

(c) The relevant part of the output is the following.

```
***** Regression Analysis *****

  Response variate: conc
      Fitted terms: Constant + airtight + uffi


*** Summary of analysis ***

                 d.f.          s.s.          m.s.       v.r.   F pr.
Regression         2          2030.9       1015.47      52.05  <.001
Residual          21           409.7         19.51
Total             23          2440.6        106.11

Change            -1          -432.5        432.50      22.17  <.001

Percentage variance accounted for 81.6
Standard error of observations is estimated to be 4.42
```

```
*MESSAGE: The following units have large standardized residuals:
        Unit      Response     Residual
          3        65.06         2.13
         10        36.37        -2.66
```

*** Estimates of parameters ***

	estimate	s.e.	t(21)	t pr.
Constant	32.00	2.05	15.64	<.001
airtight	2.808	0.308	9.13	<.001
uffi 1	8.49	1.80	4.71	<.001

The model with parallel lines, i.e. without the product variable airtight.uffi, fits the data just as well as the more general model, with the percentage of variance accounted for the same at 81.6%. Also, all the regression coefficients are associated with $SPs < 0.001$. This goes in particular for uffi, and so there seems strong evidence against a zero coefficient for uffi and hence against equal intercepts. (Confirmation of the appropriateness of separate intercepts is obtained by observing that, for the model fitted by GENSTAT using only airtight as an explanatory variable, the percentage of variance accounted for plummets to 63.9%.)

Solution 6.9
Apart from different names being used for the explanatory variables/factors, the regression results are exactly the same.

Solution 6.10
(a) The simple linear regression produces the following output.

```
***** Regression Analysis *****

Response variate: height
    Fitted terms: Constant, voice
```

*** Summary of analysis ***

	d.f.	s.s.	m.s.	v.r.	F pr.
Regression	3	81.1	27.038	3.95	0.010
Residual	103	705.7	6.851		
Total	106	786.8	7.423		

```
Percentage variance accounted for 7.7
Standard error of observations is estimated to be 2.62
*MESSAGE: The following units have large standardized residuals:
        Unit      Response     Residual
          5        76.00         2.78
```

```
*** Estimates of parameters ***

               estimate        s.e.    t(103)  t pr.
Constant         68.905        0.571    120.64  <.001
voice Tenor 2     1.000        0.808      1.24  0.219
voice Bass 1      1.813        0.708      2.56  0.012
voice Bass 2      2.480        0.768      3.23  0.002
```

The ANOVA table matches that of Solution 5.10 and shows an SP of 0.010, giving evidence against the hypothesis of no difference in mean heights between the different voice parts. The residual plots (not shown) seem good enough to continue with our analysis under the usual assumptions. Interestingly, the percentage of variance accounted for is just 7.7%. This does not seem to be due to any inappropriateness of the model and can be put down to a high degree of inherent variability (high σ^2).

(b) If μ_1 is the mean for Tenor 1 and μ_2 the mean for Tenor 2, then a suitable contrast for testing the hypothesis that there is no difference between the means is $\theta = \mu_2 - \mu_1$. Since Tenor 1 is the first level of the voice factor, it corresponds to the constant term in the model, and so an estimate of μ_1 is $\widehat{\alpha}$. From the discussion in Example 6.5, the mean μ_2 is estimated by $\widehat{\alpha} + \widehat{\beta}_2$. Thus the contrast is estimated by $(\widehat{\alpha} + \widehat{\beta}_2) - \widehat{\alpha} = \widehat{\beta}_2 =$ 1.000, and its SP is given in the Estimates of parameters as 0.219. (Likewise, β_3 and β_4 represent the differences between the mean heights for Bass 1 and Tenor 1, and Bass 2 and Tenor 1, respectively.)

There is therefore no evidence that the contrast in question is non-zero, and hence no evidence that the mean heights of the different tenors differ.

7

Solution 7.1

There is evidence that the treatments differ in their effect on the mean response ($SP = 0.023$).

The standard residual plots show nothing seriously amiss with the assumptions.

Solution 7.2

(a) Taking the sample means from the GENSTAT output generated in Exercise 7.1, the estimated contrast is

$$\widehat{\theta}_1 = \tfrac{1}{2}(83.9 + 85.9 - 79.2 - 100.0) = -4.7.$$

Its variance is $\tfrac{1}{4}(\frac{\sigma^2}{10} + \frac{\sigma^2}{10} + \frac{\sigma^2}{10} + \frac{\sigma^2}{10}) = \frac{\sigma^2}{10}$ (there being 10 rats per treatment). The error (residual) variance σ^2 is estimated by m.s.(Residual), which is given in the GENSTAT output as 223.6. Thus the estimated variance of $\widehat{\theta}_1$ is $223.6/10 = 22.36$, and the standard error of $\widehat{\theta}_1$ is estimated as $\sqrt{22.36} = 4.73$. Thus the t statistic is $-4.7/4.73 = -0.994$ on 36 degrees of freedom (d.f.(Residual)), and the corresponding SP is 0.327. You can calculate this SP in GENSTAT by using PRINT 2*cut(0.994;36). Notice that the negative sign is missing from this command. This is because we are interested in $2P(t(36) \le -0.994)$, which equals $2P(t(36) \ge 0.994)$, and cut gives probabilities with a $\ge$ sign.

There is no evidence that the average effect of changing the protein source is not zero.

(b) An appropriate contrast is

$$\theta_2 = \tfrac{1}{2}((\mu_{HC} - \mu_{LC}) + (\mu_{HB} - \mu_{LB}))$$
$$= \tfrac{1}{2}(\mu_{HC} + \mu_{HB} - \mu_{LC} - \mu_{LB}).$$

Its estimate is $\widehat{\theta}_2 = 11.4$. Its estimated standard error is also 4.73. Thus the t statistic is $11.4/4.73 = 2.410$ on 36 degrees of freedom, and the corresponding SP is 0.021. There is some evidence that changing the protein amount changes the mean response.

Solution 7.3

(a) $\widehat{\theta}_3 = -18.8$. The standard error is $\sqrt{4\sigma^2/10}$, estimated to be 9.46. The t statistic is -1.987 on 36 d.f., so that the SP is 0.055. There is only very weak evidence that the value of this contrast is non-zero.

(b) The contrast $\mu_{HB} - \mu_{LB}$ measures the change in mean response when the protein amount changes from low to high in rats receiving a beef protein diet, and the contrast $\mu_{HC} - \mu_{LC}$ measures the same change in rats receiving a cereal protein diet. The difference between these changes is $(\mu_{HC} - \mu_{LC}) - (\mu_{HB} - \mu_{LB}) = \mu_{HC} + \mu_{LB} - \mu_{LC} - \mu_{HB} = \theta_3$. Thus this contrast, which appeared to treat the two factors differently, actually does not.

Solution 7.4
This is discussed in the main text.

Solution 7.5
(a) You should have found that the residual plots exactly match those you looked at in the one-way analysis in Exercise 7.1. Again, this is because this two-way analysis of variance model is exactly equivalent to the one-way model; thus the residuals and fitted values are identical.

(b) This is discussed in the main text.

Solution 7.6
(a) The ANOVA table shows 2 d.f. for the source factor, 1 for the amount factor, 2 for the interaction, and 54 for the residual. Overall, there are $3 \times 2 = 6$ different treatments, with 10 rats on each treatment, so that makes 60 rats in all. A one-way analysis of variance on these data would thus show 5 d.f. for treatments, and 59 d.f. in total, leaving 54 for the residual. All the two-way analysis does is to split up the d.f. for the treatments, so in the two-way analysis there are still 54 for the residual. There are now three levels of the source factor, but one constraint is imposed on the parameters involved, so there are 2 d.f. for source. There are two levels of the amount factor with one constraint, so there is 1 d.f. for amount. It is rather more complicated to work out the d.f. for the interaction in this way: it is easier to note that there are 5 d.f. for treatments in all, and we have used 2 for source and 1 for amount, so there must be $5 - 2 - 1 = 2$ left for the interaction. Actually, there is another way to find this: the number of degrees of freedom for the interaction of two factors is always the product of the d.f. for the two factors involved.

The SPs given show strong evidence that the amount main effect is not zero ($SP < 0.001$), but no evidence that the source main effect is not zero ($SP = 0.541$) and only very weak evidence that the interaction is non-zero ($SP = 0.073$). Thus it is reasonable to conclude that rats on diets with different protein amounts have different mean weight gains (because the amount main effect is significant). Looking at the Tables of means, rats on a high protein amount gained 95.1 g on average, compared with 80.6 g for rats on a low protein amount. However, the size of the difference in mean weight gains of rats on different protein sources does not depend on the source of the protein (because the interaction is not statistically significant); also, on average, the source of protein makes no difference to the mean size of the weight gain (because the source main effect is not significant). Thus, within the treatments studied in this experiment, it is plausible that source of protein makes no difference at all to the response.

(b) The standard residual plots show nothing to imply that the standard ANOVA assumptions have been violated severely.

(c) The resulting plot is as shown below.

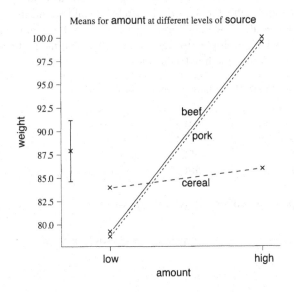

Means for amount at different levels of source

This shows clearly that the mean weight gains of rats on the two meat diets (pork and beef) are almost the same at each level of protein amount; however, the pattern for cereal protein looks different. To an extent, this impression contradicts what you found in part (a), where only the amount main effect (the difference between the mean of the three crosses on the left of the picture and the mean of the three on the right) differed significantly from zero. This means that the apparently clear difference between the treatment sources in this plot turns out to be small enough to be explained away in terms of chance variation.

Solution 7.7

(a) The variances within the treatment groups differ quite markedly, ranging from 24 to 1035. This is well beyond the rule of thumb we have been using to decide when it is reasonable to treat the variances as equal. In particular, one group (operator C with multiple pipettes) has a far smaller variance than any of the others. However, you should bear in mind that these variances are each calculated from only four values; one would expect a large amount of variability. It is probably reasonable to proceed with the analysis, bearing in mind that the assumption of equal variances is in some doubt.

(b) The analysis of variance table shows SP values of 0.58 for the main effect of the operator factor and 0.87 for the main effect of the pipette factor, but a much smaller SP (0.006) for their interaction. Thus the interaction between the two factors is highly significant, but neither main effect is.

The residual plots show nothing untoward. Perhaps it is surprising that there is no obvious pattern in the plot of residuals against fitted values, given the very different variances observed in part (a). Even the group that we know to be unusual from part (a), namely operator C with multiple pipettes, fails to show up in the fitted value plot. The reason for this failure is explained by the table of means, which shows that the mean for operator C with multiple pipettes (at 255.0) is almost equal to the mean for operator A with multiple pipettes (at 255.2), and so the plotted points for these two groups overlap on the fitted value plot.

(c) The interesting part of the table of means is shown below.

operator	pipette	single	multiple
A		273.0	255.2
B		252.8	299.5
C		288.8	255.0

The corresponding plot of means is as follows.

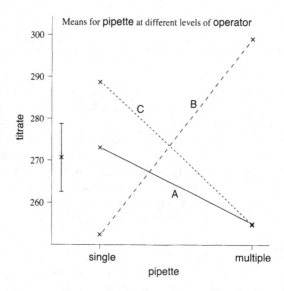

(You might have produced the means plot the other way round, with **operator** on the x-axis and **pipette** defining the groups, but the conclusions should amount to the same sort of thing.)

Operators A and C produced similar patterns of measurements, with rather higher measurements with a single pipette than with multiple pipettes. Operator B showed a different pattern, with much higher results with multiple pipettes, and overall a greater difference between the two methods than for the other two operators.

The fact that the mean response for operator B is low for a single pipette and high for multiple pipettes, while the mean responses for operators A and C go the other way, is the explanation for the lack of significance of the **operator** main effect: things have cancelled out. The same sort of thing has happened for the **pipette** main effect. (However, remember that, in terms of understanding how the measurement process works, these main effects are of no real interest, given that the interaction is significant.)

Solution 7.8

(a) The residual plots are shown below.

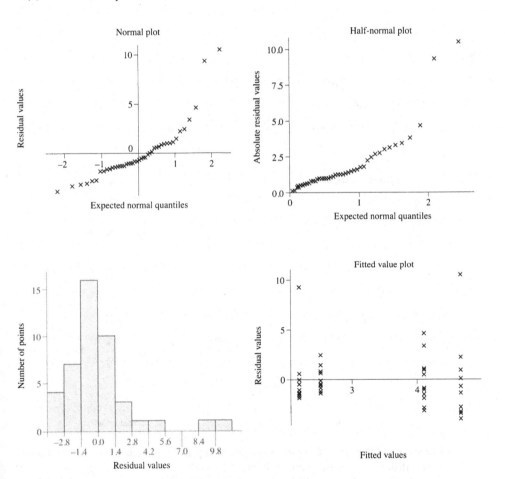

The probability plots and the histogram cast a lot of doubt on the normality assumption. There appear to be two high outliers, but more importantly the general pattern of the residuals is positively skewed, as shown by the shape of the histogram and the curvature upwards in the probability plots.

An appropriate transformation to reduce this skewness would be one that moves down the ladder of powers, such as square root, log or reciprocal.

(b) You can make the log transformation from score to lscore, say, either by using the Data|Transformations menu item or by typing CALC lscore=log (score) in the Input Log in interactive mode.

Perhaps it is most appropriate to begin the analysis of the logged responses by looking at the residual plots.

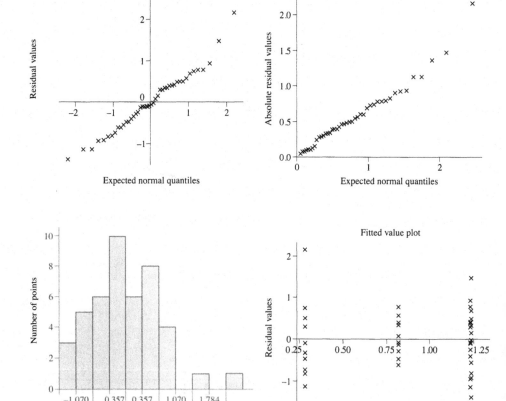

The shapes of the probability plots and the histogram are much more normal, though there still appear to be two high outliers. (GENSTAT flags one of these as a high outlier in its **Output** window.) The plot of residuals against fitted values perhaps looks a bit odd. It seems to show only three treatment groups, though there were four. This is because, after the transformation, two of the group means are almost identical. Thus it is rather difficult to assess the spread in these two groups. Overall, there is still some doubt about the appropriateness of the assumptions, but things are much better than before the data were transformed.

The plot of means is shown below. (You might have put the other factor on the x-axis; the plot below may be easier to interpret.)

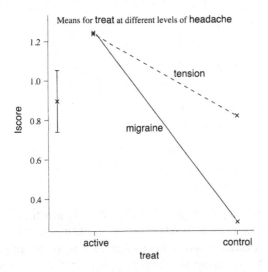

It shows that groups with both headache types have about the same average score on the active treatment, and that both groups have a lower score on the control, though the difference looks rather greater for people with migraine headaches than for people with tension headaches. The SP for the interaction of the two factors is 0.231, so there is no evidence that the interaction is non-zero. That means that it makes sense to interpret the experiment in terms of separate additive main effects of the two factors. The SPs for the main effects of headache type and of treatment type are 0.244 and 0.004 respectively. So it is plausible that the main effect of headache type is zero, but it is not plausible that the main effect of treatment type is zero. Thus we can ignore headache type in interpreting this experiment, and simply conclude that the mean logged score for those on active treatment is 1.24, significantly greater than the logged score of 0.55 for those on the control.[1] The general conclusion is that the treatment does indeed work, and that it works (on average) to the same extent for people with both types of headache.

Solution 7.9

From the analysis of variance table, the main effects of temperature and concentration are significant, as is the two-factor interaction of temperature and catalyst. The three-factor interaction is not significant, so that it can be taken as zero. The further interpretation of these results is in the main text.

The plot of means for temperature and catalyst is shown overleaf. Again, there is some discussion of its interpretation in the main text.

---

[1]The logs used throughout this book are always to base e; however, it is possible to perform such analyses using logs to base 10 (in which case you would get different numbers).

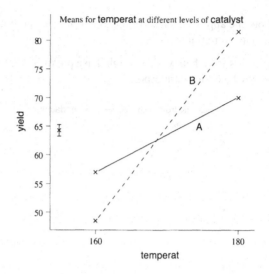

The residual plot (not shown) looks rather odd. This is partly because the amount of data is rather small, and because the residuals appear in matched pairs. The residuals for the two observations on each treatment have to be equal in magnitude and opposite in sign, because they have to add up to zero. Note that this emphasizes that residuals are not independent. It is thus rather hard to tell whether the assumptions of the method are appropriate. But at any rate the probability plots do not look grossly curved.

Solution 7.10

(a) The analysis of variance table shows very small SPs for **sex** and for the **time.vigour** interaction. There is perhaps very slight evidence that the three-factor interaction may be non-zero (its SP is 0.061).

Treating the three-factor interaction as zero, we can report (from the table of means) that the mean log faecal coliform contribution after 30 minutes for different levels of time since last bath and for different levels of bathing vigour are as follows.

time	vigour	letharg	vigorous
1hr		2.68	4.10
24hr		4.31	3.58

It might be more helpful to translate these back to the original scale (by taking exponentials using a calculator), bearing in mind that the resulting values will not give the precise mean faecal coliform contributions (though they will be reasonable estimates). This is because $E(e^X)$ is not the same as $e^{E(X)}$.

time	vigour	letharg	vigorous
1hr		14.6	60.3
24hr		74.4	35.9

Both tables show that vigorous bathing increases the bacterial concentration, on average, for people who last bathed one hour ago, but reduces it for people who last bathed 24 hours ago.

The facts that the main effect of **sex** is significantly different from zero and that none of the interactions involving it are, means that we can simply report that the mean log faecal coliform contribution after 30 minutes is 3.21 ($= 5.27 - 2.06$) higher for males than for females. In other words, on average, the contribution for males is approximately $e^{3.21} = 24.8$ times higher for males than for females. (Without being either a biologist or a water engineer, it is difficult to know why this might be.)

The residual plots (below) are not very satisfactory.

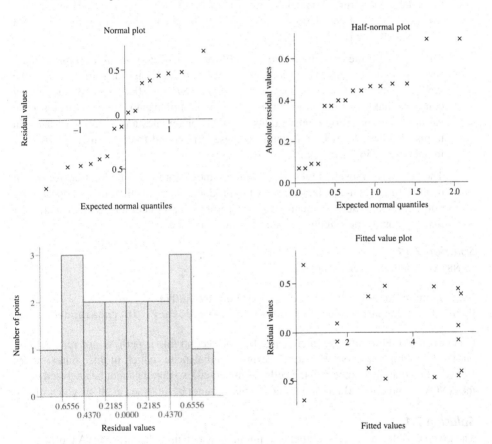

The plot of residuals against fitted values shows a reasonably even scatter (given that there are only two observations on each treatment, so that the residuals go in pairs). This indicates that the log transformation has dealt reasonably well with the problem of non-constant variance. However, the probability plots do not look at all straight.

Because of the non-constant variance, any alternative transformation needs to be broadly similar to log in its effect. The square root transformation is one such. Also, square root transformations often work well with data in the form of counts (which, essentially, we have here).

Analysing the square root of **fc30** leads to rather similar conclusions to those above, except that the evidence for a significant **time.vigour** interaction is much weaker ($SP = 0.080$) and there is no sign at all of a three-factor interaction. However, the residual plots (not shown) are still not satisfactory. The probability plots are rather straighter (though still not very straight), and the plot of residuals against fitted values now shows strong evidence of non-constant variance.

Overall, the log transformation might be preferred. You may have found something better! In any case, since the model assumptions seem not to be satisfied very well, the conclusions from the analysis must remain somewhat tentative. However, the evidence for an effect of **sex** is so strong that the problems with the model do not really undermine it.

(b) The results for the contribution to total coliform concentration after 30 minutes are somewhat clearer. After log transforming **tc30**, the only sign of significance in the ANOVA table relates to the main effect of **sex** ($SP = 0.002$). So we can simply conclude that time since last bath and bathing vigour have no effect on this response variable; the mean log contributions are 4.34 for females and 6.32 for males. (In terms of untransformed contributions, these correspond to approximately 76.7 for females and 556 for males.)

The position on residual plots is much the same as for **fc30**. The log transformation deals with the non-constant variance but produces residuals that do not appear to be normally distributed. A square root transformation produces a better distribution of residuals but some evidence of non-constant variance.

Solution 7.11

GENSTAT produces an error message:

```
Design unbalanced - cannot be analysed by ANOVA
Model term litter (non-orthogonal to term mother) is unbalanced.
```

(litter and mother may be interchanged.) The second line is perhaps not very helpful, but the first simply points out that the design is unbalanced (which in this context arises from the fact that the numbers in the cells are not equal), so that it cannot be analysed using the ANOVA command. But that does not mean it cannot be analysed at all, as you will see.

Solution 7.12

The first part of the regression output is a summary, which includes an ANOVA table testing the entire regression model. (This is similar to what you would get if you were doing a one-way ANOVA using regression.) The F test is a test of whether there are any treatment differences at all. You met this sort of thing in Chapter 5.

The next part of the output is a set of estimates of regression coefficients. This looks like nothing in the ANOVA output, so you probably did not manage to work out exactly what it meant. (See the main text after this exercise for an explanation.)

The final section of the regression output is an Accumulated analysis of variance, which (apart from a few details of layout) looks identical to the analysis of variance table at the start of the output from the two-way ANOVA. This indicates that this, key, output from a two-way analysis of variance can be produced using regression.

The residual plots give some indication why this is. You should have found that the plots you were asked to produce from the regression model look exactly like the corresponding ANOVA residual plots. This gives an indication that the fitted values and residuals from the two models (regression and ANOVA) are exactly the same – which indicates further that the two models are effectively the same, as indeed they are.

Solution 7.13

(a) The possible outlier is in row 35: it has a huge standardized residual (3.48).

On reanalysing the data with this case omitted, the accumulated ANOVA table shows strong evidence ($SP = 0.001$) of a non-zero main effect of the mother's genotype, and the interaction is also significant ($SP = 0.003$). Because the interaction is significant, it seems to be the case that the litter genotype and the mother's genotype both affect the response, and furthermore the litter genotype affects it in ways that are different for different mother's genotypes. There is more on the exact interpretation of this ANOVA table in the main text after this exercise.

The probability plot of the residuals (not shown) looks reasonably straight. The plot of the residuals against the fitted values (below) looks reasonable, but not totally satisfactory. The two very small residuals corresponding to small fitted values, at the left of the plot, correspond to the two litters where the litter genotype is I and the mother's genotype is A. The outlier that we threw away was also in this cell. It seems to be highly untypical in some way. The rest of the plot shows some evidence that the variance of the residuals decreases as the fitted values increase. Perhaps some transformation would improve matters.

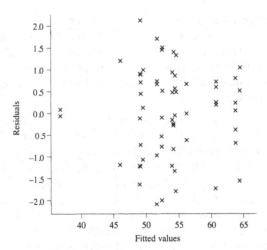

(b) The two accumulated analysis of variance tables look similar, but are not identical. Not surprisingly, the order of the two main effects is reversed. But the sums of squares for the main effects (and hence their mean squares, variance ratios and F probabilities) are not quite the same. (The reason why is given in the main text after this exercise.) The row for the interaction is identical, as are the residual plots.

Solution 7.14

(a) The analysis of variance table looks as follows.

Source of variation	d.f.(m.v.)	s.s.	m.s.	v.r.	F pr.
headache	1	2.1864	2.1864	5.79	0.021
treat	1	5.7162	5.7162	15.13	<.001
headache.treat	1	1.0228	1.0228	2.71	0.108
Residual	38(2)	14.3597	0.3779		
Total	41(2)	22.6548			

The residual and total degrees of freedom have been adjusted to allow for the missing values (which are shown in brackets). Basically, GENSTAT has estimated the missing values from the data, using the means for the data that are not missing. Interestingly, both main effects are now significant. (In the previous analysis, the headache type main effect was not.) The residual plots look reasonable.

(b) This time the analysis of variance table looks as follows.

Source of variation	d.f.	s.s.	m.s.	v.r.	F pr.
headache	1	2.0823	2.0823	5.51	0.024
treat	1	5.2388	5.2388	13.86	<.001
headache.treat	1	0.9741	0.9741	2.58	0.117
Residual	38	14.3597	0.3779		
Total	41	22.6548			

The sums of squares are slightly different, because there has been no estimation of missing data, but the overall conclusions are the same.

Solution 7.15

(a) GENSTAT performs this analysis without any error messages. There is an analysis of variance table, but there is no line for Residual, because there are no degrees of freedom left. Therefore the columns for variance ratios and F probabilities are all empty. The table of standard errors of differences between means includes no standard errors – they cannot be estimated because there is no estimate of the error variance σ^2. There are no d.f.s either.

(b) In the analysis of variance table, the row for the three-factor interaction has been replaced by a row for Residual. The degrees of freedom, sums of squares and mean squares for the main effects and two-factor interactions are, not surprisingly, the same as in part (a). The degrees of freedom, sum of squares and mean square for Residual take the same values as those in the three-factor interaction row in the analysis in part (a). This is because GENSTAT is now assuming that there are no three-factor interactions. Thus differences between treatments which, in part (a), were taken to be due to three-factor interactions are now being put down to random differences between experimental units. The residual sum of squares is precisely that due to fitting the model with the three-factor interaction missing.

Since there is now a Residual row in the table, the variance ratio and F probability columns have been completed. All three main effects are significant, as is the length.amplitude interaction. That is, all three factors affect the mean logged response, and length and amplitude do so in a non-additive way.

The rest of the output is similar to what was produced in part (a). There is now a warning about a datapoint with a large residual. In the table of means, there are no means given for specific combinations of all three of length, amplitude and load – GENSTAT assumes we are not interested in these since we are assuming no three-factor interaction. There is a similar omission in the table of standard errors of differences between means. The other standard errors are now given in this table.

The residual plots look reasonably satisfactory, though there is perhaps a hint in the plot of residuals against fitted values that the error variance increases with the mean.

8
___

Solution 8.1

The blocks in this experiment are the individual grapefruit. The experimental units within the blocks are the halves of the grapefruit.

Not all grapefruit are the same. In particular, left to their own devices, it is likely that different grapefruit contain different percentages of solids. So, there is a heterogeneity between grapefruit that it would seem advantageous to take into account. However, within grapefruit, one would expect two halves to be much the same (if they were both shaded in the same way) in terms of percentage of solids. That is, there is likely to be a much greater homogeneity between grapefruit halves (i.e. within blocks) than between different grapefruit (between blocks). As interest lies in differences between responses under exposed and shaded conditions, it seems intuitively more useful to be able to compare responses within the same grapefruit (and average these) than to collect data on a set of exposed (whole) grapefruit and another set of shaded (whole) grapefruit and compare these. The latter, the situation for an ordinary two-sample comparison, would not have separated out the between-grapefruit variability.

Solution 8.2

(a)

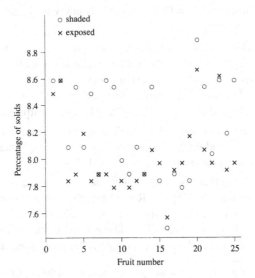

In this version of the scatterplot,[1] information on the differences between the responses for the exposed and shaded halves of each of the 25 grapefruit is clearly shown in the vertical direction. In an ordinary scatterplot, in which **exposed** is plotted on, say, the vertical axis against **shaded** on the horizontal, this information is hidden away: in such a scatterplot, fruit for which the shaded half has less solid matter than the exposed half are those whose points lie above the line $y = x$, i.e. the line at $45°$ to the axes and passing through the origin.

Looking at the plot above, there seems to be some indication that exposure to sunlight results in a smaller percentage of solids.

(b) In interactive mode in the **Input Log** window, use

```
CALC diff=exposed-shaded
```

to calculate the differences.

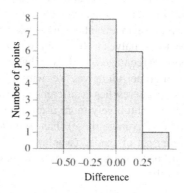

Then click on **Graphics**, click on **Histogram**, put **diff** in the **Data** field, and click on **OK**, to obtain the histogram. You may have some slight concern about the normality of the differences based on this plot, but it is not wholly unreasonable to proceed as if normality is tenable.

(c) (i)   Although the **t-test (paired samples)** option is under **Two-sample tests**, GEN-STAT actually subtracts the values in **shaded** from the values in **exposed** and performs a one-sample t test on these differences, hence the **One-sample T-test** heading. The differences are in fact stored in a temporary variate called Q[1], which is called the **Sample** in the output.

(ii) The estimates of μ_Q and σ_Q^2 are the sample mean and variance of Q[1], given as -0.1936 and 0.09403, respectively. (The sample standard deviation of the differences is therefore $\sqrt{0.09403} = 0.3066$.)

(iii) The SP is given as 0.004 when the test statistic -3.16 is referred to the t distribution on 24 d.f. (Remember that the t statistic divides the sample mean by the standard error, not the standard deviation. The standard error is $\sqrt{0.09403/25} = 0.0613$.) There is quite strong evidence that the mean percentages of solids corresponding to the exposed and shaded grapefruit halves

---

[1] Different symbols for the two sets of points are used here, but on the computer screen different colours are employed.

are different: in fact, exposure to sunlight appears to reduce the percentage of solids (if only by a small amount). This coincides with the impression given by the plot in part (a).

Solution 8.3

(a) In the GENSTAT spreadsheet, all $8 \times 5 = 40$ weight gains are given in the first column. The second column, containing the factor diet, shows the diet codes associated with each response; the third column, containing the factor litter, shows the litter codes.

In the plot, the responses are plotted against diet, which GENSTAT has recoded as $1, 2, \ldots, 5$ corresponding to diets $a, b, \ldots, e$ respectively. At first glance, it looks as though the response means for each diet might be rather similar. There also appears to be quite a lot of variability, but the most striking aspect of the plot is that much of this variability can be put down to variation between litters. To see this, notice that on your computer screen, responses from different litters are displayed in different colours. It is clear that responses from some litters are all high, and responses from other litters are all low. It certainly seems important to account for these large differences between litters (blocks) in our analysis.

(b) The ANOVA table produced is as follows.

Source of variation	d.f.	s.s.	m.s.	v.r.	F pr.
litter stratum	7	6099.47	871.35	21.44	
litter.*Units* stratum					
diet	4	346.87	86.72	2.13	0.103
Residual	28	1137.73	40.63		
Total	39	7584.07			

Taking the diet and Residual rows of the table together, it is clear that:

- in the diet row, the d.f. $= 4$ (corresponding, as in the completely randomized case, to one less than the number of treatments) divides the s.s. to give an m.s.;
- in the Residual row, m.s. also arises as s.s./d.f.;
- in the diet row, the test statistic v.r. is calculated as m.s.(diet)/m.s. (Residual), and as usual is compared with an F distribution on d.f.(diet) and d.f.(Residual) degrees of freedom to obtain an SP of 0.103.

It is not yet clear where the values of s.s. come from or what the first row of the table means.

From the diet row, there is little or no evidence for any differences in mean responses between the five rat diets.

(c) There is little untoward to be seen in the residual plots (not shown) except perhaps rather an excess of small positive residuals in the histogram. We may be disinclined to pay too much attention to this on the basis of the reasonable-looking plot of residuals against fitted values.

Solution 8.4

(a) Conclusions remain just the same (as they should) since the same test for diet effects has been performed. The ANOVA table is now as follows.

```
Source of variation    d.f.        s.s.        m.s.     v.r.   F pr.
diet                      4       346.87       86.72     2.13   0.103
litter                    7      6099.47      871.35    21.44   <.001
Residual                 28      1137.73       40.63
Total                    39      7584.07
```

The diet row, including the SP of 0.103, is just as before. Indeed, we know now that the s.s. is precisely that associated with diet when diet and litter are both treated as treatments. Because we haven't told GENSTAT anything about blocks, it also gives a test of the litter effect, which is not really of any interest to us. In fact, it reflects the very considerable variation there is between rat litters (see also the litter means given below the ANOVA table: these clearly vary a lot). It has obviously been sensible to allow for such variation between litters and to base our analysis of diets on comparisons within litters.

Notice that the ANOVA table in Solution 8.3, which is aware of the blocking structure, reminds us of this through a rather unintelligible labelling of the ANOVA table in which the block row is called litter stratum and the diet and Residual rows appear beneath a heading litter.*Units* stratum. There is more on this later.

(b) When All interactions is specified, the ANOVA table looks just like that in part (a) except that:

- the Residual is now called diet.litter;

- and the v.r. and F pr. columns are empty.

These differences demonstrate that, in part (a), we obtained a usable residual structure by assuming that there is no diet.litter interaction. (This is very similar to the no-replications situation of Section 7.4.)

Solution 8.5

(a) With One-Way ANOVA (in Randomized Blocks) in the Design field, weight in the Y-Variate field, drug in the Treatments field, and block in the Blocks field, the following table is produced.

```
Source of variation    d.f.         s.s.        m.s.      v.r.  F pr.

block stratum             7      0.056129    0.008018    1.91

block.*Units* stratum
drug                      2      0.149858    0.074929   17.84  <.001
Residual                 14      0.058808    0.004201

Total                    23      0.264796
```

The test statistic for testing whether there is any difference between treatments is v.r. = 17.84 (from the drug row), which, when compared with the F distribution on 2 and 14 d.f., yields an $SP < 0.001$. There is strong evidence of a difference between treatments.

(b) There is a hint of a slight non-normality in the residual plots (not shown), and one residual is flagged by GENSTAT as a possible outlier; but neither of these possible departures looks very serious, and we are probably justified in performing the analysis we have done.

Solution 8.6

From the previous GENSTAT output, $\overline{Y}_1 = 3.866$, $\overline{Y}_2 = 4.005$ and $\overline{Y}_3 = 4.052$. Thus,

$$\widehat{\theta}_1 = 3.866 - \tfrac{1}{2}(4.005 + 4.052) = -0.163.$$

The estimated variance, from the Residual line of the ANOVA table, is 0.004201. Thus the estimated $V(\widehat{\theta}_1)$ is

$$\widehat{V}(\widehat{\theta}_1) = \frac{3 \times 0.004201}{16} = 0.000788.$$

Using d.f. (Residual) = 14 from the ANOVA table, the required percentage point is $t_{0.975}(14)$ which, using GENSTAT's edt command, is 2.145. The 95% confidence interval for θ_1, given by

$$-0.163 \pm 2.145 \times \sqrt{0.000788},$$

is therefore

$$(-0.223, -0.103).$$

Since this interval does not include zero, and indeed it would have to be considerably expanded to do so, there is evidence that there is a difference between control and treatments (indeed, the latter do seem to increase weight).

Solution 8.7

(a) In grapeft there were three columns of length 25 each. The first column contained a fruit (block) number, the second column the responses for the shaded halves, the third column the responses for the exposed halves. The same information is imparted by grapeft1 in the following way. There are still three columns but now 50 rows, one for each response; the 50 responses are held in the first column. The second column is a factor called fruit which associates a block (grapefruit) number with each response. The third column is a factor sun denoting the 'treatment' (values: exposed and shaded). It happens that the 25 responses for the shaded halves are listed first, followed by the responses for the exposed halves.

(b) As well as One-way ANOVA (in Randomized Blocks) in the Design field, enter solid in the Y-Variate field, sun in the Treatments field and fruit in the Blocks field. Click on OK.

The F pr. entry in the ANOVA table in the Output window is indeed 0.004, just as in Solution 8.2: each has tested the hypothesis that the treatment means are different and came to the same conclusion, which is that there is strong evidence of a difference. The separate treatment means are given in the Tables of Means, and their difference (exposed – shaded) does indeed come to -0.194 as in Solution 8.2. And the standard error of the difference in means is given as 0.0613, just as in Solution 8.2.

In fact, just as in the two-treatment completely randomized case (see Section 7.1.1), the test statistic (9.96) used in the ANOVA table is precisely the square of the test statistic (3.16) used in the equivalent t test; and this squared value is referred to the $F(1, v)$ distribution (here $F(1, 24)$), which is exactly the distribution of the square of a $t(v)$ (here $t(24)$) random variable (as mentioned in Solution 4.10). In the completely randomized case, the relevant t test was the two-sample one, here it is the paired t test.

Solution 8.8

(a) An appropriate diagram is shown below.

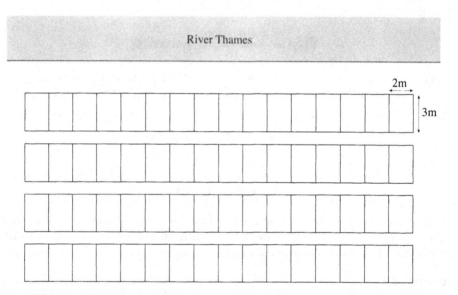

Each block has its long side parallel to the river-bank. The boundaries between the sixteen plots within a block are perpendicular to the river-bank.

(b) The flooding of the river will affect the part of the field nearest the river in a different way from the part furthest from the river. It would be expected that the flooding would have the same effect on all the plots within a block; about the same amount of each plot will get flooded, or will have been flooded in the past. Thus the blocks are made up of fairly homogeneous plots (experimental units). If the long side of the block were perpendicular to the river-bank, then the plots nearer the river would have different characteristics from the ones further away, and the plots in the block would be heterogeneous.

Solution 8.9

(a) There are four blocks and hence $4 - 1 = 3$ d.f. for blocks.

There are $16 - 1 = 15$ d.f. for treatments. Add this 15 to the 3 d.f. for blocks, and take the sum away from the d.f. (Total) $= 64 - 1 = 63$ to get $63 - 3 - 15 = 45$ for d.f. (Residual). Alternatively, remembering that the residual in a randomized block experiment is calculated in just the same way as the block by treatment interaction, there must be d.f.(block) $\times$ d.f.(treatment) $= 3 \times 15 = 45$ d.f. for the interaction, or residual. (Indeed, one *could* break the residual down into its constituent parts (with corresponding d.f.) due to block.density interaction, and block.sowing interaction, and even block.density.sowing.variety interaction, and so on. But this would not really make sense, because we are assuming that all interactions involving blocks are zero.)

(b) The SPs in the ANOVA table are rather clear about the importance of all three main effects and also of the density.sowing interaction. The sowing.variety interaction yields only rather marginal evidence for its existence, in the form of an SP of 0.057. There is no evidence for either a density.variety interaction nor a three-way (between treatments) interaction.

(c) The first bulleted question asks about the effect of variety. Since no interactions involving variety are considered significant, it is sufficient to look at the variety main effect. The appropriate entry in the Tables of Means shows that the Barkant variety gives an average yield of 6.52 kg while the Marco variety gives a lower average yield of 4.23 kg. Thus Barkant is preferred, at least in situations where turnips are to be sown in fields similar to the experimental field (in terms of other attributes such as soil type, etc.).

The second and third bulleted questions ask about sowing and density. Since they have an important interaction, the appropriate table of means is the following.

density	sowing	21/8/90	28/8/90
1Kg/Ha		1.76	2.51
2Kg/Ha		3.01	3.69
4Kg/Ha		3.91	10.74
8Kg/Ha		5.18	12.21

From this table, the biggest yield is for the highest sowing density (8 kg/ha) and the later sowing date. Not only can farmers (with conditions similar to experimental conditions) 'afford to wait an extra week before sowing' but they should be encouraged to! The interaction here pertains to there being much less to choose between sowing dates when the density of sowing is low, but it being much more important to sow at the later date when density of sowing is high. (You could also see this from a means plot, if you wished to.)

In general, looking at the main effects, yields increase with higher density and later sowing date.

(d) The residual plots (not shown) suggest two possible discrepancies from the assumptions of the randomized block model.

The most noticeable of the two, also flagged in the GENSTAT output, is the appearance of two outliers. Both refer to 'Unit 7', which is the seventh row of the data table (Table 8.4), corresponding to Barkant variety, late sowing, and the 4 kg/ha density.

The value (24.0) obtained in Block I is especially high and the value (2.6) obtained in Block IV is especially low (relative to what the model would expect of these situations).

The second slight worry is that there is some indication of increasing residual variance with increasing fitted value.

Solution 8.10

(a) The layout of the data is different from Table 8.5. There are sixteen rows in the spreadsheet, one for each experimental unit (combination of a run and a machine position). The first column is a variate called **wear**, giving the responses. The other three columns are factors giving the **run**, **position** and **material** for the experimental unit in question.

(b) The plot is as follows.

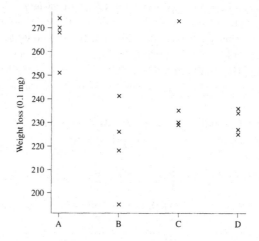

(In GENSTAT, the materials are represented by codes 1 to 4 instead of A to D.) There is some apparent evidence of systematic differences in the responses between different materials. Material B seems to wear rather better (lower response) than the others, particularly Material A. However, the pattern is not entirely clear-cut. Also, there is some evidence of differences in variability between the responses on different treatments, though we shall be better placed to judge that after we have fitted the ANOVA model and looked at the residuals.

(c) The ANOVA table is as follows.

Source of variation	d.f.	s.s.	m.s.	v.r.	F pr.
run stratum	3	986.50	328.83	5.37	
position stratum	3	1468.50	489.50	7.99	

```
run.position stratum
material                    3     4621.50     1540.50    25.15  <.001
Residual                    6      367.50       61.25

Total                      15     7444.00
```

It is laid out in much the same way as those you saw in the previous section. Really the only difference is that there are three strata rather than two. There is one stratum for each of the two types of block (run and position), and a third (labelled run.position stratum) which relates to the individual experimental units (which differ from each other in terms of run *and* position). As with the randomized block design, the only stratum that gives you information about differences between materials is the final one. The SP for a test of the hypothesis that the mean responses on all materials are equal is less than 0.001 – strong evidence that the materials differ in their resistance to wear. (One could go on to estimate relevant contrasts.)

The residual plots show nothing untoward. One residual is flagged in the ANOVA output, but it is not enormously large. The model seems adequate.

(d) The command is BLOCK run*position.

This might have reminded you of the way that factorial treatment combinations are entered in GENSTAT, with asterisks separating the factors. In contrast, in a randomized block design (with a factor called block defining the blocks), the corresponding command would be BLOCK block.

Note that, if you use the General Analysis of Variance option in the Design field, you can enter complicated block structures like this directly into the Block Structure field in the Analysis of Variance dialogue box. So another way to produce exactly the same analysis of these data would have been to use that Design option, to enter material as the Treatment Structure, and to enter run*position as the Block Structure. All the other forms of analysis of variance later in this chapter can be dealt with in a similar way, if you can work out the appropriate form for the treatment structure and the block structure.

Solution 8.11

(a) The layout of the data is different from Table 8.6. There are three rows in the spreadsheet for each washing session. Each row of the spreadsheet consists of: the washing session, the detergent used and the number of plates cleaned.

(b) You probably thought that there are systematic differences between the detergents. If you look at the spreads of values for each detergent, then you can see that the ranges of the responses for different detergents do not overlap much. This tells you that the differences between the means for the detergents are large compared with the variation within detergents. This, in turn, tells you that the differences are likely to be due to the different detergents rather than to pure chance.

From the plot, it looks as if Detergents 1, 2, 6, 7 and 8 are very similar in performance. Detergent 9 looks best, being slightly better than Detergent 5. The worst detergent, by quite a long way, is Detergent 4.

Looking at plots which indicate the different sessions does make it clear that there are reasonably large differences between the sessions, but it appears that they are mainly due to the differences in detergents used at each session. (To obtain appropriate plots, use the **Groups** field in the **Point Plot** dialogue box.)

(c) The ANOVA table is as follows.

```
Source of variation      d.f.        s.s.        m.s.      v.r.   F pr.

session stratum
detergen                   8      408.4444     51.0556    35.57  0.007
Residual                   3        4.3056      1.4352     1.74

session.*Units* stratum
detergen                   8     1086.8148    135.8519   164.85  <.001
Residual                  16       13.1852      0.8241

Total                     35     1512.7500
```

The unusual feature of this table is that detergen appears twice, once in the session stratum and once in the session.*Units* stratum (which we shall refer to, for the sake of brevity, as the units stratum). We shall concentrate first on the entry in the units stratum.

From the ANOVA table, it is clear that there is overwhelming evidence of a difference between detergents, with an SP of less than 0.001. So it is sensible to take the conclusions that we drew from the plot in part (b) seriously.

The detergen row in the session stratum is based on the block totals. A detailed discussion of how to use this information is beyond the scope of this book. GENSTAT has automatically done something called *recovery of inter-block information*. The idea is that the blocks have been treated with different combinations of detergents. So, we can use the differences between block totals to find out something about differences in detergents. There is not so much information about differences in detergents in the session stratum as in the units stratum, which is why we have concentrated on the latter. (The part of the ANOVA output labelled Information summary tells one about how the information about differences between detergents is split between the two strata, but again we shall not go into details of how it does this.)

(d) The residual plots do not look ideal. Most importantly, the normal probability plots do not look particularly straight. It seems that the assumption of normality is suspect. However, trying the obvious transformations of the data does not really improve things. Our conclusions in the previous part must therefore remain slightly suspect.

Solution 8.12

(a) There is one row in the spreadsheet for each plot. The first six rows of the spreadsheet correspond to the first row of Table 8.7, the next six to the second row, and so on. The columns called **row** and **column** in the spreadsheet indicate the row and column that the plot appears in in Table 8.7. The column called **variety** is the variety of turnip. The column called **weight** is the fresh weight, in pounds, of the turnips in each plot.

(b) From the plot (not shown), it looks as if there are systematic differences in the fresh weights for different varieties. The variety which produces the lowest weight of turnips is Variety C. The varieties which produce the biggest fresh weights are Varieties A and D.

(c) The output produced by trying to carry out the ANOVA for a latin square design should be something like the following.

```
******** Fault (Code AN 1). Statement 1 on Line 163
Command: ANOVA [PRINT=aovtable,information,mean; FPROB=yes; PSE=diff] weight
Design unbalanced - cannot be analysed by ANOVA
Model term column (non-orthogonal to term row) is unbalanced, in the
row.column stratum.
```

The output tells you that something went wrong. The words that tell you what went wrong are `Design unbalanced` on the third line of the output.

(d) The ANOVA produced using **General Linear Regression** is as follows.

```
*** Accumulated analysis of variance ***
```

Change	d.f.	s.s.	m.s.	v.r.	F pr.
+ row	5	119.15	23.83	0.75	0.596
+ column	5	270.50	54.10	1.71	0.187
+ variety	5	1140.81	228.16	7.20	<.001
Residual	17	538.45	31.67		
Total	32	2068.91	64.65		

The ANOVA tells us that the extra sum of squares due to adding **variety** to a model which already contains **row** and **column** is 1140.81 on 5 degrees of freedom. When this is combined with the residual sum of squares of 538.45 on 17 degrees of freedom, it gives a v.r. of 7.20. This corresponds to an $SP < 0.001$. There is thus overwhelming evidence that there are systematic differences in the yields for different varieties.

(e) The ANOVA table has told us that there are differences between yields for different varieties. The table of regression coefficients tell us what the differences are.

For example, the estimated coefficient is -6.12 for Variety B and -18.00 for Variety C. The difference between these coefficients is 11.88. This means that the mean weight of turnips per plot is 11.88 pounds less for Variety C than for Variety B.

There is no value for Variety A because its mean weight has been constrained to be zero. So, the coefficients for the other varieties are all measured relative to Variety A. For example, if we want to compare Variety A and Variety C, instead of Variety B and Variety C, we just look at the coefficient for Variety C. This is -18.00, which means that the mean weight of turnips per plot is 18.00 pounds less for Variety C than for Variety A.

Solution 8.13

(a) The layout of the data is different from Table 8.8. Each row in the spreadsheet corresponds to an experimental unit (i.e. a subplot). The first column is a variate called yield, giving the responses. The other five columns are factors defining the block and plot structure (block, wholplot and subplot) and the treatments (variety and nitrogen).

(b) The yield clearly increases with nitrogen level (and indeed the increase looks linear). However, there is a lot of variability at each level. It remains to be seen how much of this is taken account of by the block structure. There is little evidence of a pattern of different yields for different varieties from this plot. (A plot with variety on the horizontal axis is a lot clearer for the purpose of investigating this, but again shows no particular sign of a relationship between variety and yield.)

(c) The ANOVA table is as follows.

Source of variation	d.f.	s.s.	m.s.	v.r.	F pr.
block stratum	5	15875.3	3175.1	5.28	
block.wholplot stratum					
variety	2	1786.4	893.2	1.49	0.272
Residual	10	6013.3	601.3	3.40	
block.wholplot.subplot stratum					
nitrogen	3	20020.5	6673.5	37.69	<.001
variety.nitrogen	6	321.8	53.6	0.30	0.932
Residual	45	7968.8	177.1		
Total	71	51985.9			

It is rather more complicated than most of those you have met before. As you might have expected, there are three strata, corresponding to blocks, whole plots and subplots respectively. But in this case, the variance ratio for the main effect of variety is calculated in the whole plot (block.wholplot) stratum. You might have expected this. Varieties were planted on whole plots, so in a sense the 'experimental unit' for the variety factor is a whole plot, and comparisons between varieties need to be made between average yields for whole plots. The SP for varieties is rather large at 0.272, so there is no evidence that yields differ for these different varieties.

The other aspects of treatment – the nitrogen main effect and the interaction – are tested in the subplot stratum. The SP for the interaction is huge (0.932), showing no evidence of interaction. The SP for the nitrogen main effect, on the other hand, is less than 0.001, so there is strong evidence that nitrogen fertilizer level affects yield, and that it affects it in the same way for each of the varieties. The plot of means (opposite) shows this too.

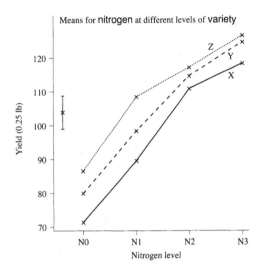

The residual plots show nothing untoward. The ANOVA output seems to flag one residual, but in fact it is the value of the average yield for Block I. (This is indeed very high compared with the other blocks, assuming that the block effects are normally distributed, but this has no effect on our conclusions.) The model seems adequate.

(d) The command is BLOCK block/wholplot/subplot.

This is different from the corresponding command in the latin square design, in that the different blocking factors are separated by slashes rather than by asterisks. The reason is as follows. In a latin square design, the units corresponding to row 2, say, in each column, have something in common in that they are all in the same row. In a split plot experiment, the units corresponding to whole plot 1 in different blocks have nothing particular in common in terms of what is going on in the field – they just happen to be those labelled with a 1. Similarly the units corresponding to subplot 3 in different whole plots have nothing important in common. If the command had been BLOCK block*wholplot*subplot then GENSTAT would have assumed that all the subplots numbered 3, say, had something in common, which would be incorrect.

(e) Apart from a few changes of labels of strata, the outputs are the same. The BLOCK command takes the form BLOCK block/variety. Here /variety has taken the place of /wholplot, as you would expect, but /subplot has been replaced by nothing. The analysis still works in the same way because GENSTAT *assumes* by default that the experimental units are the rows in the spreadsheet, and that these are arranged within the last level of blocking given in the BLOCK command.

Solution 8.14

(a) There are sixteen rows in the spreadsheet, each corresponding to one response in Table 8.9. The first four rows of the spreadsheet correspond to the first block, the next four to the second block, and so on. The columns clqualty, clamount and produnit correspond to ammonium chloride quality, ammonium chloride amount and production unit, respectively. The treat column indicates the treatment combination. The meanings of block and yield are the obvious ones.

(b) The plot is shown here.

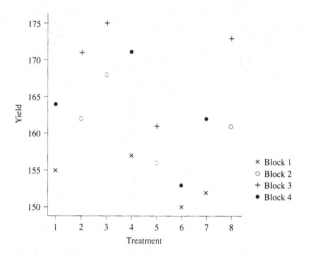

You can see that Block 3 gives higher yields than Block 2. This is not a consequence of the fact that different blocks contain different treatments, because Block 2 and Block 3 involved the same treatments. Similarly, you can see that Block 4 gives higher yields than Block 1.

It is very difficult to make any sense of the graph in terms of differences between treatments. The main problem is that you cannot tell whether the patterns that you can see are due to differences between blocks or to differences between treatments. For example, Treatment 3 produces higher yields than Treatment 6. There are (at least) two possible reasons for this:

- Treatment 3 is better than Treatment 6;
- the input materials used for Blocks 1 and 4 were not as good as the materials used for Blocks 2 and 3.

The fact that these two possibilities can't be distinguished is a side-effect of the confounding.

(c) Perhaps the clearest point made by this graph is that Production Unit 1 (Treatments 1 to 4) gives higher yields than Production Unit 2.

The next thing that can be seen is that using extra ammonium chloride produces a higher yield than using the usual amount. This can be seen because the adjusted yields for Treatments 3 and 4 (extra) are higher than those for Treatments 1 and 2 (usual amount), and there is a similar pattern for Treatments 5 to 8.

The question of whether one quality (fine or coarse) of ammonium chloride gives a higher yield does not have an obvious answer. Looking closely, one can see that Treatment 1 (coarse) gives adjusted yields which are slightly higher than those of Treatment 2 (fine), that the adjusted yields for Treatment 3 (coarse) are bracketed by those for Treatment 4 (fine), and that there is a similar pattern for Treatments 5 and 6. There seems to be no consistent difference here.

In summary, there is some evidence from the graphs that Production Unit 1 produces higher yields than Production Unit 2, and that using more ammonium chloride increases the yield.

(d) The ANOVA table (not shown) confirms that the confounding has worked as planned. All the main effects and two-factor interactions are indeed tested in the units stratum, and the three-factor interaction is tested in the blocks stratum.

Look at the units stratum first. The ANOVA tells us that there is very strong evidence that the amount of ammonium chloride and the production unit both alter the mean yield from the process. The SPs are 0.005 and 0.002, respectively. The yields are consistent with there being no difference in yield between the fine and coarse ammonium chloride. There is no evidence for claiming that there are any two-way interactions.

Look now at the three-way interaction, tested in the blocks stratum; there is no evidence for claiming that it is non-zero ($SP = 0.326$). However, the residual mean square in the blocks stratum is far bigger than that in the units stratum. This reflects the fact that comparisons between blocks are far less precise than comparisons within blocks. We just cannot estimate the three-factor interaction accurately from this experiment, because of the confounding.

(e) You would tell the manager that there is no evidence that the quality (fine or coarse) of ammonium chloride affects the yield.

One can calculate a 95% confidence interval for the difference in mean yields between the usual amount and the extra amount of ammonium chloride. The calculation is

$$(164.87 - 159.00) \pm t_{0.975}(6) \times 1.360 = 5.87 \pm 2.447 \times 1.360,$$

giving the 95% confidence interval

$$(2.54, 9.20).$$

The numbers 164.87, 159.00 and 1.360 are all taken from the Tables of means. The 6 d.f. in $t(6)$ and the standard error (1.360) come from the table of Standard errors of differences of means. The s.e.d. in this part of the output means *standard error of difference*. Alternatively, you could work the standard error out as follows. The contrast for the difference in mean yields that we require is the difference between the average yield on the eight runs where extra ammonium chloride was used and the average yield on the eight runs where the usual amount was used. Its variance is thus $2\sigma^2/8$. From the ANOVA table, σ^2 is estimated by the residual mean square, 7.396. Thus the standard error of the difference is $\sqrt{7.396/4} = 1.360$.

If you were talking to the manager, you might simply say that, on average, between 3 and 9 extra units of chemical are produced by using extra ammonium chloride.

The same approach as above gives a 95% confidence interval for mean difference in yield between the production units of

$$(-10.20, -3.54).$$

So, you might tell the manager that, on average, between 4 and 10 extra units of chemical are produced by Production Unit 1 over Production Unit 2.

Solution 9.1

From a purely computational point of view, there is no difficulty in obtaining a least squares fit to the data. You could load the dataset into any package, such as GENSTAT, choose the usual multiple linear regression command and obtain some least squares estimates for the regression coefficients, as well as t statistics, analysis of variance table, F statistic, etc.

However, if you used the fitted equation to estimate the probabilities, or to predict probabilities at new values of the explanatory variable, there is no guarantee that the estimated probabilities would lie between 0 and 1. The least squares estimates of the regression coefficients can take any negative or positive values and the linear combination of them, calculated for any combination of the explanatory variables, could in principle lie anywhere in the range $(-\infty, \infty)$, which might be difficult to explain! Direct prediction of a Y value itself is even more problematic: we cannot expect a linear formula to result in values of exactly 0 or 1. The most sensible target for prediction is to predict the probability that Y is 1.

The normality assumption of the usual multiple linear regression model is unrealistic when we know that Y_i is binary and has a Bernoulli distribution. The assumption of constant variance is also untenable when we know that $V(Y_i) = p_i(1 - p_i)$. Without these assumptions, the distributional theory underlying the F and t tests is lost. We also lose the result that the least squares estimates are identical to the maximum likelihood estimates. All told, none of our current methods is justified!

An assumption that is still reasonable is independence of observations. And you will soon see that it is possible to retain a linear component to the relationship between the response variable and the explanatory variables.

Solution 9.2

After loading fasten into GENSTAT, make the Input Log window active and enter interactive mode. Give GENSTAT the command

```
CALC propn=failed/samsize
```

to compute a new variate propn containing the proportions. Choose the Graphics|Point Plot menu item. Enter propn as the Y Coordinates and load as the X Coordinates, and click on OK. (As a shortcut, you could enter failed/samsize in the Y Coordinates field without obtaining propn.)

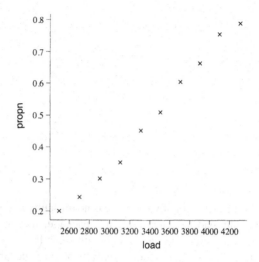

The resulting figure, like Figure 9.2, suggests an elongated S-shaped curve, fairly linear in the middle but with flattening at either end. However, in this case the curve is increasing rather than decreasing (i.e. it is forwards rather than backwards S-shaped).

Solution 9.3

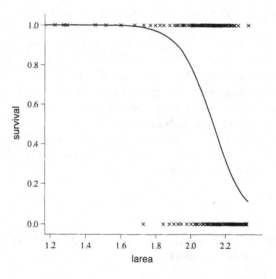

The diagram shows the binary regression data together with a fitted logistic regression curve, i.e. $\log(p(x)/1 - p(x)) = \alpha + \beta x$ for appropriate α and β where $p(x)$ is $P(Y = 1)$ when the explanatory variable **larea** takes the value x. The fitted curve is close to 1 for small values of **larea**, plunges smoothly down towards zero as **larea** increases from about 1.8 to 2.2 and beyond (with a fairly linear decrease in the central part of that range) and starts to slow its rate of decrease towards zero at the right-hand end. The curve does not stray outside [0, 1]. It has a (backwards) elongated S-shape, albeit cut off at the right-hand end. It appears to be just the right kind of model for the survival probability suggested by the points plotted in Figure 9.2.

Solution 9.4
In the summary table for the logistic regression, the d.f. column is as expected. The s.s. and m.s. columns have been renamed deviance and mean deviance. The deviance ratio replaces the v.r., but it is *not* the same thing; you may have noticed that the value here is in fact equal to the mean deviance and is not what you might have expected, namely the ratio of the Regression and Residual entries in the mean deviance column. There is an 'approx chi pr' column replacing the F pr. column.

There is no longer anything corresponding to the Percentage variance accounted for. Nor is there an estimated standard deviation, but this makes sense because there is no σ parameter to estimate as there was in normal regression.

The GENSTAT messages also differ. The first GENSTAT message for the burns data is that ratios are based on dispersion parameter with value 1. This will be the case throughout almost all the rest of the book: the 'dispersion parameter', discussed briefly in Sections 10.5 and 13.4, allows a more general modelling component that need not concern us now. The other messages are, in this case, to do with residuals, and we shall come back to these later in the section.

The estimates of regression coefficients (parameters) are laid out in much the same way except for an additional 'antilog of estimate' column which you can ignore. (Ignore also the final message, repeating the dispersion parameter information.)

Solution 9.5
The value of the regression deviance is 188.4. The SP is $P(Z \geq 188.4)$ where $Z \sim \chi^2(1)$. To find this, use the command line

    PRINT cuchisquare(188.4;1)

in the Input Log in interactive mode.

You should obtain the answer 0, i.e. a very small SP. Clearly, there is very strong evidence of a regression effect (an effect of larea on $P(\text{survival} = 1)$) with these data.

This same approximate χ^2 probability is in fact given in the GENSTAT Summary of analysis table, under approx chi pr, except that there GENSTAT simply writes <.001 for any probability less than 0.001.

Solution 9.6
The fitted regression line is

$$\log\left(\frac{p}{1-p}\right) = 22.22 - 10.45\ \text{larea}$$

where $p = P(Y = 1)$ is the probability of survival.
For a patient with larea $= 2$,

$$\log\left(\frac{p}{1-p}\right) = 22.22 - 10.45 \times 2 = 1.32;$$

therefore

$$\frac{p}{1-p} = e^{1.32} = 3.743$$

and

$$p = \frac{3.743}{1 + 3.743} = 0.789.$$

An increase of 0.1 in larea results in an odds multiplier of $e^{0.1\beta} = e^{-1.045} = 0.352$. That is, the odds in favour of survival for the larger larea are only 0.352 of the odds associated with the smaller larea. (The odds in favour of survival go down with increased area, in correspondence with the backwards S-shape of the fitted curve.)

Solution 9.7

Choose the Stats|Regression Analysis|Generalized Linear menu item. In the Generalized Linear Models dialogue box, set the Analysis field to Modelling of binomial proportions (e.g. by logits), the Numbers of Subjects field to 1, the Numbers of Successes field to gvhd, and put all six explanatory variables in the Model to be Fitted field, separated by commas (or, if you prefer, plus signs). Then click on OK. The GENSTAT output is as follows.

```
***** Regression Analysis *****


 Response variate: gvhd
  Binomial totals: 1
     Distribution: Binomial
    Link function: Logit
     Fitted terms: Constant, recage, recsex, donage, donmfp,
                   type, lindx



*** Summary of analysis ***

                                      mean  deviance approx
                 d.f.    deviance  deviance   ratio chi pr
Regression          8       27.55    3.4432    3.44  <0.001
Residual           28       23.50    0.8394
Total              36       51.05    1.4180
* MESSAGE: ratios are based on dispersion parameter with value 1


* MESSAGE: The following units have large standardized residuals:
          Unit     Response     Residual
           20         0.00        -2.18
           29         1.00         2.35
* MESSAGE: The residuals do not appear to be random;
           for example, fitted values in the range 0.00 to 0.11
           are consistently larger  than observed values
           and fitted values in the range 0.82 to 1.00
           are consistently smaller than observed values
* MESSAGE: The error variance does not appear to be constant:
           intermediate responses are more variable than small or
           large responses
```

```
* MESSAGE: The following units have high leverage:
        Unit      Response      Leverage
         13        0.00          0.53
         22        1.00          0.51
         28        1.00          0.68
```

```
*** Estimates of parameters ***
```

	estimate	s.e.	t(*)	t pr.	antilog of estimate
Constant	-7.98	4.42	-1.81	0.071	0.0003417
recage	0.0459	0.0949	0.48	0.629	1.047
recsex 1	-1.39	1.32	-1.06	0.291	0.2481
donage	0.176	0.122	1.44	0.149	1.193
donmfp 1	2.00	2.25	0.89	0.376	7.366
donmfp 2	1.54	1.35	1.14	0.255	4.645
type 2	-0.69	1.49	-0.46	0.645	0.5024
type 3	2.37	1.64	1.45	0.148	10.72
lindx	2.146	0.994	2.16	0.031	8.549

```
* MESSAGE: s.e.s are based on dispersion parameter with value 1
```

The d.f. (Regression) is eight and not six because there are (i) three quantitative explanatory variables giving one d.f. each, (ii) one binary explanatory variable giving a further one d.f., and (iii) two nominal explanatory variables, with three classes each, giving two d.f. each.

The approx chi pr is < 0.001; there seems to be strong evidence of a regression effect here, i.e. that not all the β_js are zero.

The deviance (Regression) – the test statistic – of 27.55 has been compared with the χ^2 distribution on d.f. (Regression) $= 8$ degrees of freedom. You could have used PRINT cuchisquare (27.55;8) to find the specific value for the SP, which is 0.00057.

You can use the t (*) values output by GENSTAT to set the explanatory variables in rough order of importance (in the presence of the other explanatory variables). Using the rule of thumb that 'significant' t values are those greater than 2 in absolute value (but recalling that this argument may be rather dubious) shows that none of the explanatory variables seems especially important (at least in the presence of the others) except for lindx ($t = 2.16$).

Solution 9.8

After fitting the model with lindx, donage and recage in the Model to be Fitted field, click on Change Model, and in the resulting dialogue box, insert recage in the Terms field, and then click on Drop.

This does indeed give the same information. There is again a Change line underneath the Summary of analysis table containing the same d.f., deviance (difference) and approx chi pr values, the only difference being that the (unimportant) minus signs have disappeared.

Solution 9.9

Fit donage, lindx and the factor donmfp to gvhd. You should find that the regression coefficients give you (1.1). This model has regression deviance 22.30 on 4 d.f. An assessment of the worth of the addition of donmfp to the successful model incorporating donage and lindx is the increase in deviance, from 19.98 to 22.30, i.e. by 2.32, using up $4 - 2 = 2$ extra d.f. (because of the three levels of factor donmfp); this has an SP of 0.314. You will probably have obtained these values by adding donmfp to the model containing donage and lindx using the Change Model button.

The obvious conclusion is to return to the model (9.6) involving just donage and lindx.

Solution 9.10

(a) The relevant part of the correlation matrix is as follows.

```
tht1    -0.604
tht2    -0.542    0.936
tht3    -0.428    0.770    0.882
tht4    -0.349    0.699    0.798    0.943
cfor1    0.180   -0.319   -0.327   -0.322   -0.383
cfor2    0.070   -0.298   -0.329   -0.416   -0.496    0.928
cfor3    0.066   -0.326   -0.320   -0.389   -0.452    0.803    0.914
cfor4    0.070   -0.353   -0.305   -0.370   -0.444    0.703    0.814
grav1   -0.157    0.075    0.129    0.088    0.099    0.220    0.193
grav2   -0.285    0.080    0.145    0.137    0.150    0.022   -0.027
grav3   -0.403    0.132    0.201    0.060    0.081    0.128    0.149
grav4   -0.248   -0.067   -0.021   -0.141   -0.098    0.186    0.170

         larea    tht1     tht2     tht3     tht4     cfor1    cfor2


cfor4    0.949
grav1    0.156    0.099
grav2   -0.008   -0.004    0.833
grav3    0.157    0.166    0.744    0.826
grav4    0.171    0.160    0.703    0.838    0.914

         cfor3    cfor4    grav1    grav2    grav3    grav4
```

All of tht1, tht2, tht3, tht4 have correlations of 0.699 and (considerably) higher with each other. Much the same goes for the four cfor variables and for the four grav variables. This makes sense since each variable is measuring much the same thing as each other member of its group. It is clear that there should be need for (at most) one 'representative' explanatory from each group.

There are at most weak relationships between different groups and/or larea. The largest (negative) correlations are between the tht variables and larea, but they are only around 0.3–0.6. These weak relationships are reflected in the upper parts of the scatterplot matrix opposite. The authors chose to take tht4, cfor4 and grav4, along with larea, as 'representative' explanatory variables, choosing the fourth ones in each case because they encompassed the others in their group; but you might just as well have chosen other representatives.

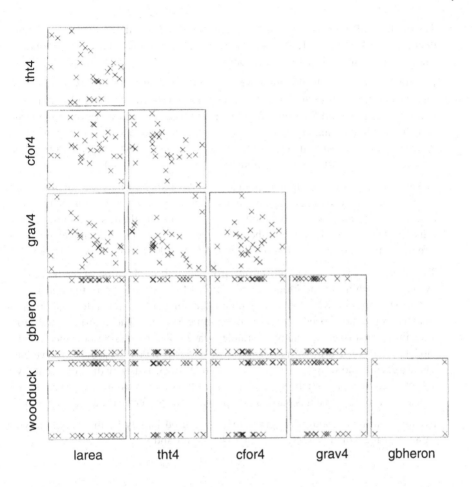

Looking at the row for **gbheron**, it is not easy to see much in the way of dependencies of **gbheron** on the four representative explanatory variables. Perhaps the herons tend to be present more for large **larea** and, in particular, for small **grav4**. For **woodduck**, dependence on **larea** may be less strong, but that on **grav4** seems a little clearer; also, there seems to be a greater wood duck presence for larger **cfor4**. (Ignore the meaningless plot of **woodduck** against **gbheron**.)

(b) The following table summarizes the results for the four individual logistic regressions with **woodduck** as the response variable.

	d.f.	Regression deviance	SP
larea	1	0.27	0.604
tht4	1	0.92	0.337
cfor4	1	2.52	0.112
grav4	1	4.87	0.027

Impressions from the scatterplot matrix are confirmed: there seems to be an effect of **grav4**, and perhaps a hint of an effect of **cfor4**.

The full multiple regression model does not fit especially well. The overall regression deviance, on 13 d.f., is 21.74: $P(\chi^2(13) \geq 21.74) = 0.060$. The largest t statistic (in absolute value) is just 1.03 (for cfor1).

In addition, we see for the first time an additional GENSTAT warning message concerning 'near collinearity', which is GENSTAT's way of referring to the multicollinearity that was mentioned in Section 6.2. This in turn is a formal way of referring to the fact that many explanatory variables are highly correlated with other explanatory variables, a fact we are already aware of from part (a). (You need not worry about the details of the 'variance inflation factors' given by GENSTAT.)

(c) Starting from the full 13-variable model, GENSTAT issues various warnings and fault messages that indicate that its results are unreliable. None the less, it takes eight steps before stopping at a six-variable model; however, one should be suspicious, because it includes (for example) three of the correlated cfor variables. The model does have a regression deviance of 17.70 on 6 d.f. (an SP of 0.007). But can we do better?

To perform stepwise regression starting from the null model, first fit the null model by putting nothing in the **Model to be Fitted** field. Starting from the null model, GENSTAT comes up with something more interesting. (This illustrates that different stepwise regressions need not arrive at the same answer.) The algorithm now stops after fitting just one explanatory variable, grav2. The regression deviance for this model is 6.76 on 1 d.f., so its (approximate!) SP is 0.009. It seems that we have found a useful, and simple, model. Recalling that grav4, as used in the 'representative' subset of explanatory variables, was a rather arbitrary choice from the grav variables, it is just as reasonable that it is grav2 which provides the preferred (one-variable) model.

(d) It seems that this 'good fit' to grav2 may be caused largely by the single outlying value of grav2 which is a little over 20. This is wetland 9.

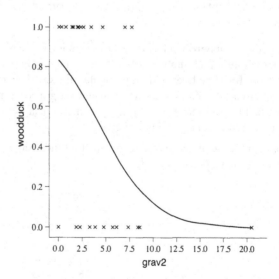

(e) GENSTAT gives no error messages when starting from the full 13-variable model this time, but still stops at the same six-variable model. Starting from the null model, GENSTAT still comes up with just the one explanatory variable, grav2. The regression deviance for the one-variable model is reduced to 5.10, but this now gives an SP of 0.024. The fitted model is drawn below.

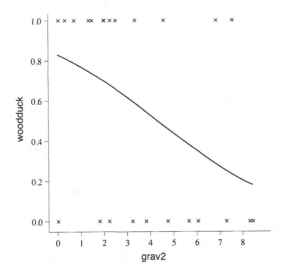

So we now have more confidence in the regression model relating **woodduck** to grav2. (Indeed the fitted model has hardly changed at all.)

(f) The fitted model (with wetland 9 included) is

$$\log\left(\frac{p}{1-p}\right) = 1.545 - 0.361 \text{ grav2}$$

where p is the probability that wood ducks are present. As can be seen from the diagram above, over the range of **grav2** values in the data (apart from that for wetland 9), p declines gently from about 0.8 to about 0.2. (Indeed, over this range, the response is essentially linear.) A unit increase in **grav2** corresponds to the odds in favour of wood ducks being present being multiplied by a factor of $e^{-0.361} = 0.697$.

Not only might it be interesting to the researcher that it is one of the **grav** variables (density of gravel roads) that best predicts wood duck presence or absence, but that it is **grav2**, the density up to 500 m from the edge of the wetland, that seems most important.

(g) The analysis with **gbheron** as the response variable leads to quite different results. There is little indication from individual regressions that any explanatory variables are important. The stepwise regression starting from the full model removes every explanatory variable, and suggests the null model. Unsurprisingly, this is confirmed by the stepwise regression starting from the null model, which remains at the null model.

Let us call a halt to our investigation there, with a strong suspicion that, for green-backed herons, we can do little better than the null model. This says that the logit of p is -0.074 (the estimated value of the constant term) regardless of the values of the explanatory variables. The value -0.074 corresponds to a probability of 0.482, which in turn arises from there being 13 presences and 14 absences of green-backed herons in the dataset.

Solution 9.11

(a) The information given in GENSTAT's tables is summarized as follows.

BD

		1	2
death	0	21	30
	1	26	23

WV

		1	2
death	0	20	31
	1	6	43

FV

		1	2
death	0	35	16
	1	34	15

VS

		1	2
death	0	32	19
	1	17	32

V2

		1	2
death	0	43	8
	1	43	6

MS

		1	2
death	0	46	5
	1	42	7

YV

		1	2
death	0	46	5
	1	46	3

AC

		1	2	3	4	5	6	7
death	0	4	13	10	14	5	3	2
	1	1	4	14	15	12	3	0

Patterns of counts seem very similar for different values of the explanatory variables for FV, V2, MS and YV. For example, for FV = 1, there were 34 death sentences out of a total of 69 and, for FV = 2, there were 15 out of a total of 31 (proportions 0.49 and 0.48, respectively). The patterns seem very different for different values of BD, WV, VS and AC. On this basis, one might expect one or more of BD, WV, VS and AC to be in the final model.

(b) Individual correlations in the correlation matrix measure the strength of the relationships between pairs of explanatory variables. There is a striking lack of high correlations, the greatest (in absolute value) being the -0.467 between BD and WV. So there is no indication from the correlation matrix of which variables may be omitted from the model. By the way, there are better ways to measure the strength of the relationship between two binary variables, but correlations are convenient to calculate in GENSTAT.

(c) For each of the eight individual logistic regressions, the deviances and SPs are as follows.

	d.f.	Regression deviance	SP
BD	1	1.4	0.233
WV	1	9.9	0.002
AC	1	3.0	0.083
FV	1	0.0	0.934
VS	1	8.0	0.005
V2	1	0.2	0.619
MS	1	0.5	0.490
YV	1	0.5	0.495

Impressions from this analysis are similar but not identical to impressions we got from the contingency tables. WV seems very important, as well as VS. AC is somewhat marginal, and BD does not figure strongly.

The eight-variable fit yields a regression deviance of 26.4 on 8 d.f. Looking at the t statistics in the Estimates of parameters, it is again VS and WV that show up strongly (and there are t statistics just under 2 for AC and MS).

(d) Starting from either the full model or the null model, GENSTAT's stepwise regression arrives at the same model. It has two explanatory variables and these are WV and VS. The regression deviance for this model is 19.5 on 2 d.f. No surprises here: these are just the two explanatory variables that showed up strongly both in the individual and in the eight-variable regressions.

It seems that the probability of getting the death sentence instead of life is strongly influenced (increased) by there being one or more white victims and by the victim being a stranger.

(e) The regression deviance for the three-variable model is 19.9 on 3 d.f. The increase in regression deviance in adding BD to the model is $19.9 - 19.5 = 0.4$ on 1 d.f., $SP = 0.524$. The neatest way to get this information in GENSTAT is, after ensuring that the WV.VS model was the last one fitted, to add BD to the model by using the Change Model button. The required information is in the Change line under the Summary of analysis table. There is no evidence in favour of adding BD to the model involving WV and VS.

(f) The deviance associated with the WV, VS, BD model is 19.9. The deviance differences associated with WV, VS, BD plus interaction are as follows.

Interaction	Deviance difference	SP
WV.VS	0.8	0.361
WV.BD	0.1	0.755
VS.BD	0.0	0.880

Clearly there is no support for adding any of these interactions.

(g) We return to the fitted model interpreted qualitatively in part (d):

$$\text{logit}(p) = -2.057 + 1.745 \text{ WV} + 1.355 \text{ VS}.$$

If there are no white victims and the victim was not a stranger, then WV = VS = 0 in the model. (You might object that WV and VS take values 1 and 2; but recall that they have been made factors, which effectively recodes 1 and 2 as 0 and 1 (by use of indicator variables).) In the case WV = VS = 0, the odds in favour of death are $e^{-2.057} = 0.128$ (i.e. a probability of 0.113, so quite likely to get life). If there are no white victims and the victim was a stranger, the odds ratio is increased to $e^{-2.057+1.355} = 0.496$; if there was a white victim and the victim was not a stranger, the odds ratio is $e^{-2.057+1.745} = 0.732$; and if the victim was a white stranger, the odds ratio becomes $e^{-2.057+1.745+1.355} = 2.838$, a whopping increase in favour of the death penalty.

(In fact, you can easily find the probabilities of getting the death sentence in each situation by rerunning the regression model with Fitted Values switched on in the Generalized Linear Model Options dialogue box. The four different fitted values given in the resulting table are 0.11, 0.33, 0.42 and 0.74. As for normal regression models, the fitted values are the mean responses, according to the fitted model, corresponding to each observed individual. For a Bernoulli random variable with probability of success p, the mean is just p. Thus we can deduce that these four fitted values are the fitted probabilities for $P(\text{death} = 1)$ in each of the four cases. You might like to check that the probabilities and odds ratios are consistent with one another.)

Solution 10.1

(a) Taking logs of the explanatory variable should make the picture clearer. However, three of the values of z are zero, and $\log 0$ is undefined. A way out of this problem is to add a small positive quantity to all the z values before taking logs. A common version of this idea is to add one-half to each value to be transformed, i.e. use

$$x = \log\left(z + \tfrac{1}{2}\right).$$

This gives the following plot of y against x.

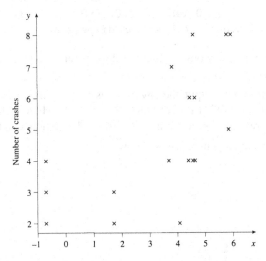

(b) The response variable takes integer values ranging from 2 up to 8. The normal distribution is a model for values on a continuous scale. It seems unlikely that the normal distribution would provide anything other than the very crudest of approximations if applied to data on such a limited integer scale. The Bernoulli distribution, on the other hand, models the behaviour of data taking just two distinct values, so how to apply it to responses taking seven distinct values is not at all clear!

Solution 10.2

(a) The two tables should be essentially the same.

(b) The GENSTAT output is as follows.

```
***** Regression Analysis *****

Response variate: crashes
   Distribution: Poisson
   Link function: Log
   Fitted terms: Constant, lcover

*** Summary of analysis ***
                                       mean   deviance approx
               d.f.      deviance    deviance    ratio chi pr
Regression        1        6.614      6.6141     6.61  0.010
Residual         15        8.681      0.5787
Total            16       15.295      0.9560
* MESSAGE: ratios are based on dispersion parameter with
           value 1


*** Estimates of parameters ***
                                                     antilog of
            estimate         s.e.      t(*)  t pr.    estimate
Constant       1.009        0.265      3.81  <.001      2.743
lcover         0.1450       0.0599     2.42  0.016      1.156
* MESSAGE: s.e.s are based on dispersion parameter with value 1
```

The null hypothesis would be that the slope parameter, the regression coefficient for lcover, is zero.

From the GENSTAT output, the deviance is 6.614, compared with the χ^2 distribution on 1 d.f., giving an SP (approx chi pr) of 0.010. There seems to be some evidence that the regression coefficient is non-zero. (Take no notice of the t(*) value here.)

(c) The scatterplot and curve are as follows.

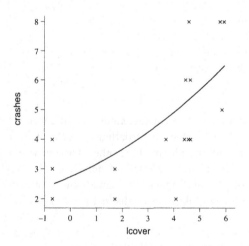

Broadly speaking, one might be happy with this model. The main worry might be the fit near $x = \log\left(5 + \frac{1}{2}\right) = 1.70$; here both responses are below the curve.

The residual plots are shown next.

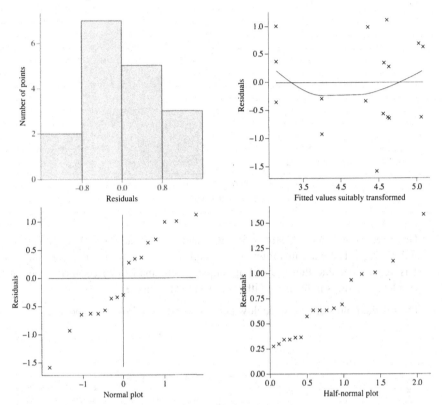

The message of the plot of residuals against fitted values is much the same as for the scatterplot. One extra point to notice is that no standardized residual is anywhere near as great in absolute value as 2, lending credence to a reasonable fit of the model. In the normal and half-normal plots, points occur in short 'patches' with gaps between them. This is how the discrete nature of the response variable is manifested in these plots.

(d) The fitted model is that, given a value for the index of newspaper coverage z, the number of crashes Y is distributed as a Poisson random variable with

$$\log(E(Y)) = 1.009 + 0.1450 \log \left(z + \tfrac{1}{2} \right).$$

Equivalently,

$$E(Y) = e^{1.009} e^{0.1450 \log(z+1/2)} = 2.743 \left(z + \tfrac{1}{2} \right)^{0.145},$$

which is an interesting 'power law' interpretation of the model. This power law simplification happens to arise in this case only because it turned out to be appropriate to take logs of the explanatory variable as well as of the mean of the response variable.

Solution 10.3

(a) To obtain the scatterplot overleaf, choose the **Graphics|Point Plot** menu item, and in the resulting dialogue box fill in the **Y Coordinates** field as **crew**, the **X Coordinates** field as **tonnage**, the **Groups** field as **power**, and click on **OK**.

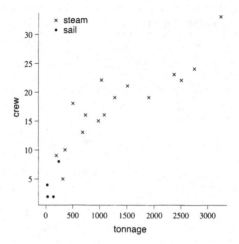

There seems to be a fairly strong, if non-linear (in x) relationship between crew size and tonnage. There are just four sailing ships as opposed to sixteen steam ships, and it is very noticeable that the sailing ships are all small (they comprise four of the smallest five ships in terms of either crew size or tonnage).

(b) The residual plots are shown below and the fitted model plot is opposite.

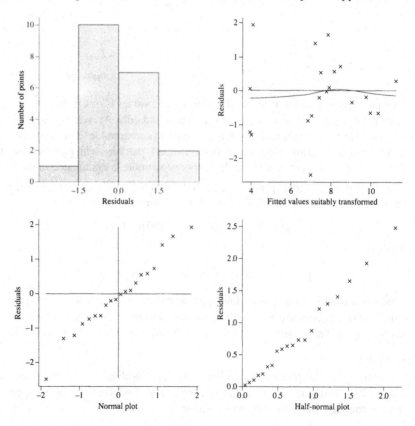

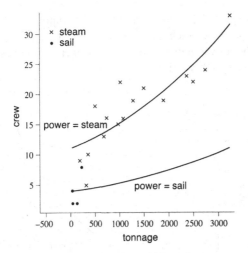

Many people's first impression from the residual plots would be that there is little amiss with the model. There are no very extreme outliers and the residuals appear to be fairly well distributed about zero, with no very obvious pattern. (The four residuals at the extreme left of the fitted value plot correspond to the four sailing ships.)

Note that the fitted model plot given here uses different symbols for **steam** and **sail** rather than the different labels used by GENSTAT. The fitted model plot shows the data together with different curves for the response means associated with the different methods of powering the ships. There are certain disquieting features. The most obvious is the enormous degree of extrapolation going on in producing the whole of the lower curve from just the four observations on sailing ships which are grouped together towards the left-hand end. (Actually, this is only because GENSTAT has plotted both curves for the entire range of steam and sail tonnages together. You don't have to go on and use the curve outside a sensible range!) Look too at the steam ship results. While the fitted response curve increases its slope as you move to the right, you could well argue that the steam ship data (with the possible exception of the very largest ship) follows a curve that is 'the other way up'. Having seen this, you can see the pattern reflected in the plot of residuals against fitted values also. (You might have observed this pattern in the residual plot before looking at the fitted model.) In particular, as the eye moves from the middle of the plot towards the right, there is a patch of positive residuals followed by a patch of negative ones; but the pattern is not *very* clear.

A better model is called for.

(c) A log transformation would pull the smaller tonnages apart and bring the larger tonnages closer together, and this would seem to work 'in the right direction'. (Notice that, although one *might* have thought of doing this at the outset, because of the variation in tonnages, the desirability of a log transformation is nothing like as obvious as in Exercise 10.2.)

(d) There seems now to be relatively little untoward in the residual plots (not shown); certainly, the pattern mentioned before has gone. You might have felt that the normal plots are less satisfactory, but bear in mind that we are not fitting a normal model here anyway.

The fitted model plot, shown below, is also much improved. Certainly, the fit for the steam ships seems to be much improved. The fit for the sailing ships also seems not unreasonable, but extrapolation problems, together with the fact that there are only four datapoints, inevitably remain.

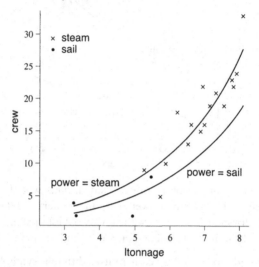

It is noticeable that the two response curves are rather similar and quite close together. Perhaps there is a case for unifying steam and sailing ships and providing just a single response curve (this is addressed in the next part of the question).

(e) With ltonnage, power as the last model to be fitted, click on **Change Model**, put power in the **Terms** field of the resulting dialogue box, and click on **Drop**. The Change line of the resulting output shows us that the difference in deviances between the two models is 1.42, to be compared with the χ^2 distribution on $2 - 1 = 1$ d.f.; the resulting SP is 0.234. There is no evidence for including **power** in the final model.

The residual plot (not shown) and fitted model plot (shown below) seem satisfactory.

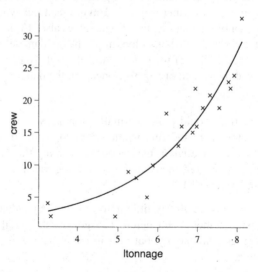

Note that although ltonnage *and* power still show in the Model to be Fitted field of the Generalized Linear Models dialogue box, the Further Output button leads to the right residual and fitted model plots. This is because the model with only ltonnage was the last to be fitted (when you clicked on Drop). (Notice also that, because only ltonnage is being fitted, GENSTAT will not allow you to specify power as a grouping variable in producing the fitted model plot. However, you know which of the points are for sailing ships from the previous fitted model plot.)

(f) The finally fitted model is that the crew size, Y, has a Poisson distribution given the tonnage, x, with its mean being related to the tonnage via the equation

$$\log(E(Y)) = -0.578 + 0.4875 \log x.$$

There do not seem to be different relationships for steam and sailing ships (at least not given the very small amount of data on sailing ships that we have).

Exponentiating both sides above gives

$$E(Y) = e^{-0.578} x^{0.4875} = 0.561 x^{0.4875}.$$

This is, again, a power law relationship. It is particularly tempting to approximate this equation further and state that, approximately, the mean crew size (for ships in 1907) is $\frac{1}{2}\sqrt{\text{tonnage}}$.

Solution 10.4

For each group of insects, you *could* expand the data into 30 (or 29) rows, each repeating the same values of explanatory variables for that group, and each with a zero or one in the equivalent of the ndead column, signifying whether or not any particular insect had been killed (nins would always be one and could be removed). Instead of the current neat data summary in 20 rows, the dataset would be $17 \times 30 + 3 \times 29 = 597$ rows long! But binary regression could then be performed. This is not, however, recommended!

Solution 10.5

(a) The scatterplot is as follows.

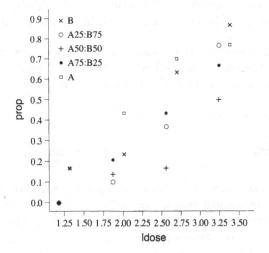

The proportion dead tends to increase with (log) dose. It appears that the single treatments (A and B) tend to kill more insects than the mixed treatments. It is not too clear what to say about different versions of treatments within the single-treatment and mixed-treatment classes.

(b) (i) The test for zero regression effect compares the regression deviance, 215.51, with the χ^2 distribution on d.f. (Regression) = 5 degrees of freedom, giving $SP < 0.001$. There is overwhelming evidence of a regression effect.

(ii) In performing this regression, GENSTAT has, of course, split the factor mix, which takes five levels, into four separate binary (indicator) variables (one each for A25:B75, A50:B50, A75:B25 and A, each relative to treatment B alone). A look at the t statistics suggests that some of these binary variables are more important than others. Perhaps there is scope for removing one or more of these.

(iii) The residual plots show nothing particularly untoward. The smooth blue curve on the plot of residuals against fitted values seems to indicate a curved pattern, but if you look at the actual points there is not really much evidence of curvature. The other plots, particularly the histogram, show some evidence of non-normality – but we are not fitting a normal distribution!

(c) Variable m13 has zeros for all but its fifth, sixth, seventh and eight values, which are all one. This is the binary (indicator) variable associated with the treatment A25:B75.

Likewise

```
CALC m22=mix.EQ.3
CALC m31=mix.EQ.4
CALC ma=mix.EQ.5
```

produce the binary variables corresponding to A50:B50, A75:B25 and A, respectively.

We have followed GENSTAT's choice and made these variables reflect effects relative to treatment B alone. This is why there is no binary variable corresponding to B.

(d) The binomial regression is done exactly as in part (b), except that the Model to be Fitted field is filled in as ldose,m13,m22,m31,ma. Apart from the different names for the binary explanatory variables, the two sets of output are exactly the same.

Stepwise regression obtained by entering

```
STEP [MAX=10;IN=4;OUT=4] ldose,m13,m22,m31,ma
```

yields a four-explanatory-variable model in which ma has been removed. If this is reasonable, it means that there appears to be no effect of using treatment A alone relative to using treatment B alone. In other words, insecticides A and B, when used on their own, are about equally effective.

That the reduction to the four-variable model is justified is confirmed by taking the difference between 215.51, which is the regression deviance under the full model, and 214.90, which is the regression deviance under the reduced model. This difference is 0.61. The loss of d.f. is 1. Comparing with the $\chi^2(1)$ distribution using PRINT cuchisquare(0.61;1) gives an SP of 0.4348, and this confirms that there is no need for the ma term. (Alternatively, you might perform this test by using the Change Model button.)

(e) The coefficients of m13, m22 and m31 are all negative, suggesting (more strongly than in part (b)) that each of these mixtures of insecticides is less effective at killing the tobacco budworm than is either of the individual insecticides alone.

The regression coefficients of m13 and m31, the two 25%–75% mixtures, are rather similar. Could it be that they have the same effect? Treatment m22 may be different; and, if so, even worse.

(f) Variables m13 and m31 have ones wherever treatments A25:B75 and A75:B25 are applied, respectively, and zeros elsewhere. Adding them together gives ones wherever one of treatments A25:B75 and A75:B25 are applied, so m1331 signifies any 25%–75% mixture of A and B.

The regression of ndead out of nins on Idose, m22 and m1331 yields a regression deviance of 214.77 on 3 d.f. compared with the four-variable deviance of 214.90 on 4 d.f. The difference is just 0.13 on 1 d.f. and the SP is 0.7184. Again, there is absolutely no evidence to prefer the more complicated model. We should be prepared to accept that $\beta_{13} = \beta_{31}$.

Next, set m123 = m22 + m1331. Thus, m123 signifies application of any of the three insecticide mixtures. Regression on just Idose and m123 yields a deviance of 207.36 on 2 d.f. The difference is 214.77 − 207.36 = 7.41 on 1 d.f., yielding an SP of 0.0065. There is strong evidence that the more complicated model is necessary; m22 and m1331 do not have the same effect.

(g) In the final model, there is a (not unexpected) strong dependence on (log) dose (note the t statistic of 11.24). Individual insecticides were the most potent, and about equally effective to one another. 25%–75% mixtures were the next most potent, and also about equally effective. The 50–50 mixture of A and B was least effective of all. These insecticides appear to act antagonistically, in that combinations of the two are less effective than either on its own.

(Residual plots, which you were not asked to produce, support the appropriateness of this model.)

Solution 10.6

(a) You may have noticed a great variation in these values. For instance, in many cities only a few individuals were tested (in four cases, just one); on the other hand, in some cities many people were tested, e.g. 54 in city 18 and 82 in city 27.

(b) Your scatterplot should look like that overleaf, except that all the symbols should be the same. There seems to be no very clear pattern.

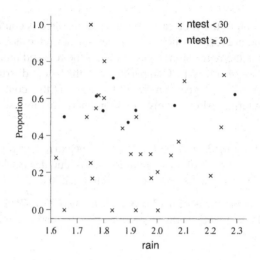

In part (a), it was noted that some of these points are based on just a few individuals while others are based on much larger numbers. The latter should afford better estimates of proportions than the former. On the figure above, the points corresponding to ntest $\geq$ 30 are represented differently from those corresponding to ntest < 30. Looking just at those for ntest $\geq$ 30, perhaps there is a clearer structure here: might a simple constant fit suffice?

(c) Choose Stats|Regression Analysis|Generalized Linear. In the dialogue box: change the Analysis field to Modelling of binomial proportions. (e.g. by logits); fill in the Number(s) of Subjects as ntest, the Numbers of Successes as npos, and the Model to be Fitted as rain; click on OK.

The regression deviance for the model is just 0.12. The SP associated with this is 0.724. This enormous SP shows that there is no evidence whatsoever that the single regression coefficient is non-zero, i.e. there is no evidence that a linear model (on the logistic scale) should be preferred to a constant model.

(d) As many as five points (out of 34) have very large standardized residuals. This is a first indication that the model we have – be it constant or linear on the logistic scale – may not be adequate.

(e) The regression deviance is now 11.58 on 3 d.f. The SP is 0.009. There is quite strong evidence that not all three regression coefficients (linear, quadratic and cubic) are zero. The cubic model (on the logistic scale) seems to be better than the linear. However, you should notice the continued presence of four large standardized residuals, which may be indicating continued lack of fit to the data of even this model.

(f) One way to fit quadratic and quartic models is by entering pol(rain;2) and pol(rain;4), respectively, in the Model to be Fitted field of the Generalized Linear Models dialogue box.

From the various bits of GENSTAT output, the regression deviances for the linear, quadratic, cubic and quartic models are 0.12, 0.12, 11.58 and 11.77, respectively. So, for example, the extra deviance due to the cubic term (over the quadratic) is $11.58 - 0.12 = 11.46$. The complete table is as follows.

Polynomial	d.f.	Extra deviance
linear	1	0.12
quadratic	1	0.00
cubic	1	11.46
quartic	1	0.19

The SPs associated with each of the linear, quadratic and quartic terms are all very high. The linear term is, therefore, not helpful; nor are the quadratic and quartic terms. But the cubic term gives rise to a very small SP (in fact, 0.0007). There is a clear indication that a logistic-cubic curve is much the best.

(g) The diagram produced is shown below.

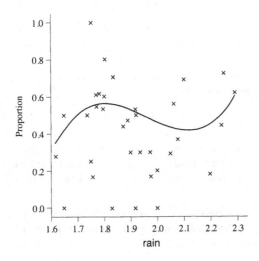

This logistic-cubic curve may be the best of the logistic-polynomials, but it is still not explaining much of what is going on. A very wiggly curve might be dreamt up to fit the data better, but this hardly simplifies things, and would almost certainly not fit well if we acquired more data from the same situation. Indeed, one might ask whether this cubic curve would really work better, in a predictive sense, than the constant model mentioned in part (b).

Solution 10.7

If μ is related to the linear predictor $\eta_x = \sum_j \beta_j x_{i,j}$ through $1/\mu = \eta_x$, then the parameter λ is related through $\lambda = \eta_x$. That is, λ is taken to be a linear function of explanatory variables.

Since $\lambda > 0$, so too is $\mu > 0$. However, the linear predictor η_x, which can take any positive or negative value, may not respect this constraint. None the less, such modelling may remain useful because a (positive) linear function throughout the range of the data may still provide a useful description even if we know that it will break down for a sufficient degree of extrapolation.

Solution 10.8

(a) The first plot below has most of its points 'scrunched up' towards the left-hand end. Those points can be spread apart and the points towards the right-hand end brought closer together by a log transformation. Making lwbc = log(wbc) and obtaining a new plot gives the second plot below.

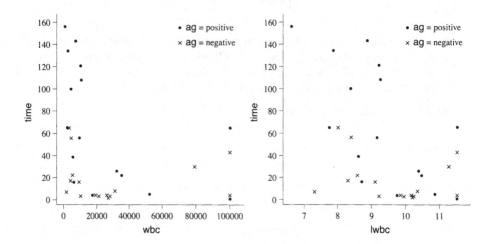

The points are at least now much more clearly separated in the x direction. Survival times for some members of the AG positive group are much bigger than those for the AG negative group. But it is difficult to say much more than this.

(b) From the GENSTAT output, the deviance is 18.36 on 3 d.f. yielding an $SP < 0.001$. There is strong evidence of some regression effect, i.e. that some of the regression coefficients are non-zero.

(c) A way to fit a model without the **ag.lwbc** term is to click on the **Change Model** button, enter **ag.lwbc** in the **Terms** field, and then click on **Drop**. The Change line in the resulting output gives an SP of 0.607 (corresponding to a deviance difference of 0.27 on 1 d.f.). There is no evidence that the coefficient of **ag.lwbc** is not zero, and so the model without **ag.lwbc** is adequate.

(d) The **Drop** button in the **Change Model** dialogue box is useful here too. However, don't forget to add **ag** back into the model before dropping **lwbc** (if you drop **ag** first and **lwbc** second). The result is this. When **lwbc** only is present, the SP is 0.004, strong evidence that the **ag** term is needed. The model with **ag** only present has an SP of 0.013. There is some evidence of a need for the **lwbc** term. The most suitable model seems to include **ag** and **lwbc** but not **ag.lwbc**.

The fitted model is as follows. (This diagram results when, in the **Graph of Fitted Model** dialogue box, **lwbc** is in the **Explanatory Variable** field and **ag** is in the **Grouping Factor** field.)

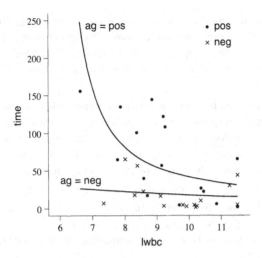

(e) The fitted equations are, for AG negative,

$$\frac{1}{E(Y)} = -0.0020 + 0.0061 \log(\textbf{wbc}),$$

and, for AG positive,

$$\frac{1}{E(Y)} = -0.0364 + 0.0061 \log(\textbf{wbc}).$$

It is equally valid to write, e.g., $E(Y) = 1/(-0.0020 + 0.0061\log(\textbf{wbc}))$. These equations arise from the Estimates of parameters part of the GENSTAT output: because it is a coefficient for ag pos that is given, the Constant is the constant in the AG negative case, and it is Constant + ag pos $= -0.0020 + (-0.0344)$ that forms the constant in the AG positive case.

In this model, the constants are different but the 'slope' coefficients are the same. Had ag not been in this model, there would have been a single equation in terms of log(wbc) only, equivalent to equal constants and slopes. Had lwbc, ag and ag.lwbc all been in the model, there would have been different constants and different slopes for each AG value. (See Example 6.4 for why this works in the normal linear regression case; the more general GLM framework makes no difference to this kind of argument.)

Solution 10.9

(a) The model fitted is a gamma response distribution with an identity link function and a straight-line predictor, i.e. of the form

$$E(Y) = \beta_0 + \beta_1 x.$$

The fitted model does not seem too bad except towards the right-hand end, where the fitted model is beneath the entire group of five observations with the largest values of density. It appears that a model that behaves much like a straight line over most of the range of the data but curves upwards towards the right-hand end would be preferable.

Approaches to making this kind of change are: (i) to change the link function; (ii) to change the linear predictor; or (iii) both. We could change to the canonical link function, which is the reciprocal, but we shall in fact continue to work with the identity link function and see if we can obtain a satisfactory fit by changing the linear predictor alone.

(b) The identity link function does not automatically make the fitted model respect the constraint that the responses must be positive. Care should therefore be taken that the linear predictor does not predict negative values. However, note that all values of **hardness** in the dataset are considerably greater than zero and so there is unlikely to be any difficulty within the range of densities of interest.

(c) Having performed the transformation, and fitted the model with both explanatory variables by the procedure given in the question, we see from the Change line in the GENSTAT output, that the approx F pr. is less than 0.001. This makes it clear that it is not appropriate to prefer the model containing **density** alone to the one containing **density** and **rden**.

Having fitted **density** and **rden**, comparison can be made with the model containing **rden** only by clicking on **Change Model**, putting **density** in the **Terms** field, and clicking on **Drop**. The resulting Change line also has an approx F pr. that is less than 0.001. It appears inappropriate to leave either variable out of the model.

(d) The model involving both **density** and **rden** is much better at fitting the 'difficult' group of five points towards the right-hand end of the plot. This fit seems much preferable, on these grounds, to that obtained in part (a). In producing the fitted model plot, GENSTAT adjusts the observed and fitted values of **hardness** to allow for the fact that the model contains **rden** as well as **density**, but in this case the adjustment has little effect.

(e) The diagram obtained is as follows. The two models give very similar fits indeed.

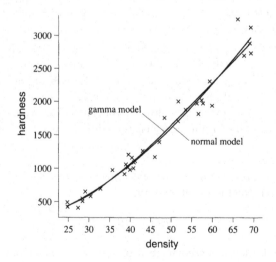

11

Solution 11.1

Looking at Figure 11.1, most people would probably say that the high-leverage point looks as if it matches the pattern of the other data. If a regression line were fitted to the rest of the data, it would go close to the position of the high-leverage point.

Solution 11.2

(a) The plot produced is shown below.

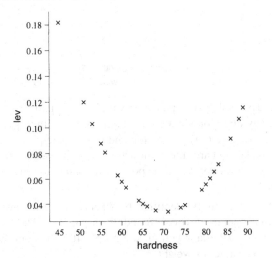

The leverages are low towards the middle of the range of values for **hardness**. High leverages are found towards the extremes of the range of **hardness** values. The left-most point (hardness 45) has the highest leverage, simply because it is a relatively long way from the other points in terms of **hardness**. The relationship looks as if it is quadratic (and, indeed, it can be shown that it is).

(b) You should not be able to see a clear relationship between **lev** and **loss**. If you could see a relationship, it would mean that **loss** and **hardness** were related very strongly, because the leverages are calculated solely from the values of the explanatory variable(s) (**hardness** in this case).

(c) The plot of leverages against fitted values looks generally like a mirror image (left and right reversed) of the plot you produced in part (a). This is simply because the fitted values, in this case where there is only one explanatory variable, are a linear function of the explanatory variable. The reversal occurs because the slope of the line is negative.

The index plot of leverages shows no particular pattern. It does emphasize, however, that unit 1 has the highest leverage.

Solution 11.3

(a) The plot is as shown.

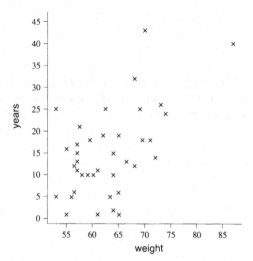

Since this is a plot solely of explanatory variables, points a long way from the 'average' position in any direction should have high leverage. In this case, the most obvious such points are those towards the top and right of the plot. The specific points you might have picked out are: the point with a value for **weight** of over 85; the point with the highest value for **years**. These points correspond to rows 39 and 38 respectively in the spreadsheet.

(b) The high-leverage points identified by GENSTAT are in rows 8, 38 and 39. We identified those points in rows 38 and 39 in part (a). The point in row 8 was not mentioned there. (Perhaps you picked it up!) It is the point with a value of 25 for **years** and a low value for **weight**.

(c) The regression coefficients when row 39 is included are as follows.

	estimate	s.e.	t(35)	t pr.
Constant	62.6	15.1	4.15	<.001
years	-0.383	0.184	-2.08	0.045
weight	1.104	0.260	4.25	<.001

Those produced when row 39 is excluded are as follows.

	estimate	s.e.	t(34)	t pr.
Constant	72.1	17.1	4.21	<.001
years	-0.416	0.186	-2.24	0.031
weight	0.955	0.289	3.31	0.002

These tables show that the regression coefficients change, but it is hard to say if the change is large or important. All we can say, without looking back at the data, is that none of the coefficients has changed by more than one standard error, and that there are no qualitative changes in interpretation associated with individual regression coefficients. Thus the impact of removing row 39 is certainly not immense.

Solution 11.4

(a) The resulting plot is as shown. Though the first point does have the largest Cook statistic value, perhaps it does not have an exceptionally large value when compared with the Cook statistics for the other points.

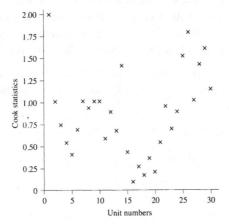

What is odd about this plot is that there should be no structure to the pattern of the Cook statistics, but there seems to be one. Remember that this plot shows the Cook statistics simply in order of the points in the datafile. If the points really are a random sample, this order should not matter at all. However, on the right-hand side of the plot, from about unit 15 upwards, the Cook statistics seem to get bigger as the unit number increases.

(b) The index plot for the residuals is as shown. (Ignore the curved line which is not at all helpful.)

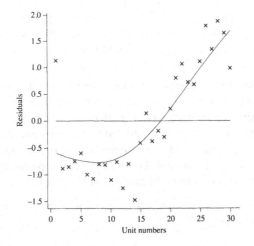

With the exception of point 1, the residuals for the first 14 points are all below -0.5. The other residuals seem to get bigger as the unit number increases. Perhaps the most likely explanation for this structure is that the unit numbers correspond to the order in which the data were collected and that something happened to change the conditions of the experiment after the first 14 points were collected.

When you looked at the data in the spreadsheet, one thing that might well have caught your eye is that the values for hardness seemed to be getting bigger up to point 14 and then jumped to a lower value for point 15. If you look more closely, it becomes clear that there are several sequences of points with increasing hardness values. Specifically, these sequences are: rows 1 to 8; rows 9 to 14; rows 15 to 19; rows 20 to 24; rows 25 to 30. You might well suspect that the experiment was carried out in several runs and that the conditions changed between different runs, with later runs tending to result in higher abrasion losses for some reason.

(c) The index plot of deletion residuals looks practically identical to the index plot of standardized residuals. In this case, as in many others, the two sets of residuals are practically the same.

Solution 11.5

(a) The resulting plot is as shown.

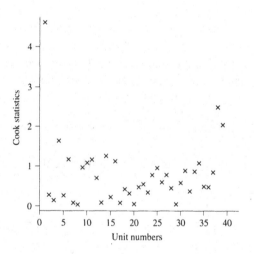

The Cook statistic for the first point is large compared with those for the other points. Most of the Cook statistics are less than 2, but point 1 has a Cook statistic of well over 4. This means that point 1 has a big effect on the regression. The main cause of this is that point 1 has a large standardized residual (essentially because of its very large value for **sbp**) and a moderately high leverage (which you can see in an index plot of leverages, if you want to). The high value of **sbp** was the original reason for the exclusion of this point from the analysis in Chapter 6.

To see the effect of point 1, compare the regression coefficients in the **Output** window with those in Solution 11.3.

(b) The resulting plot is as shown. There are three points with Cook statistics that are considerably higher than the others: points 4, 38 and 39. The Cook statistics for these three points are all greater than 2, whereas the others are less than 1.5. (Notice that the exclusion of the first datapoint alters the Cook statistics of the remaining points.)

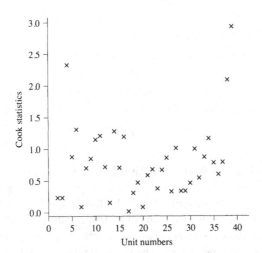

The three points with large Cook statistics are not quite the same as the points with high leverage. The points with high leverage are 8, 38 and 39. Point 8 has high leverage, but does not make much difference to the regression coefficients (essentially because its response follows the pattern of the responses for the other points). Point 4 makes quite a big difference to the regression coefficients, but does not have a particularly high leverage.

Solution 11.6

(a) The regression output flags two points, numbers 11 and 29, as having large standardized (Pearson) residuals. A composite plot of Pearson residuals (or deletion residuals, which are practically identical) shows nothing untoward except, possibly, the high residuals that have already been flagged. An index plot of the residuals shows no particular pattern. (There is no reason why it should; the states are listed in alphabetical order. Only 47 states are included. The omitted states are Alaska, Hawaii and New Jersey.)

No points are flagged as having high leverage, and an index plot of leverages confirms that no points have leverages that are well out of line with the majority. An index plot of Cook statistics (overleaf) shows that state 29 has the largest influence; state 11 also has a large Cook statistic as, arguably, does state 36.

An appropriate conclusion would be that the model of Exercise 6.5 fits reasonably well, but that a few of the states (particularly number 29) have had considerable influence on the regression coefficients and, through them, on the choice of which explanatory variables to include in the model.

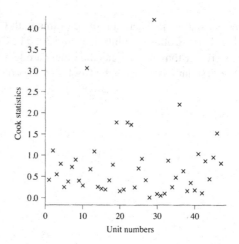

(b) State 29 has a large (negative) standardized residual and the largest leverage. The relatively large leverage arises because it is relatively extreme in terms of the explanatory variables included in the model. (Indeed this point, which corresponds to New York State, has the lowest value of **malyth** and the highest value of **pol60** in the whole dataset.) The large leverage and the relatively low response, leading to a large negative residual, are the reasons why this point has a large Cook statistic.

Point 11 (corresponding to Illinois) does not have a notably high leverage. Its large Cook statistic has more to do with its large positive residual. It has a large influence on the regression coefficients largely because its crime rate is untypically high.

(c) After excluding point 29, the stepwise regression procedure (whether you start from the full model or the null model) ends up with a model including just three explanatory variables, **pol60, poor** and **school**. This is certainly different from the model in Exercise 6.5, because it leaves out the explanatory variables **malyth** and **unemid** (and the coefficients of the variables that remain have changed a bit). The percentage of variance accounted for has gone up slightly compared with the model of Exercise 6.5, from 69.7% to 71.1%. The regression flags two different points as having large residuals, and this time flags two points as having high leverage (but overall, plots of residuals, Cook statistics and leverages look reasonable). Interestingly, point 11, which came up as having a very large residual and a large Cook statistic in the previous analysis, no longer appears to be particularly unusual (though it still has the largest positive residual, just).

We have come up with a different model by excluding the influential point for New York State. Is this new model better? Arguably not. In some respects (pattern of residuals, percentage of variance explained) it is slightly better than the original model, but the improvement is pretty tiny and has come at the price of excluding a perfectly good state from the analysis. On the other hand, the new model is simpler than the original. However, the reanalysis does emphasize again the difference that one point can make to an analysis.

Solution 11.7

(a) The index plot is as follows.

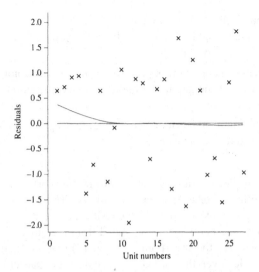

There is an interesting feature. There is a gap in the middle of the plot. Only one residual takes a value between about −0.5 and about 0.5. A gap somewhere in the middle of the pattern is a common (but not universal) feature of such residual plots for binary regression, and it does not usually tell one anything useful. The one small residual corresponds to Wetland 9, which has an outlying value of **grav2**, as we saw in Exercise 9.10.

There is perhaps a tendency for there to be more positive residuals towards the start of the data sequence. For binary regression, the observed values are all 1 or 0 while the fitted values are somewhere between. Thus a positive residual can occur only if the observed value is 1. For some reason, datapoints near the top of the spreadsheet seem to have a slight tendency to result in one (i.e. wood ducks present) rather more frequently than the model would indicate. Without further information we cannot tell why this might be; but in any case it is not marked enough to worry about.

(b) The half-normal plot is as follows.

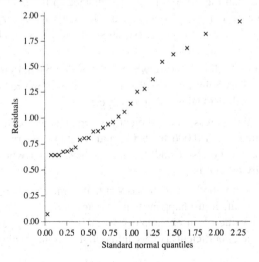

It is smooth, but with one point (again Wetland 9) clearly out of line at the bottom left. The line is not straight, but since we have not assumed normality, we should not expect it necessarily to be straight.

(In this case, the normal plot (as opposed to the half-normal plot) looks very strange, essentially because of the gap in the middle of the distribution of the residuals that we noted in part (a).)

(c) The plots of deletion residuals look very similar to the plots of standardized deviance residuals, and lead to the same conclusions.

Solution 11.8

Neither of the residual plots (index plot and half-normal plot) shows any major reason to doubt the fit of the model or to suspect any outliers. There is no pattern in the index plot. The half-normal plot looks curved, but the residuals are not normal in an exponential model anyway. Two residuals do appear separated from the others at the top right of the half-normal plot. These are the two (points 14 and 15) that are flagged by GENSTAT in the regression output. However, they are not huge and are not particularly out of line with the other values. (Both these patients survived for a very short time only, recorded as 1 week. The survival times are all rounded to whole weeks, yet the exponential distribution, which we used to model the times, is continuous. It could be that the rather bad fit for these patients is a consequence of the rounding.)

Solution 11.9

(a) No points are flagged as having high leverage. This is not surprising when you look at the index plot of leverages (not shown), which shows a reasonably uniform pattern. You may have found this strange in view of the fact that we found Wetland 9 to have a very outlying value of **grav2** in Exercise 9.10, though in fact it seemed to have no strong influence on the model (which changed very little when this wetland was removed). In fact, the leverage plot shows Wetland 9 as having the lowest leverage. There is no very clear pattern in the Cook statistic plot either. We return to the question of leverage and Cook statistics in binary data in the next exercise.

(b) This time, one point (unit 2) shows up in the index plot of leverages as having far higher leverage than the others. (It is also flagged in the main output.) The same point has a huge value for the Cook statistic. It clearly has a very large influence on the fitted model. Interestingly, the two points (14 and 15) which have large residuals do not turn out to be influential.

(c) This point has the smallest value of **wbc** by a considerable margin. This is why it has high leverage. (This point also has the highest value of the response variable **time**, but its standardized residual is not at all large.)

On omitting it, the regression coefficient for **lwbc** changes considerably relative to its previous value, from 0.006 to 0.009, and its standard error has also increased markedly; also, two points (1 and 4) with small values of **wbc** are now flagged as being potentially influential.

Unless there is some good reason to suspect that point 2 is in error, there is no good reason to leave it out. It just happens that, in many exponential regressions, the point with the lowest value of an explanatory variable (or the highest value, if it has a negative regression coefficient) is very influential, and this cannot really be avoided.

Solution 11.10

(a) Both index plots show a reasonably clear pattern – low towards the beginning and end of the data sequence and high in the middle. The pattern is more marked for the Cook statistics plot (shown below) than for the leverage plot (not shown). (The Cook statistics plot also shows that point 5, which was flagged as having high leverage, is not in fact all that influential.)

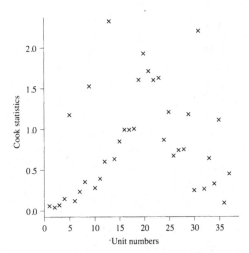

Looking at the original data, you can see that they are listed in two groups, first those with gvhd $= 0$ and then those with gvhd $= 1$. Within each of these groups they are listed in increasing order of indx. You might have thought that the patterns of influence and leverage had something to do with this, and you would be right.

Let us consider leverage first. Patients with a high value of indx have a higher probability of exhibiting graft-versus-host disease (gvhd $= 1$). Thus patients near the ends of the range of indx tend to have responses with a fitted probability $\hat{p}$ of graft-versus-host disease near to 0 or to 1. Also (since the model fits reasonably well) the patients in the first group, for whom gvhd $= 0$, tend to have lower values of $\hat{p}$ than do patients in the second group. Thus, because of the way the data are sorted in order of gvhd and indx, those near the beginning of the list have low values of $\hat{p}$ and those near the end have high values of $\hat{p}$, with the ones near the middle of the list somewhere in between in terms of $\hat{p}$. The variance of a Bernoulli (binary) random variable is $p(1 - p)$, which is smallest for values of p near 0 or 1. Because of the way the maximum likelihood estimation process works in generalized linear models, this means that values of the explanatory variables that lead to values of $\hat{p}$ near 0 or 1 tend to have low leverage compared with those that lead to values of $\hat{p}$ near 0.5. Hence the strange pattern of leverages is a consequence of the facts that the data are sorted into a particular order and that the variance of the Bernoulli distribution is small for p near 0 and 1.

Why is the pattern more marked for the Cook statistics? In this dataset (as often happens in logistic regression), points with $\hat{p}$ right at the ends of the range tend to have small standardized residuals. Since they have small leverage as well this means that many points with values of $\hat{p}$ near 0 or 1 have relatively very small Cook statistics.

(b) If this were a normal linear regression, these points would have high leverage because they are a long way away from the general pattern of the others. However, this is not normal linear regression, and you may have suspected from part (a) that they would not have high leverage. In fact, these two points have the lowest leverages of all!

(c) The resulting plot is as follows.

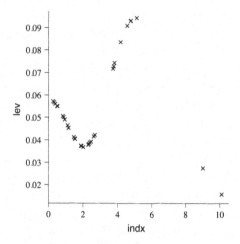

It has a rather bizarre shape compared with the quadratic plot of leverage against the explanatory variable that you saw in Solution 11.2. The leverages are relatively low for values of indx in the middle of the main group of points, and rise as the values of indx move away from there. But they do not keep on rising. They start to fall off with increasing distance from the 'centre', and the two points (36 and 37) with very extreme values of indx have, again, the lowest leverage. If you do the same exercise using lindx as the explanatory variable rather than indx, the plot looks similar in general shape.

The explanation is related to what was said in part (a) of this solution. For the same reasons as in a normal linear model, points near the 'centre' of the set of values of the explanatory variable have relatively low leverage, and the leverage initially increases as the explanatory variable moves away from the centre. But after it has moved a certain distance, the effect of the variance of the Bernoulli random variable comes into play, and the leverage begins to decrease again because the response variance is so low. For a point near the 'centre', like point 14 with indx = 2.01, $\widehat{p} = 0.394$ and the estimated response variance is $0.394 \times (1 - 0.394) = 0.239$. Even for a point some way from the 'centre', say point 35 with indx = 5.07 and $\widehat{p} = 0.874$, the estimated response variance is $0.874 \times (1 - 0.874) = 0.110$, not a great deal less. But for point 37, with indx = 10.11 and $\widehat{p} = 0.997$ the estimated response variance is only $0.997 \times (1 - 0.997) = 0.003$. This is why point 37 has such low leverage.

Solution 11.11

Index and half-normal plots of the residuals do not provide any strong evidence that there is anything wrong with the model. The residual for point 16 is rather large and negative, is flagged in the output, and shows up in the index plot. But its value is only −2.1. It does not look particularly out of line on the half-normal plot. Since this is Poisson regression, you may have produced a normal plot too. It is reasonably straight, though at the left-hand end it shows some evidence of the 'patchy' patterns, due to the discreteness of the distribution involved, that was mentioned in Solution 10.2(c).

An index plot of leverages shows one point (9) that has rather higher leverage than the others. But this point does not have a high Cook statistic. The most notable aspect of the index plot of the Cook statistics is that point 16 has a rather higher influence than all the others. A look at the data shows that this point corresponds to a (sailing) ship of 138 tons with a crew of only 2. This crew size does appear rather low, and should be checked if possible. (Since you do not have the luxury of checking it here, you could try to refit the model excluding this point to see how much difference it actually makes.)

12

Solution 12.1

(a) The spreadsheet gives all four cell counts in a variate called count, with factors imp-sulph and toxicity representing the rows and columns respectively. The arrangement is different from that in Table 12.1; but, apart from the row and column totals, the information is all there in the spreadsheet.

(b) GENSTAT reports the following.

```
Pearson chi-square value is 15.90  with    1 df.
```

```
Probability level (under null hypothesis) p < 0.001
```

The chi-squared test statistic is large, and the resulting SP is small. There is strong evidence that the row and column factors are associated.

The maximum likelihood test statistic has a slightly different value (16.43), but again the SP is very small. (More precise SPs could be obtained from the cuchisquare command. They are 0.000067 and 0.000050, respectively for the chi-squared and maximum likelihood test statistics.)

Either way, there is strong evidence of an association between toxicity and impaired sulphoxidation capacity.

(c) GENSTAT gives a great deal of output for Fisher's exact test, as follows.

```
***** Fisher's exact test *****

One-tailed significance level =    0.00007867
Mid-P value =    0.00004321

Two-tailed significance level
     Two times one-tailed significance level =    0.0001573
     Mid-P value =    0.00008642
     Sum of all outcomes with Prob<=Observed =    0.0001000
     Mid-P value =    0.00006456
```

This is essentially because there is more than one way of calculating the two-sided SP for this test. But, by whatever means it is calculated, the SP is tiny and the conclusions are as in the previous part. (The SP labelled Sum of all outcomes with Prob<=Observed may well be what you have seen before.)

(d) In the analysis of deviance table for the model with the interaction term, the residual deviance is given as 0.00 with no degrees of freedom. This means that, with this model, the data are fitted exactly (i.e. the fitted values are equal to the observed values).

The deviance difference, given in the Change line of the GENSTAT output, is 16.43 on 1 d.f. Note that this corresponds to the residual deviance for the first model (with no interaction); this is because the total deviance is the same in both models, and the residual deviance in the second model is zero. The SP is less than 0.001. Thus there is strong evidence that the interaction term should be in the model. In fact, this amounts to the same thing as you saw in the previous two parts; the row and column factors are not independent. (You will see why in Section 12.2.)

The Likelihood chi-square value of 16.43 with 1 d.f. from part (b) is the same as the residual deviance for the model with no interaction. It is also the same as the deviance difference we have just tested. (This is no coincidence, as you will see later in this section.)

(e) The fitted values from the model with an interaction term are exactly the same as the data values given in Table 12.1. This corresponds to the fact that this model fits the data exactly, with zero residual deviance.

The fitted values from the model with no interaction term are as follows.

	toxtab	
toxicity	yes	no
impsulph		
yes	22.20	16.80
no	14.80	11.20

This time they do not match the data exactly. But they do match chi-squared expected values. Using the marginal totals in Table 12.1, the expected value in the top left cell, for instance, is $(39 \times 37)/65 = 22.2$, as given. The expected value in the bottom left cell is $(26 \times 37)/65 = 14.8$. The others check out similarly.

Solution 12.2

(a) GENSTAT reports the Pearson chi-squared value as 27.68 and the likelihood chi-squared value as 28.60, both with 3 d.f. In both cases the SP is less than 0.001. There is strong evidence that beetles that come out in different seasons tend to have different colours.

(b) According to the Change line in the GENSTAT output, obtained when colour.season is added to the model colour+season, the deviance difference is 28.6 on 3 d.f., and the SP is less than 0.001. The conclusion is the same as in part (a).

(c) In a loglinear model for a two-way contingency table with an interaction term included, as you saw in Exercise 12.1, the fitted values are exactly equivalent to the observed data. You saw this there for a 2×2 table, but the result holds for general $r \times c$ tables. Thus the residual deviance in the model with an interaction term must be zero. The total deviance in this model and in the model without interaction must be the same, and so the regression deviance in the model with interaction must be equal to the total deviance. The difference between the regression deviances in the interaction model and the no-interaction model is thus the same as the difference between the total deviance and the regression deviance in the no-interaction model. This difference is equal to the *residual* deviance in the model with no interaction term. (This was also true of the example on toxicity.) Thus we can check for interaction by comparing the residual deviance in the no-interaction model with a chi-squared distribution. The degrees of freedom add up in the same way (there are no degrees of freedom for the residual in the interaction model), so the appropriate number of degrees of freedom for the chi-squared distribution is just that given in the residual row of the analysis of deviance table for the no-interaction model.

In this case, the residual deviance associated with the **colour+season** model is indeed 28.60 on 3 d.f. You would have to calculate the SP for yourself. The command PRINT cuchisquare(28.60;3) gives 0.000 003. The conclusion is again the same.

Solution 12.3

The joint probability mass function, found by putting $k = 2$ in Box 12.1, is as follows.

$$P(N_1 = n_1, N_2 = n_2) = \begin{cases} \dfrac{n!}{n_1!n_2!} p_1^{n_1} p_2^{n_2} & \text{if } n_1 + n_2 = n \\ 0 & \text{otherwise} \end{cases}$$

Now $\dfrac{n!}{n_1!n_2!} = \binom{n}{n_1} = \binom{n}{n_2}$ when $n_1 + n_2 = n$. In general, we would find $P(N_1 = n_1)$ by summing the expression for $P(N_1 = n_1, N_2 = n_2)$ over all the possible values of n_2. But, since the probability mass function is zero unless $n_1 + n_2 = n$, there is only one value to sum, and

$$P(N_1 = n_1) = \binom{n}{n_1} p_1^{n_1} (1 - p_1)^{n - n_1}$$

(since $n_1 + n_2 = n$ and $p_1 + p_2 = 1$). This is the probability mass function of the binomial distribution $B(n, p_1)$. The result for N_2 is found in a similar way.

Solution 12.4

(a) The mean counts in all four cells come directly from the model, as described just above the exercise, and are shown boxed below. The mean total for the first row is the sum of mean counts in the first row, i.e. $e^\mu + e^{(\mu + \beta_c)} = e^\mu(1 + e^{\beta_c})$. The other mean row and column totals are as shown below. The mean total count for the whole table is the sum of the two mean row totals or, equivalently, the sum of the two mean column totals. The complete collection of mean counts is as follows.

e^μ	$e^{\mu + \beta_c}$	$e^\mu(1 + e^{\beta_c})$
$e^{\mu + \beta_r}$	$e^{\mu + \beta_r + \beta_c}$	$e^{\mu + \beta_r}(1 + e^{\beta_c})$
$e^\mu(1 + e^{\beta_r})$	$e^{\mu + \beta_c}(1 + e^{\beta_r})$	$e^\mu(1 + e^{\beta_r})(1 + e^{\beta_c})$.

(b) The cell probabilities are as follows.

$$
\begin{array}{|cc|}
\hline
\dfrac{1}{(1+e^{\beta_r})(1+e^{\beta_c})} & \dfrac{e^{\beta_c}}{(1+e^{\beta_r})(1+e^{\beta_c})} \\[3mm]
\dfrac{e^{\beta_r}}{(1+e^{\beta_r})(1+e^{\beta_c})} & \dfrac{e^{\beta_r+\beta_c}}{(1+e^{\beta_r})(1+e^{\beta_c})} \\
\hline
\end{array}
$$

(c) The probability of falling in the first row is

$$
\frac{1}{(1+e^{\beta_r})(1+e^{\beta_c})}+\frac{e^{\beta_c}}{(1+e^{\beta_r})(1+e^{\beta_c})}=\frac{1+e^{\beta_c}}{(1+e^{\beta_r})(1+e^{\beta_c})}
$$

$$
=\frac{1}{1+e^{\beta_r}}.
$$

In a similar way, the probability of falling in the second column is

$$
\frac{e^{\beta_c}(1+e^{\beta_r})}{(1+e^{\beta_r})(1+e^{\beta_c})}=\frac{e^{\beta_c}}{1+e^{\beta_c}}.
$$

The product of these two probabilities is $\dfrac{e^{\beta_c}}{(1+e^{\beta_r})(1+e^{\beta_c})}$, which is indeed the probability of falling in the cell in the first row and the second column.

Solution 12.5

(a) The layout of the data in the spreadsheet is considerably different from that in Table 12.3. There is a variate called, again, count, which gives all the cell counts, and then three factors called class, vote and gender defining the classifying variables.

The table produced from the Summary by Groups dialogue box looks much like Table 12.3, except that the blocks for Conservative and Labour voters appear above one another rather than side by side.

(b) The residual deviance for this model is 133.0 on 4 d.f. This is so large that it isn't really worth calculating the SP using cuchisquare; the SP is clearly going to be very small. (If you do calculate it, GENSTAT gives the value as 0.) There is ample evidence that this model does not fit. The three factors are *not* all independent of one another. (Because the model fits so badly, seven of the eight counts are flagged as having large residuals.)

(c) This time the residual deviance is 2.393 on 1 d.f. Using cuchisquare (2.393;1), the SP is 0.1219. There is very little evidence in favour of rejecting the model; this one appears to fit.

(d) The deviance differences, d.f.s and SPs for comparing the models with one two-factor interaction omitted with the model with them all included are as follows, taken from the Change line in the output generated by using the Change Model button.

Model	Deviance difference	d.f.	SP
class.vote omitted	120.0	1	< 0.001
class.gender omitted	0.032	1	0.857
vote.gender omitted	7.77	1	0.005

Clearly it is not on to drop either **class.vote** or **vote.gender**, but **class.gender** can be dropped from the model. (Strictly speaking, you should go back and check that **class.vote** and **vote.gender** remain essential to the model after **class.gender** has been dropped. It is possible in principle that they are no longer essential, though since their SPs were so small it would be highly unlikely. In fact, if you do this check, you will find that both are indeed still essential.) There is no evidence of a relationship between class and gender, but the two other pairs of variables are not independent.

(e) The table of fitted values, classified by all three factors, is as follows.

		Total	
	gender	male	female
vote	class		
con	nonman	135.4	156.6
	manual	113.6	131.4
lab	nonman	55.9	44.1
	manual	209.1	164.9

The table of totals of the original data, classified by **vote** and **class** (and summed over gender), is as follows.

	Total	
class	nonman	manual
vote		
con	292.0	245.0
lab	100.0	374.0

The corresponding table of totals for the fitted values looks absolutely identical. (The two tables are identical not only because the totals are exactly the same but also because GENSTAT does not label tables like this with the name of the variate from which they were calculated.)

For classification by **gender** and **class**, the table of totals for the original data is as follows.

	Total	
gender	male	female
class		
nonman	190.0	202.0
manual	324.0	295.0

The corresponding table for fitted values is as follows.

	Total	
gender	male	female
class		
nonman	191.3	200.7
manual	322.7	296.3

The two tables differ, though not by much. (If they had been very different, a model omitting the **class.gender** interaction would not have fitted well.)

(f) From the table of fitted values in part (e), the fitted value for $P(\text{class} = \text{manual}, \text{vote} = \text{con}, \text{gender} = \text{female})$ is $131.4/1011 = 0.1300$.

(g) The fitted value of $o = p/(1 - p)$ is $0.1549/0.1300 = 1.192$. The fitted value of the conditional probability p is then given by

$$p = \frac{o}{1+o} = \frac{1.192}{1+1.192} = 0.544.$$

(There is actually a rather quicker, but less instructive, way to find p. Working with fitted values, there is a total of $156.6 + 131.4 = 288$ female Conservative voters, of which 156.6 are in the non-manual occupational class. Thus the probability that a voter is in the non-manual class given that she is a female Conservative voter is $156.6/288 = 0.544$.)

Solution 12.6

For the chi-squared test, the test statistic is 5.66 on 1 d.f., and the SP is 0.017. For Fisher's exact test, the two-sided SP (or, at any rate, the version calculated as a Sum of all outcomes with Prob<=Observed) is a little larger at 0.01890. For the loglinear model, an appropriate test statistic is the residual deviance in the model with no interaction term. This is 5.717 on 1 d.f., corresponding to an SP of 0.017. Each test shows a reasonable amount of evidence that the probability of being offered help is not the same for male students as for female students (and it is clear from the original table that female students are more likely to be offered help).

Solution 12.7

(a) The five models are as follows.

1 The saturated model, **depth*seedtype*mortalty**.

2 The null model, i.e. the minimum model that we can fit to these data, namely **depth*seedtype+mortalty**.

3 The model with all the two-factor interactions but not the three-factor interaction, namely **depth*seedtype+mortalty+mortalty.depth +mortalty.seedtype**.

4 **depth*seedtype+mortalty+mortalty.depth**.

5 **depth*seedtype+mortalty+mortalty.seedtype**.

(b) The residual deviances and degrees of freedom for models 2–5 in part (a) are as follows.

Model	Residual deviance	d.f.
2	50.10	3
3	1.284	1
4	25.03	2
5	27.79	2

It is plain, without bothering to do any detailed chi-squared calculations, that models 2, 4 and 5 do not fit. However, it looks as if model 3 fits reasonably well, and indeed cuchisquare(1.284;1) gives the SP of a test comparing it with the saturated model as 0.2572. Model 3 is the winner, with the others nowhere!

Therefore, a good description of the mortality of pine seedlings requires the presence of all three two-way interactions (in addition to all the main effects) but not the three-way interaction. This implies that mortality does depend both on depth of planting and on type of seedling, but there is no interaction between the two in their effect on mortality.

Solution 12.8

(a) This model has a residual deviance of 1038 on 64 d.f. Again, the deviance is so big compared with the degrees of freedom (or, if you like, the mean residual deviance is so much bigger than 1) that it is not really worth bothering with a chi-squared calculation (but, if you did, GENSTAT gives the SP as 0). This model clearly does not fit.

This lack of fit is reflected in the huge list (52 cases out of a total of 72) of residuals flagged by GENSTAT as being large. They are large because the model does not fit, not because there is anything wrong with any of the data.

(b) The model with all the main effects and two-factor interactions has a residual deviance of 99.91 on 45 d.f. It is not quite so obviously a bad fit; cuchisquare(99.91;45) gives its SP as 0.000005, so its fit is, indeed, inadequate.

One way to fit the model with all main effects and interactions up to third order is to tack on another .(age+work+tenure+acctype+resp) at the end of the Model to be Fitted. This model has a residual deviance of 26.87 on 20 d.f. Its SP is given by cuchisquare (26.87;20) as 0.1390. This is large enough to accept the fit as adequate.

(c) GENSTAT takes nine steps and finishes up with a model including all the two-factor interactions but only two three-factor interactions, namely age.work.acctype and age.tenure.acctype. The model has a residual deviance of 38.59 on 39 d.f. (corresponding to an SP of 0.49 – though, since they do not really take account of the stepwise method of choosing the model, SP values resulting from a stepwise fit are not always to be taken too seriously). It certainly seems to fit well.

(You may have noticed, in the output, a message saying that FACTORIAL limit for expansion of formula = 3. This means that, because of the way GENSTAT is set up, the STEP command did not attempt to fit any interactions of order higher than three. In fact we did not want it to fit any such interactions, so no problem!)

(d) This time GENSTAT prints many messages about not being able to add a certain term because another term 'is marginal to [the first term] and is not in the model', and similar messages about not being able to drop terms. Here GENSTAT is refusing to fit a non-hierarchical model; but since it does this automatically, you need not worry about it.

It eventually finishes up with almost the same model as in part (c), but not quite. The model includes all the main effects, all the two-factor interactions *except* acctype.resp, and the two three-factor interactions (**age.work.acctype** and **age.tenure.acctype**) that appeared in the model of part (c). The model has a residual deviance of 38.59 on 40 d.f. Again this appears to be a good fit.

Solution 12.9

The model resulting from the stepwise process includes only the main effects of **age1**, **work1** and **tenure1**. The other explanatory factor, **acctype1**, does not come into it. There are three datapoints flagged as having high leverage, but generally this model appears to fit well.

You might have noticed that the t value corresponding to **age1** 31–45 is small at -0.83. This indicates that we might well be able to produce a simplified model that still fits well by not fitting a separate term for this group, thus effectively treating them the same as those aged under 30. (In fact this is true, but we shall not pursue it here as it rather obscures the relationship between logistic regression and loglinear modelling.)

Solution 12.10

(a) The interaction terms involving **resp** in the model in Exercise 12.8(d) were **age.resp**, **work.resp** and **tenure.resp**. In terms of the logistic regression model, these interactions correspond to the main effects of **age**, **work** and **tenure**. The fact that there are no higher-order interactions involving **resp** in the loglinear model corresponds to the absence of interactions in the logistic regression model. This model thus matches the final model in Exercise 12.9 exactly, in terms of the effects it includes.

(b) The model including **resp+age*work*tenure*acctype** indeed fits badly. Its residual deviance is 158.9 on 35 d.f. This should hardly come as a surprise – the corresponding logistic regression model is one that just fits a constant, because there are no interactions with **resp**, so that the response variable does not depend on the explanatory variables at all. (You probably noticed that GENSTAT gives a message about 'near collinearity', which you can ignore in this context.)

The model arrived at by the stepwise fitting includes the main effect of **resp** and a saturated model for the other four factors. (It has to, since these were all in the model we started with, and the STEP command was constructed so that it would not change them.) In addition to this, it contains just three terms, **resp.tenure**, **resp.work** and **resp.age**. In logistic regression terms, these again correspond to the main effects of **age**, **work** and **tenure**, so that this model again corresponds to that of Exercise 12.9.

But, because this loglinear model satisfies the proviso, it is essentially *the same model* as the logistic regression model. This is reflected in the fact that both models have the same residual deviance (29.67 on 30 d.f.). In addition, their regression coefficient estimates match. The estimates from the logistic regression model are given by GENSTAT as follows.

	estimate	s.e.	t(*)	t pr.	antilog of estimate
Constant	0.305	0.135	2.26	0.024	1.356
age1 31–45	−0.113	0.137	−0.83	0.409	0.8933
age1 46+	−0.436	0.140	−3.12	0.002	0.6464
work1 unskill	−0.763	0.152	−5.02	<.001	0.4664
work1 office	−0.305	0.141	−2.17	0.030	0.7369
tenure1 own	1.014	0.114	8.87	<.001	2.758

The loglinear model produces far more estimated regression coefficients, because it has to estimate those describing how the explanatory variables are related to one another. The estimates for the coefficients of terms involving resp are as follows.

	estimate	s.e.	t(*)	t pr.	antilog of estimate
resp no	-0.305	0.135	-2.26	0.024	0.7373
resp no .tenure own	-1.014	0.114	-8.87	<.001	0.3626
resp no .work unskill	0.763	0.152	5.02	<.001	2.144
resp no .work office	0.305	0.141	2.17	0.030	1.357
resp no .age 31-45	0.113	0.137	0.83	0.409	1.120
resp no .age 46+	0.436	0.140	3.12	0.002	1.547

You can see that, apart from a sign change, the corresponding estimates are equal, and their standard errors match too. This is because, in these respects, the loglinear and logistic models are identical. The sign change occurs because we fitted a logistic regression model to the yes responses, whereas in the danish.gsh file the yes value of resp is coded as 1 and the no value as 2. Thus GENSTAT fits a parameter for no instead of yes, and the sign switches.

Either way, the model fitted says that a 'yes' to a greater amount of do-it-yourself home maintenance is more likely if the respondent owns the accommodation (rather than rents) and less likely if he is older (the older the respondent is, the less likely to do more home maintenance), if he works in an office and even more so if he is unskilled (relative to being in the skilled employment class). There is no dependence òn type of accommodation. Nor is there any more complicated dependence on interactions between age, work and tenure.

Solution 12.11

(a) The explanatory variables in the logistic regression model would be depth and seedtype. The proviso says that, for a loglinear model to match the corresponding logistic regression model, the loglinear model must include the main effect of the response variable and a saturated model for the explanatory variables. In this case, the loglinear model has therefore to include mortalty+depth*seedtype. Because of the fixed marginal totals in this dataset, this was the null model, and every model we fitted included these terms at least.

(b) This model corresponds to the logistic regression model depth+seedtype, since mortalty is the response variable and since the only interactions involving mortalty in the loglinear model are mortalty.depth and mortalty.seedtype. (The logistic regression model does not include the interaction depth.seedtype since the three-way interaction depth.mortalty.seedtype does not appear in the loglinear model.)

Because all the models we considered in Exercise 12.7 satisfied the proviso, they all match the corresponding logistic regression models exactly in terms of parameter estimates and residual deviance. Thus the modelling decisions we made in Exercise 12.7 would all have been made in exactly the same way if we had used logistic regression instead, and we would have ended up with the same model.

13

Solution 13.1

All variables in Table 13.1 can be treated as continuous. As the response variable is one of these continuous variables, this appears to be a task for normal multiple linear regression, as described in Chapter 6.

The main body of the scatterplot includes a number of striking patterns, most notably made by a few points that do not, in some respect or other, conform with the rest of the data. For example, there is one particularly high value of **solover** and a few particularly low values of **ratover**. The variable **ratpreg** is particularly odd for having almost all values tightly clustered around 0.6, but one value much smaller and one other rather greater.

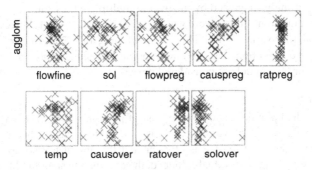

The bottom row of the scatterplot matrix is reproduced above.[1] First, note that there are no particularly extreme values of the response variable **agglom**. But in trying to judge which of the **agglom**/explanatory variable relationships might, disregarding other variables (i.e. in a simple linear regression sense), be important, it is hard not to be led to one's conclusions by the most extreme datapoints in the x directions (i.e. those points which potentially – and certainly in one dimension – have high leverage). It seems that **agglom** might depend on **causpreg** and **causover**. There may be some increase with increasing **ratover** ... or is that simply suggested by the four points furthest left? You just can't see what might be happening in the main body of the data with regard to **ratpreg** or **solover** because of the extreme x points.

Solution 13.2

The correlation matrix, obtained by entering all explanatory variables (but not **agglom**) in the **Data** field of the **Correlations** dialogue box (under **Stats|Summary Statistics| Correlations**),

---

[1] The row of plots will not be split like this on your computer.

as just after Exercise 6.4, yields only one correlation greater than 0.5 (in absolute value). This is the 0.779 correlation between the two caustic concentration variables causpreg and causover. Looking at the scatterplot of causover against causpreg (which you might redraw using the Graphics|Point Plot menu item), there seems to be a strong (linear) relationship between the two. The outlying point on this plot makes the correlation less than it would otherwise be. It seems that there may not be a need for both these variables in the final model.

The multiple regression of agglom on all nine explanatory variables is obtainable in the usual way using the Stats|Regression Analysis|Linear option. Those variables with small SPs are ratpreg, with $SP < 0.001$, causpreg, with $SP = 0.007$, ratover, with $SP = 0.023$, and flowpreg, with $SP = 0.026$ (flowfine with $SP = 0.080$ might be noted too).

The STEP command, with MAX equal to something sensible (e.g. 18) and IN = OUT = 4 (as in Example 6.2), yields a four-variable model when started from the full nine-variable model. Reassuringly, the very same model is produced when the same STEP command is used starting from the null model (see near the end of Example 6.2).

This chosen model contains flowpreg, causpreg, ratpreg and ratover. The scatterplots (in Solution 13.1) suggested causpreg and maybe ratpreg and ratover, although these last two appear to have leverage problems. In retrospect, looking again at the scatterplots, perhaps a negative dependence on flowpreg, not mentioned in Solution 13.1, is not unreasonable. In light of the correlation values, it is gratifying that only one of the caustic concentration variables is in this selected model. These four variables are the same four as were most strongly suggested by the overall multiple regression.

We should note some quite severe warnings issued by GENSTAT, however, in association with this final four-variable fit. There is a warning that The error variance does not appear to be constant and four points are flagged as having high leverage. (Diagnostics for this model are considered next.)

Solution 13.3

(a) The residual plot does show some kind of pattern, which GENSTAT flags as demonstrating non-constant variance. It is not entirely clear whether this is the best interpretation, or at least it isn't clear that this might not be driven by just a few of the more extreme residuals. Indeed, the eye and the smooth blue curve tend rather to suggest some evidence of curvature in the residual plot.

The four points flagged by GENSTAT as having high leverage (units 12, 32, 42 and 43) show up as being clearly greater than the others in the leverage plot. But are these points having a great influence on the fit? In the plot of the Cook statistics, there are large values, but they are not so clearly separated from the rest. The largest Cook statistic is given by unit 32. The second largest is given by unit 43. (You have to take care in identifying particular units on the plot.) Both units 32 and 43 are points of high leverage, but the third biggest Cook statistic belongs to point 37, which isn't. (The other high leverage points, numbers 12 and 42, have small values of the Cook statistic.)

It seems that points 32 and 43 are having a particularly big influence on the current fit, and it is sensible to investigate them further. Further consideration of the residual plot can be put in abeyance until after considering them.

(b) Units 32 and 43 have the lowest two values of ratover. (Recall that ratover is the variable on which three (or maybe four) units have particularly low values.) The third lowest value of ratover belongs to unit 42, which also has the (single) very low value of **ratpreg**; so, with respect to these two explanatory variables, unit 42 is particularly unusual. The single particularly high value of **ratpreg** belongs to unit 12. So their values for **ratover** and **ratpreg** are what seem to determine the high leverage of these four datapoints.

(c) Repeating the stepwise regression using all nine variables but the restricted number of datapoints, starting from both the full nine-variable model and the null model, results in the same two-variable model: the two variables chosen are **causpreg** and **ratpreg**. Perhaps unsurprisingly, **ratover** has disappeared from the model. It seems that units 32 and 43, with their very small values of **ratover**, were influencing the procedure to select that variable, possibly wrongly. Interestingly, the relatively weak effect of **flow-preg** has also no longer been chosen, but the rather stronger effect of **causpreg** remains.

When regressing on just **causpreg** and **ratpreg**, GENSTAT flags units 12 and 42, once more, as points of high leverage. You can see this clearly on a plot of leverage values too; yet when Cook statistics are plotted, no values are particularly large (although unit 12's is the largest). (GENSTAT does not now flag the shape of the residual plot, although you might think the variance is larger for medium fitted values.)

(d) Without units 12, 32, 42 and 43, the stepwise procedure still comes up with the two-variable model involving **causpreg** and **ratpreg**. Thus, the importance of **ratpreg** in the model-building exercise was not a function purely of the unusual values taken by units 12 and 42. The plot of residuals against fitted values now looks a little better than before; there is just one flagged leverage value (unit 16), which has the largest Cook statistic.

Solution 13.4

(a) The fitted model is

$$E(\text{agglom}) = -545 + 1.357 \,\text{causpreg} + 445 \,\text{ratpreg}.$$

This model has been fitted ignoring datapoints 12, 32, 42 and 43 which have particularly unusual values of **ratpreg** and/or **ratover**. The fitted model therefore seems good over a limited range of values for **ratpreg** (all but one of those remaining in the dataset being 0.59, 0.60 or 0.61) and for **ratover** (all but one being 0.52, 0.53, 0.54 or 0.55) (see Figure 13.1 and Table 13.1). For points with values of **ratpreg** outside the values mentioned here, indications are that a similar model continues to hold; for points with smaller values of **ratover**, there may be a greater dependence on those values, and hence perhaps a change of appropriate model. But these indications are based on very few datapoints, and much further data of these types would have to be collected to make firmer conclusions about the model appropriate over a wider range of explanatory-variable values.

(b) Notice first that the **ratpreg** value at which the confidence interval is required is one of those for which the fitted model in part (a) is considered to be valid. Therefore, perform the multiple regression of **agglom** on **causpreg** and **ratpreg** again, if it was not the last analysis you ran. Then, use the following.

```
PREDICT [pred=yhat;se=se] causpreg,ratpreg;244,0.60
CALC lower=yhat-se*edt(0.975;44)
CALC upper=yhat+se*edt(0.975;44)
PRINT lower,upper
```

(The 44 is d.f. (Residual).) This gives the following 95% confidence interval:

$$(50.87, 55.60).$$

The source of variability not taken into account here is that induced by uncertainty over the appropriate form of model (remember that the two-variable model was selected from the data).

Solution 13.5

The response variable is nodal, the explanatory variables are age, acid, xray, tgrade and tsize. Since nodal is a binary response variable and not all the explanatories are categorical, the appropriate approach is logistic regression. (If all the explanatories had been categorical, logistic regression would still have been appropriate, but a possible alternative would have been loglinear modelling.)

Solution 13.6

The Stats|Regression Analysis|Generalized Linear menu item allows the appropriate logistic regression when the Analysis field is changed to Modelling of binomial proportions. (e.g. by logits), the Number(s) of Subjects field is filled in as 1, the Numbers of Successes field is nodal, the Model to be Fitted field contains the five explanatory variables acid, age, xray, tgrade and tsize, separated by commas, and the Transformation (link) field remains at Logit.

GENSTAT warns about three points with high leverage, one of them (point 24) having much greater leverage than the others, as confirmed by an index plot of leverage values (under Further Output|Model Checking). (The message given about the error variance and indeed the residual plots are uninteresting because we are doing binary regression.) Importantly, the same point shows up strongly in the index plot of Cook statistics too, implying that its Cook statistic is large, and thus that this point is having a strong influence on the results of the regression analysis.

The outstanding thing about point 24 is that this patient had much the largest acid level and yet, slightly surprisingly given an apparent tendency for nodal involvement to increase with increasing acid level (see the scatterplot below), has no nodal involvement.

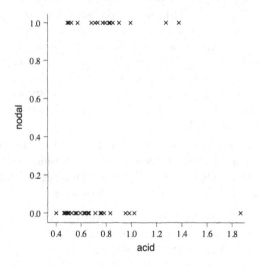

Solution 13.7

(a) The first attempt at the regression allows all interactions up to three-way interactions; this is because GENSTAT's default for the level of interactions to include is 3 (as is indicated by FACTORIAL limit for expansion of formula = 3 in the GENSTAT output). GENSTAT hits trouble in fitting this 25-parameter model. Adding FACTORIAL=2 restricts GENSTAT to fitting just the main effects and two-way interactions (15 of these in all), which it manages successfully. Given that we didn't really have three-way interactions in mind, because they would not be very interpretable in this kind of context, the restriction is reasonable.

(b) This stepwise regression may take a while on some computers. GENSTAT decides on a model with seven explanatory variables, namely the four main effects acid, xray, tgrade and tsize and the three interactions acid.tgrade, acid.tsize and tgrade.tsize. The regression deviance is 38.24 on 7 d.f.

(c) Moving forwards, GENSTAT settles on the three-variable model with just the acid, xray and tsize main effects. The regression deviance is 22.10 on 3 d.f.

(d) The difference between the deviances is $38.24 - 22.10 = 16.14$ which has to be compared with the χ^2 distribution on $7 - 3 = 4$ d.f. The SP is 0.0028. There is strong evidence against the null hypothesis – which is for the model with fewer variables – and so it seems that the larger model is the better of the two.

This demonstrates the fallibility of the stepwise approach: going forwards, GENSTAT was unable to find a four-variable model containing the three-variable model that was better; yet we know of a seven-variable model that is better. Part of the problem is that GENSTAT will not include an interaction term unless both of the corresponding main effects are already in the model.

(e) GENSTAT warns about five high-leverage points, although the index plot of leverages does not show them up as being particularly extreme. However, the index plot of Cook statistics does show up two of these as being strongly influential, namely points 20 and 34, which are in fact the two points with highest leverage. Perhaps the fit of the model should be investigated further.

Solution 13.8

(a) The fitted model is

$$\text{logit}(P(\text{nodal} = 1)) = -12.57 + 12.67\,\text{acid} + 2.45\,\text{xray} - 5.67\,\text{tgrade} + 13.56\,\text{tsize}$$
$$+ 21.3\,\text{acid.tgrade} - 14.23\,\text{acid.tsize} - 8.58\,\text{tgrade.tsize}$$

where $\text{logit}(p) = \log(p/(1-p))$.

(b) The given patient has $\text{acid} = 0.55$ and $\text{xray} = \text{tgrade} = \text{tsize} = 0$. This means that the interaction terms, which are just products of the variables, are all zero. Thus, simply, in this case,

$$\text{logit}(P(\text{nodal} = 1)) = -12.57 + 12.67 \times 0.55 = -5.6015,$$

so that the odds in favour of $\text{nodal} = 1$ are $e^{-5.6015}$ ($= 0.00369$) and

$$P(\text{nodal} = 1) = \frac{e^{-5.6015}}{1 + e^{-5.6015}} = 0.00368,$$

as also given by the following GENSTAT command.

```
PREDICT acid,xray,tgrade,tsize;0.55,0,0,0
```

The output from PREDICT shows that the standard error for this estimated probability is quite large, so that the estimate is not very precise. (The same goes for the estimated probabilities in the next two parts.) Notice that the patient's age, although given, doesn't come into it because it isn't in the model.

(c) Now, **tsize** = 1. This increases the log odds of nodal involvement by $13.56 - 14.23 \times 0.55 = 5.7335$ (i.e. tsize.acid is $1 \times 0.55 = 0.55$ and the other interaction terms remain zero). Thus the odds increase by a factor of $e^{5.7335} = 309.049$. The resulting probability of nodal involvement rises dramatically to $0.00369 \times 309.049/(1 + 0.00369 \times 309.049) = 0.533$, as readily confirmed by GENSTAT (to within rounding error).

(d) When **tgrade** = 1 as well, the additional terms in the model are $-5.67 + 21.3 \times 0.55 - 8.58$, since acid.tgrade = 0.55 and tgrade.tsize = 1. To cut a long story short (use PREDICT), the model estimate of $P(\text{nodal} = 1)$ now reduces to 0.083.

That having a 'more serious' tumour rather than a 'less serious' tumour lessens the probability that the cancer has spread may or may not be surprising to you on intuitive grounds. You might have expected an increased probability because 'one bad aspect implies another'; but perhaps if the 'badness' is concentrated in the tumour it has not spread to the lymph nodes. Clearly, specialist knowledge is needed here. This behaviour is a consequence of the negative signs attached to **tgrade** and **tgrade.tsize** in the model (outweighing the positive sign on **acid.tgrade**).

Solution 13.9

(a) The response variable is clearly the yield of apples from each plot during the experiment, i.e. the variable labelled y in Table 13.3. There are three different explanatory variables. One is the yield from the previous cropping records, i.e. the variable labelled x in Table 13.3. It is continuous. There are also two categorical explanatory variables, one for the blocks and one for the ground-cover treatments.

(b) With a continuous response variable and three explanatory variables, one might start with a normal multiple linear regression model.

Solution 13.10

(a) In the spreadsheet, each experimental unit has one row (as usual). There are two variates, the response variable (**weight**) and the covariate (**before**), and factors for the block and the treatment.

The mean response ranges from 267.8 on Treatment B to 284.5 on Treatment A. Four of the experimental treatments appear worse than the control and one appears better. The variances differ quite widely for different treatments. Not surprisingly, the same sort of thing shows up in the boxplots, where probably the most noticeable feature is that the experimental units on Treatment A are much less variable, and on average have higher yields, than the others. (The 'whiskers' on all the boxes look rather short, too.) There is some indication that different ground-cover treatments may lead to different mean yields, but one cannot be certain because of all the variability.

(b) Again the means and variances differ between treatments, with considerably higher yields before the experiment on the units that got the control treatment (O), and relatively low yields before the experiment for Treatments D and E. The variance of the **before** yields is greatest for Treatment D and least for Treatment A, just as for **weight**. The experimental treatments cannot be responsible for these differences in the covariate, because they had not even been applied when the covariate was recorded and because they were allocated at random. It seems just to have happened by chance that the control units had high-yielding trees.

The scatterplot shows reasonably clearly that there is a positive relationship between the covariate and the response variable, overall. Within each treatment group, there are not many experimental units, so the relationship between covariate and response variable is not quite so clear within groups, but overall an assumption of a linear relationship seems not too wide of the mark.

(c)   The accumulated ANOVA table shows a v.r. of 3.14 for the ground-cover treatments (**treat**), and a corresponding SP of 0.042. There is thus some evidence (though it is far from overwhelming) to reject the null hypothesis that the ground-cover treatments make no difference. No outliers or high-leverage points are flagged. In the table of estimates of regression coefficients, we are not particularly interested in the rows for the blocks and the covariate. The estimates for the treatments are all positive (though, judging by the t values, some of them do not differ significantly from zero). The **treat** factor has been set up so that the control treatment (O) corresponds to its first level (we can deduce this since there is no row for this treatment in the table of estimates). Therefore these estimates correspond to differences between the corresponding ground-cover treatment and the control. Thus the fact they are all positive indicates that (on average) all the experimental ground-cover treatments show higher average yields than does the control treatment, after the covariate has been taken into account, though for some of the treatments this difference may be attributable to chance. Note that the control treatment did *not* have the lowest mean response overall, as you saw in part (a). We thus needed to take account of the covariate in the analysis for the suggestion that all the experimental treatments improve the yields to become apparent.

(d)   The plot of the fitted model is as follows.

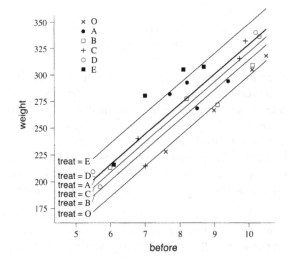

The lines fit the data reasonably, though there remains a lot of scatter. They are parallel because the model we fitted (block+before+treat) constrains them to be parallel. For an experimental unit in block j, with treatment l, and with a value x for the covariate, the mean response according to the model is $\mu + \gamma_j + \tau_l + \beta x$. What GENSTAT does in plotting the fitted model is to average over the values of the block parameters (the γ_js) so that the resulting line for treatment l is (constant) $+ \tau_l + \beta x$. The slopes of these lines are the same for different treatments, though their intercepts differ. To fit non-parallel lines, we would have to add a term before.treat to the model. This would allow different slopes for different treatment groups. (If you do add such a term, its SP turns out to be 0.152, so the model is not significantly improved.)

(e) Plots of residuals against fitted values and index plots of leverage and of Cook statistics show nothing untoward. A normal probability plot of the residuals (shown below) is less encouraging. It looks rather bent. The more extreme residuals are somewhat less extreme than one would expect under the assumption of normality. The histogram is not very symmetric either. We should thus treat the conclusions from the analysis with a little circumspection.

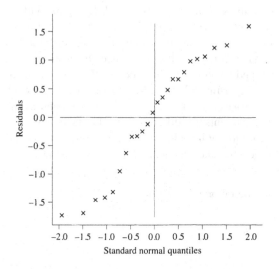

(f) The appropriate mean increase (because the control treatment is the first level of treat) is just the regression coefficient corresponding to level D of the treat factor. From the table of parameter estimates, this quantity is estimated to be 29.8, with a standard error of 12.7. To calculate the confidence interval, we need also the 0.975 quantile of a t distribution with 14 degrees of freedom (since this is the d.f.(Residual)). The quantile can be found using PRINT edt(0.975;14). It is 2.145, so the confidence interval is given by $29.8 \pm 2.145 \times 12.7$, which results in $(2.6, 57.0)$ (which is still pretty wide, even after taking the covariate into account).

(g) The output from the PREDICT commands begins with a lot of descriptive messages explaining how GENSTAT has averaged over the values of the other variables. The means are then given as follows (mean followed by its standard error).

treat		
0	251.34	8.98
A	280.48	8.34
B	266.57	8.33
C	274.07	8.33
D	281.14	8.43
E	300.92	8.79

Some of the means are reasonably close to those you found in part (a) without taking the covariate into account. However, the mean for Treatment O (the control) is much smaller than the observed mean (which was 279.5). It is now the smallest mean, while it was the second highest mean before the covariate was taken into account. The means for Treatments D and particularly E, taking the covariate into account, are considerably higher than they were without accounting for the covariate. This reflects the fact that (as you found in part (b)) the units that got Treatment O happened to contain trees that produced high yields before the experiment, and the units that got Treatments D and E happened to contain low-yielding trees. The means produced by the PREDICT command adjust for these differences by using the same value of the covariate (its mean) for all treatment groups.

(h) Without the covariate in the model, the variance ratio for **treat** is only 0.1, and the corresponding SP is huge at 0.991. There is absolutely no evidence of treatment differences. This is essentially because the treatment differences that do exist (on the basis of our previous analysis) have been hidden by the huge variability in yields between different trees.

The residual mean square for the analysis without the covariate is 1562; with the covariate it was only 277.5. Putting the covariate into the model has taken account of much of the variability in yields, allowing us to distinguish the treatment effects much more clearly.

Solution 13.11

The output looks rather different both from the regression output and from the output from previous analyses of variance. Comparing the ANOVA table here with the accumulated analysis of variance table in the regression output, you can see that the Residual and treat rows (in the units stratum here) are the same, apart from an extra entry in the column headed cov.ef., which we shall come to. This is hardly surprising, since we have fitted the same model. The SP for the test of no treatment differences is exactly the same, so the conclusions are the same.

The usual table of response means appears in the ANOVA output, but it says that it has been adjusted for covariate. The figures given are exactly those you found in part (g) of Exercise 13.10 (apart from being given to one fewer place of decimals). The 'adjustment' consists of fixing the covariate at its mean level, as we did in the previous exercise.

There is a section in the output headed Covariate regressions. GENSTAT has essentially fitted a regression model of the response variable on the covariate (as you did in Exercise 13.10). Within the units stratum, the regression is identical to what we did before (and indeed the regression coefficient, 28.4, with a standard error of 3.38, is exactly what we found before). GENSTAT also does regressions in other strata, which can be important in more complicated designs. (Here, in the block stratum, the response means for each block are regressed on the covariate means.) We shall not pursue the details.

We have already mentioned that there is an additional column in the ANOVA table, headed cov.ef. which stands for 'covariate efficiency'. We shall not go into the details of what this means, except to note that the entry of 5.63 in the Residual row in the units stratum is just the ratio of the residual mean square in the model with the covariate omitted to the residual mean square in the model with the covariate included. (See part (h) of Exercise 13.10.)

The residual plots are very similar to those you found in Exercise 13.10, and show nothing untoward. The slight differences are due to the fact that the ANOVA residual plots use simple (unstandardized) residuals whereas the regression residual plots use Pearson (standardized) residuals.

Solution 13.12

The number of seizures is the response variable and the treatment and period indicators are obvious explanatory variables. Patient number is often ignorable as a potential explanatory variable because it just identifies individuals, but not so in this case. This is because there are two observations per patient, and so the patient number reflects part of the design of the study.

A straightforward scatterplot of **seizures** against **patient** is dominated by the huge numbers of seizures of patient 11, so you can't really see what is going on with the other patients. There are various possibilities for getting round this problem. In the plot below, the response variable is $\log(\textbf{seizures} + \frac{1}{2})$, where the $\frac{1}{2}$ is added to avoid the problem with the one zero count in the data. Treatment information is added using the **Groups** field in the **Point Plot** dialogue box. The full set of GENSTAT instructions to produce the colour version of the plot below is to choose the **Graphics|Point Plot** menu item, fill in the Y Coordinates field either as **lseiz** (if **lseiz** has already been set to log(seizures+1/2)) or as **log(seizures+1/2)** itself, fill in the X Coordinates field as **patient**, fill in the Groups field as **treat**, and click on **OK**.

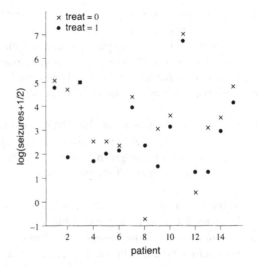

There is much variability in **seizures** for different patients (remember, this plot is on a log scale). We should account for this major source of variability by including **patient** as an explanatory variable and thinking of it as a blocking factor, i.e. we should treat it as an important explanatory variable whose effect is not of major interest per se.

The new treatment seems to result in a lower number of seizures for most of the patients, and you might guess that the new treatment is beneficial overall. (But remember that this plot takes no account of **period** or **exposure**.)

Your first recourse should be to a Poisson regression model since the responses are counts. There are three basic explanatory variables, all factors, namely **treat, period** and **patient**.

Solution 13.13
Choose the **Stats|Regression Analysis|Generalized Linear** menu item. In the resulting dialogue box, make sure the **Analysis** field reads **General Model**, put **seizures** in the **Response Variate** field, patient+treat+period in the **Model to be Fitted** field, **Poisson** in the **Distribution** field, and leave the **Link Function** field as **Canonical**. You can either make a new variate equal to log(**exposure**) before entering this in the **Offset** field obtained through the **Options** button, or simply write log(**exposure**) directly in the **Offset** field. Click on **OK**.

As many as 14 datapoints (out of 30) are flagged by GENSTAT as having large standardized residuals (and several of these are very large indeed). Something is seriously amiss with our model.

Solution 13.14
(a) The interesting effects are those of **period** and **treat**. Those due to **patient** (blocks) are just 'soaking up' the between-patient variability, but the patient effect is not of interest in its own right.

(b) The (approximate) SP for the removal of **treat** is 0.028. For the removal of **period**, the SP is 0.737. There is clearly no evidence to include **period** in the model containing **treat**. The result for **treat** is more equivocal – particularly in the light of the approximate nature of the F test – but you would probably wish to retain it in the model already containing **period**. Combining these two conclusions suggests a model containing **treat** but not **period**.

(c) Dropping **treat** from the Poisson model with **treat** and **patient** (and the offset) gives an SP of 0.004. There is strong evidence for the inclusion of **treat** in the model.

So, the best model includes just a treatment effect in addition to the patient blocking factor and the exposure offset term.

Solution 13.15
Both these interactions are of the form of block by treatment interactions. Following the usual practice for blocking (Chapter 8), and a little thought to check that this practice makes sense in this particular case, it seems appropriate to set such interactions to zero.

Solution 13.16
(a) Units 9 and 24 both refer to patient number 2, who displayed a major improvement from 110 seizures (in 56 days) when using the old drug (in the first period) to just 6 seizures (in 56 days) using the new drug (in the second period). (Unit 9 therefore

has a large positive residual corresponding to a larger than expected response under treat $= 0$ and Unit 24 a large negative residual corresponding to a smaller than expected response under treat $= 1$, measured relative to the average difference between responses for other individuals.) The removal of the pair of units 9 and 24 therefore makes sense as it removes the single unusual patient (number 2) from the analysis.

(b) The main consequence of omitting units 9 and 24 is to make much the same analysis occur, but with some strengthening of the evidence for including treat and not period in the model.

(c) The majority of the data are well fitted by a GLM based on a Poisson response distribution but with an additional (estimated) dispersion parameter. For the model containing just the patient and treat main effects, and with patient 2 omitted, this dispersion parameter takes the value 4.28. The mean response behaviour has, as its point estimate,

$$\log E(Y_i) = \log t_i + 1.055 - 0.3253 \, \text{treat}_i + \widehat{\beta}_{3,j(i)} \, \text{patient}_{j(i)}$$

where Y_i is the number of seizures for the ith patient and t_i is the ith exposure time. Here we have not put in any values for the patient effect because they are not of direct interest, but the $\widehat{\beta}_{3,j(i)}$s are available in the Estimates of parameters part of the output.

The treatment effect cannot be removed from the model, suggesting that its effect is real: the new drug appears to lessen the seizure rate relative to the old drug. (For two identical patients with identical exposure times, the rate of seizures under the new treatment is estimated to be only $\exp(-0.3253) = 0.72$ times what it was under the old treatment.)

It is important to note also that the one patient out of the fifteen who did not follow the modelled pattern actually made a much more dramatic improvement under the new drug, from 110 seizures to 6. However patient 11, with large number of seizures under both treatments, *is* fitted by this model.

Solution 13.17
When ϕ is estimated, $\widehat{\beta}_1 = -0.3253$ with estimated standard error 0.0747. When $\phi = 1$, $\widehat{\beta}_1$ is still -0.3253 but its estimated standard error is now 0.0361. Taking the overdispersion into account has greatly increased the standard error, as it would with other standard errors and confidence (or prediction) intervals too. But in the process the result has been made much more realistic; without accounting for the dispersion, our confidence in our point estimates is much too great.

In fact, the ratio of the two estimated standard errors, $0.0747/0.0361 = 2.07$, equals (to within rounding error) $\sqrt{\widehat{\phi}} = \sqrt{4.28} = 2.07$ (the square root being appropriate because ϕ is a parameter on the variance rather than standard deviation scale).

Index of datasets

Subject index